Two Manuals Combined into One Book



# Military Police Operations

# And

# Physical Security

Department of Defense

Pentagon
Publishing

Field Manual
No. 3-19.1

Headquarters
Department of the Army
Washington, DC, 22 March 2001

# Military Police Operations

## Contents



**Distribution Restriction:** Approved for public release; distribution is unlimited.

*This publication supersedes FM 19-1, 23 May 1988.

Page

# Preface

The Military Police (MP) Corps supports the commander across the full spectrum of military operations. This manual is the foundation for all MP doctrine as it relates to this support. It communicates to all levels of leadership and staffs how the MP provide a flexible and lethal force capable of operating across this full spectrum. As the keystone manual, it identifies what the MP train on and how their forces are organized and equipped in support of all Army echelons. Additionally, this manual provides guidance that can be used by United States (US) Army service schools, students, sister services, and federal agencies.

This manual is based on the purpose, organization, responsibilities, and goals of the US Army as set forth in Field Manuals (FMs) 100-1 and 3-0, as well as corps, division, and brigade manuals. Additionally, this manual is fully compatible with current joint, multinational, and interagency doctrine.

Appendix A contains a metric conversion chart.

The proponent of this publication is HQ TRADOC. Send comments and recommendations on Department of the Army (DA) Form 2028 directly to Commandant, US Army Military Police School (USAMPS), ATTN: ATSJ-MP-TD, 401 Engineer Loop, Suite 2060, Fort Leonard Wood, Missouri 65473-8926.

Unless this publication states otherwise, masculine nouns or pronouns do not refer exclusively to men.

# Chapter 1

# Introduction

*The MP Corps has a strong history evolving over the past five decades. We, as a corps, continue to transform our organizations and doctrine as we have in the past to support the Army in the active defense of the 1970s, the AirLand battle of the 1980s, and now the force-projection doctrine of the 1990s. Our five MP functions clearly articulate the diverse role the MP play across the full spectrum of military operations. We cannot bask in our successes, nor reflect or celebrate. Our charter is to continue our legacy of stellar performance and strive to perfect it.*

*BG Donald J. Ryder*

When the Army developed the Active Defense strategy in 1976, the US was facing the Cold War scenario of central Europe. Military strategy and doctrine were related to a single, focused threat that revolved around the countries in the Warsaw Pact. We were an outnumbered and technically inferior force facing an armor-dominated European battlefield. The MP Corps supported the Active Defense strategy by tailoring its forces to meet the threat. In 1982, when the AirLand Battle strategy was developed, US forces were still outnumbered, but were no longer technically inferior. Still threat-based and focused on a central European conflict, the AirLand Battle strategy used a relatively fixed framework suited to the echeloned attack of soviet-style forces. It delineated and clarified the levels of war; emphasized closed, concerted operations of airpower and ground forces; balanced the offense and the defense; and highlighted the synchronization of close, deep, and rear operations. MP doctrine kept pace with the Army's AirLand Battle strategy by supporting the battlefield commander through four basic missions—battlefield circulation and control, area security (AS), enemy prisoner of war (EPW), and law and order (L&O).

## OVERVIEW

1-1. In October 1983, MP capabilities in the AirLand Battle strategy were tested during operation Urgent Fury in Grenada. The MP performed missions that ranged from assisting the infantry in building-clearance operations to assisting Caribbean peacekeeping forces in restoring L&O. These actions secured the MP's place in the combat-support (CS) role, demonstrating the professional knowledge and flexibility necessary for rapid transition from combat to CS to peacetime missions. The changing battlefield conditions of operation Urgent Fury set the stage for the demand of MP units today.

1-2. Evolving simultaneously with the changing definition of the modern battlefield, MP performance in Operations Hawkeye, Just Cause, and Desert Shield/Storm galvanized their ability to perform at any point along the operational continuum. With the publication of FM 100-5 in 1993, the Army adopted the doctrine of full-dimensional operations, relying on the art of battle command to apply those principles and to shift the focus from AirLand Battle to force-projection doctrine. This new doctrine was based on recent combat experience in a multipolar world with new technological advances. Already trained and expected to perform in this new strategy, MP support was already in place and fully operational. The MP continued to perform their basic battlefield missions and to refine their capabilities while supporting the battlefield commander as he deployed to contingency operations throughout the world.

1-3. In 1996, the MP Corps went through a doctrinal review process to determine if it was properly articulating its multiple performance capabilities in support of US forces deployed worldwide (see Appendix B). The review process identified the need to restructure and expand the EPW mission to include handling US military prisoners and all dislocated civilians. This new emphasis transformed the EPW mission into the internment and resettlement (I/R) function. The review process also identified the need to shift from missions to functions. In the past, the four battlefield missions adequately described MP capabilities in a mature theater against a predictable, echeloned threat. However, that landscape is no longer valid. Accordingly, the four MP battlefield missions have become the following five MP functions:

- Maneuver and mobility support (MMS).
- AS.
- L&O.
- I/R.
- Police intelligence operations (PIO).

1-4. These new MP functions are shaped by the following factors:

- The application of stability and support operations where the integration of joint, multinational, and interagency capabilities are common occurrence.
- The lack of traditional linear battlefields, requiring theater commanders in chief (CINCs) to request forces that meet a specific function to accomplish operational requirements.
- The impact of asymmetric threats (such as drug traffickers and terrorist factions) and the effects of man-made and natural disasters.
- The impact of advances in information and communication technologies and specifically in understanding the increased vulnerabilities presented by these technologies.

1-5. Articulating MP capabilities along functional lines benefits the MP and the Army echelon commander as well as the combatant commander. Since there is a multinational, interagency, and sister-service overlap of security services, the importance of including MP leaders and staffs early in the operational planning process cannot be overemphasized. This means before units are designated, unit boundaries are drawn, and unit missions are assigned. Early involvement ensures the proper development of common security responsibilities, communication and connectivity, liaisons, processes, and the rules of interaction between all forces. The ultimate goal should be the optimal, phased employment of MP forces in support of a commander's operational plan. MP functions not only reflect and capture current capabilities, they define the MP Corps in the twenty-first century.

1-6. As the Army reshapes and focuses its resources on transformation, Force XXI, and other redesign efforts, the MP Corps stands proud and ready to support this progress and reiterate its commitment to assist, protect, and defend.

## OPERATIONAL FRAMEWORK

1-7. The operational framework consists of the arrangement of friendly forces and resources in time, space, and purpose with respect to each other, the enemy, or the situation (see Figure 1-1). The operational framework for Army forces (ARFOR) rests within the combatant commander's theater organization. Each combatant commander has an assigned geographical area of responsibility (AOR), also called a theater, within which he has the authority to plan and conduct operations. Within the theater, joint-force commanders at all levels may establish subordinate operational areas such as areas of operation (AOs), joint-operations areas (JOAs) and joint rear areas

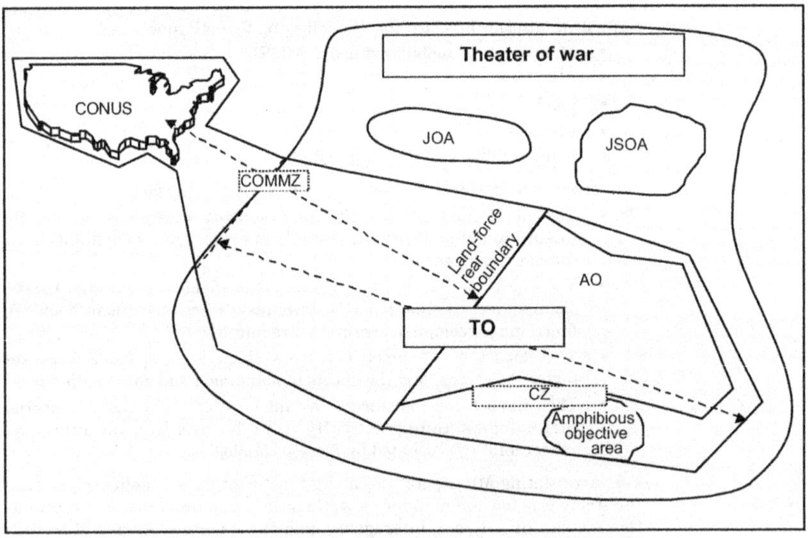


Figure 1-1. Operational Framework

(JRAs). The JRAs facilitate the protection and operation of bases, installations, and forces that support combat operations. When warranted, combatant commanders may designate theaters of war, theaters of operations (TOs), combat zones (CZs), and communications zones (COMMZs).

1-8. A theater of war is that area of air, land, or water that is, or may become, directly involved in the conduct of the war. A theater of war may contain more than one TO. It does not normally encompass the geographic combatant commander's entire AOR. A TO is a subarea (defined by a geographic combatant commander) within a theater of war in which specific combat operations are conducted or supported.

1-9. A CZ is the area required by combat forces for conducting operations. It normally extends forward from the land force's rear boundary. The COMMZ is the rear part of the TO (behind but contiguous to the CZ) that contains the lines of communication (LOC) and provides supply and evacuation support. Other agencies required for the immediate support and maintenance of field forces may also be located in the COMMZ. The COMMZ spans back to the continental US (CONUS) base, to a supporting combatant commander's AOR, or both.

1-10. An AO is an operational area defined by the joint-force commander for land and naval forces. An AO does not typically encompass the entire operational area of the joint-force command (JFC), but it should be large enough for component commanders to accomplish their mission and protect their forces. Army commanders use control measures to describe AOs and to design them to fit the situation and take advantage of the joint force's capabilities. Commanders typically subdivide the assigned AO by assigning subordinate-unit areas. These subordinate-unit areas may be contiguous or noncontiguous (see Figure 1-2). When friendly forces are contiguous, a boundary separates them. When friendly forces are noncontiguous, the concept of operations links the force's elements, but the AOs do not share a boundary. The intervening area between noncontiguous AOs remains the responsibility of the higher headquarters.

## BATTLEFIELD ORGANIZATION

1-11. Battlefield organization is the arrangement of forces according to purpose, time, and space to accomplish a mission. Battlefield organization has both a purpose- and spatial-based framework. The purpose-based framework centers on decisive, shaping, and sustaining (DSS) operations. Purpose unifies all elements of the battlefield organization by providing the common focus for all actions. However, forces act in time and space to accomplish a purpose. The spatial-based framework includes close, deep, and rear areas. Despite the increasing nonlinear nature of operations, there may be situations where commanders describe DSS operations in spatial terms. Typically, linear operations involve conventional combat and concentrated maneuver forces. Ground forces share boundaries and orient against a similarly organized enemy force. In such situations, commanders direct and focus simultaneous DSS operations in deep, close, and rear areas, respectively (see FM 3-0).

1-12. The MP battlefield organization supports every Army echelon, from the Army service component command (ASCC) and the theater-support command

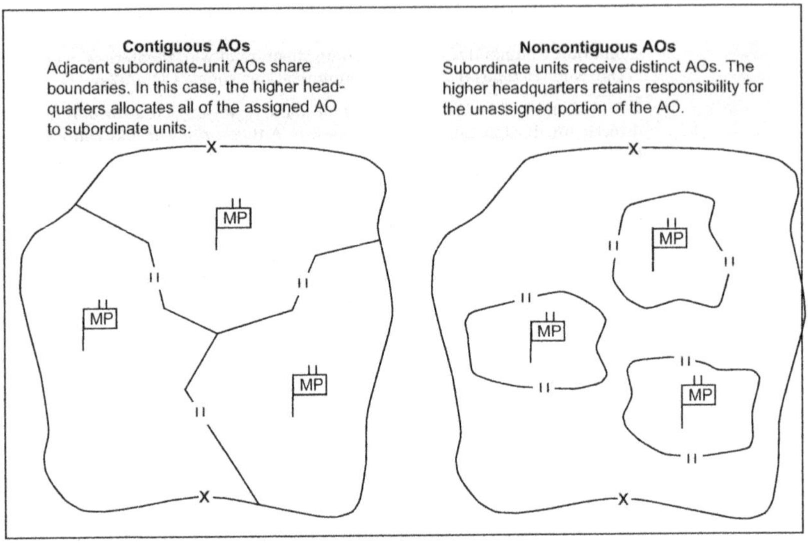


**Figure 1-2. Contiguous and Noncontiguous AOs**

(TSC) to the maneuver brigade. Regardless of the battlefield organization (purpose or spatial based), MP support to the Army commander is based on available resources and mission, enemy, terrain, troops, time available, and civilian considerations (METT-TC).

1-13. MP support throughout the theater of war may include MP units in the JOA and in the TO. If the combatant commander designates a COMMZ and a CZ within his TO, MP support will come from the established MP modified table of organization and equipment (MTOE) at the subordinate-command echelon. MP support to the JOA is also provided based on METT-TC and available MP assets. Typical MP support may include an I/R brigade liaison detachment (BLD), MP brigades and battalions, a division MP company, a military-working-dog (MWD) team, an L&O team, and a customs team. Figure 1-3, page 1-6, depicts a typical MP organization throughout the TO. In the COMMZ, Figure 1-3 depicts the different types of MP units that are assigned to echelons above corps (EAC) (the ASCC or the TSC). In the CZ, Figure 1-3 depicts the different types of MP units that are assigned to corps, division, and the separate brigades.

## SUPPORT IN THE COMMZ

1-14. MP support in the COMMZ is provided by an array of multifunctional MP units. The following MP units provide MP support to EAC:

- The MP brigade (I/R). The MP brigade (I/R) may augment the ASCC or the TSC during wartime. Its mission is to provide command, staff

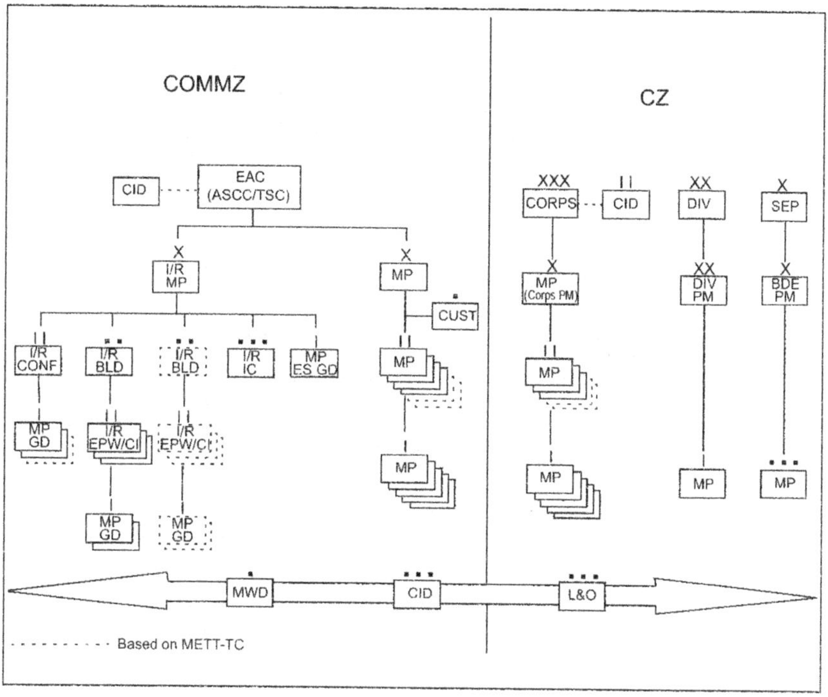


**Figure 1-3. MP Structure in the TO**

planning, and supervision of I/R operations. This includes coordination with joint and host-nation (HN) agencies, civilian police authority, nongovernmental organizations (NGOs), and US federal agencies.

- The MP brigade (CS). The MP brigade (CS) is assigned to the ASCC or the TSC during wartime (based on METT-TC). The MP brigade (CS) is capable of performing all five MP functions.
- The Criminal Investigation Division (CID) group. The CID group is a stovepipe organization that reports directly to the Commander, US Army Criminal Investigation Command (USACIDC). The CID group provides support to the ASCC and subordinate commands (TSC, corps, or division). See Chapter 9 for further discussion of CID support.

1-15. MP support to other EAC subordinate commands is performed only if MP resources are available. See Chapter 5 for further discussion of MP support to EAC.

### SUPPORT IN THE CZ

1-16. MP support is provided in the CZ to each corps, division, and brigade (separate teams or initial/interim brigade combat teams [IBCTs]). An MP brigade (CS) is assigned to each corps, and the MP brigade commander is the corps's provost marshal (PM). A PM and his section, along with an organic division MP company, are assigned to each division. A PM cell and an MP platoon are organic to a separate brigade. A two-person PM cell is organic to the IBCT. The MP units assigned to corps, divisions, and separate brigades are capable of performing all five MP functions. They provide combat, CS, and combat-service-support (CSS) operations within their command's AO.

## TYPES OF MILITARY POLICE UNITS

1-17. Most MP units supporting a TO and a JOA are capable of performing all five MP functions. However, the functions must be prioritized based on METT-TC and the availability of MP assets. Current MP structures are designed and tailored to better support the level of command deployed. For example, at the division level, division MP companies are organized as light, heavy, airborne, or air assault and are organic to their respective divisions. The EAC and corps MP brigades and battalions are equally designed to command and control a force mix of up to six battalions or companies. An MP escort-guard and guard company are designed to transport, guard, and provide security to EPWs, civilian internees (CIs), or dislocated civilians. The MP escort-guard company is assigned to the MP brigade (I/R), and the MP guard company is assigned to the MP battalion (I/R).

1-18. MP units can also be tailored and augmented to accomplish multiple, diverse, or specific missions. Customs, L&O, and MWD teams are examples of MP capabilities and flexible responses to a combatant commander's operational requirements. (See Table 1-1, pages 1-9 through 1-19, for a more complete description of MP units. See FM 19-10 and Army Regulation (AR) 190-12 for further information.) The battlefield workload analysis (BWA) is a tool used to determine the number of MP units required to perform some of these multiple missions (see Appendix C).

## JOINT, MULTINATIONAL, AND INTERAGENCY OPERATIONS

1-19. In today's environment, the Army will rarely operate or fight alone. The high probability that the Army will operate in concert with its sister services, in an alliance with the forces of foreign nations, or in support of United Nations (UN) operations (when it is committed) is fully reflected in joint doctrine. In such operations, protecting LOC, key facilities, and command and control ($C^2$) centers will be a shared responsibility. Under this framework, MP units can expect to share the AO with joint, combined, multinational, or interagency resources. MP forces must be prepared to conduct a number of full-spectrum operations with a variety of government and nongovernmental agencies, other services, allied nations, and international agencies.

1-20. Corps and division commanders and staffs must plan (in advance) the transition from a single-service headquarters with joint representation to a joint headquarters capable of functioning as a joint task force (JTF) headquarters. When tasked to form a JTF headquarters, the corps or division must ensure that all of the staff sections and agencies have joint representation (see FMs 100-15 and 71-100). To this end, MP planners must ensure that the JTF is augmented with the appropriate MP forces and with the appropriate echeloned $C^2$.

1-21. Regardless of the force mix, the MP provide the force with unparalleled, multifunctional capabilities. Among these capabilities is the MP's ability to generate firepower or to handle populations such as EPWs/CIs, dislocated civilians, and refugees. Additionally, MP expertise in investigations and law enforcement enhances the capabilities of other joint, multinational, and interagency police and security forces.

1-22. MP security plans must reflect the joint synergy derived from combining the multiple and diverse capabilities of all participants. To capitalize upon that synergy, MP leaders must keep an open line of communication and coordination to offset the challenges presented by interoperability. Some of these interoperability challenges include—

- Differing political objectives.
- Differing capabilities.
- Cultural/language differences.
- Legal and policy constraints.
- Media impacts.
- Compromise of sensitive processes, procedures, and equipment.
- $C^2$.
- Communications (digital- and analog-equipment differences).

1-23. MP plans must also accommodate differences in planning capabilities, as well as differences in doctrine, training, and equipment. The intent is to match security missions with force capabilities. MP leaders must understand that operations will often involve multinational teams. While US forces routinely task-organize, this may be more difficult to accomplish with some multinational security forces. This kind of orchestration requires employing standardized procedures, communications, equipment, and liaison within the constraints of operations security (OPSEC).

1-24. Coordination is the key to mission accomplishment in multinational and interagency operations. A military coordination center or a civil-military operations center (CMOC) may meet this coordination requirement. The CMOC provides access for nonmilitary agencies desiring military (to include MP/CID) assistance and coordination. These nonmilitary agencies may include—

- Government organizations (GOs)
- NGOs.
- International organizations (IOs).
- International humanitarian organizations (IHOs).
- HN authorities and agencies.

1-25. The introduction of US Army MP forces in any joint, multinational, or interagency operation is based on METT-TC and the capabilities they bring to the operation. Effective integration of MP forces with other security forces reduces redundant functions, clarifies responsibilities, and conserves resources.

## Table 1-1. Unit Descriptions

| MP Unit | Mission | Action Elements | Support Role | Capabilities (Include But Are Not Limited to [See TOEs for Details]) | Augmentation Needed for Mission Accomplishment |
|---|---|---|---|---|---|
| **DIVISION MP UNITS** | | | | | |
| MP Company (Airborne Division) | Perform the 5 MP functions. | A PM cell and 4 platoons. Each platoon has 3 squads and each squad has 2 three-man teams.<br>Total: 24 three-man teams. | 1 platoon deploys with each division brigade and provides DS during the assault phase. Then platoons revert to GS. | During the assault phase, 6 mounted, mobile or fixed-position teams deploy to provide DS for each brigade and 6 teams deploy to provide GS near the division main CP<br>OR<br>after the assault phase, 1 division EPW collection point (6 teams) and 18 mounted, mobile patrols or fixed-position teams provide GS<br>OR<br>24 mounted, mobile or fixed-position teams are used in any combination. | The division band to provide close-in security for the division CP or to guard EPWs<br>AND<br>the corps CS MP to conduct combat operations within the division rear and provide sustained MMS, area security, special operations support, and escort for evacuating EPWs/CIs between division collection points and to the corps holding area<br>AND<br>the corps L&O MP and CID to conduct sustained L&O operations and criminal investigations<br>AND<br>division/corps transportation assets for evacuation of EPWs<br>**NOTE: Both heavy division companies require corps CS MP augmentation for each forward support platoon (1 corps MP squad per forward support platoon).** |
| MP Company (Light Infantry Division) | Perform the 5 MP functions. | A PM cell and 3 platoons. Each platoon has 3 squads and each squad has 2 three-man teams.<br>Total: 18 three-man teams | Provide GS for all units in the AO. | Provide screening security for the division main CP (6 teams), the division EPW collection point (6 teams), and 6 mounted, mobile or fixed-position teams. | |
| MP Company (Heavy Division)*<br>*19333F000 | Perform the 5 MP functions. | A PM cell and 2 division support platoons. Each platoon has 3 squads and each squad has a squad leader and 3 three-man teams<br>AND<br>3 forward support platoons. Each platoon has 2 squads and each squad has a squad leader and 3 three-man teams.<br>Total: 36 three-man teams | 2 platoons provide GS and 3 platoons provide DS. | Provide screening security for the division main CP (6 teams), the division EPW collection point (6 teams), and 6 mounted, mobile or fixed-position teams<br>AND EITHER<br>3 DS platoons providing their brigade with either 1 forward EPW collection point (3 teams) and 3 mounted, mobile or fixed-position teams or 6 teams in any combination<br>OR<br>18 mounted, mobile or fixed-position teams are used in any combination. | |
| MP Company (Heavy Division)*<br>*19333L000 | Perform the 5 MP functions. | A PM cell and 5 platoons. Each platoon has 2 squads and each squad has 3 three-man teams.<br>Total: 36 three-man teams | 3 platoons provide GS and 3 platoons provide DS. | Provide screening security for the division main CP (6 teams), the division EPW collection point (6 teams), and 6 mounted, mobile or fixed-position teams<br>AND EITHER<br>3 DS platoons providing their brigade with either 1 forward EPW collection point (3 teams) and 3 mounted, mobile or fixed-position teams or 6 teams in any combination<br>OR<br>18 mounted, mobile or fixed-position teams are used in any combination. | |

Table 1-1. Unit Descriptions (continued)

| MP Unit | Mission | Action Elements | Support Role | Capabilities (Include But Are Not Limited to [See TOEs for Details]) | Augmentation Needed for Mission Accomplishment |
|---|---|---|---|---|---|
| MP Company (Air Assault Division) | Perform the 5 MP functions. | A PM cell and 4 platoons. Each platoon has 3 squads and each squad has 2 three-man teams. Total: 24 three-man teams | Provide GS for all units in the AO. | 24 mounted, mobile or fixed-position teams are used in any combination. | (See augmentation needs on previous page.) |
| **MP COMBAT SUPPORT UNITS** | | | | | |
| HHC MP Brigade | Provide C2 and coordinate the combat, CS, and CSS of all MP elements assigned or attached. | Command and staff elements and long-range planning section | Provide command, control, planning, and supervision for up to 6 MP battalions and other assigned or attached subordinate HQ. | Provide continuous command, control, coordination, planning, and supervision for subordinate units. | None |
| HHD MP Battalion (Combat Support) | Provide C2 and coordinate the combat, CS, and CSS of all MP elements assigned or attached. | Command and staff elements plus a support element | Provide command, control, planning, and supervision for up to 6 MP companies and other assigned or attached subordinate HQ. | Provide continuous command, control, coordination, planning, and supervision for subordinate units. | None |
| MP Company (Combat Support) | Perform the 5 MP functions. | 4 platoons with 3 squads. Each squad has a squad leader and 3 three-man teams. Total: 36 three-man teams | Provide GS for all units in the AO. | Provide security for a unified or combined HQ or higher HQ or 1 main CP and 1 tactical CP OR operate 1 EPW holding area (9 teams) and 27 mounted, mobile or fixed-position teams for use in any combination. | Corps band assets to guard EPWs AND corps transportation for evacuation of EPWs AND external MP to escort EPWs |
| MP Company (Combat Support) | Perform the 5 MP functions. | 4 platoons with 3 squads. Each squad has a squad leader and 3 three-man teams. Total: 36 three-man teams | Provide GS for all units in the AO. | Provide security for a unified or combined HQ or higher HQ or 1 main CP and 1 tactical CP OR operate 1 EPW holding area (9 teams) and 27 mounted, mobile or fixed-position teams for use in any combination. | Corps band assets to guard EPWs AND corps transportation for evacuation of EPWs AND external MP to escort EPWs |
| MP Company Arctic Support Augmentation Detachment | Augment MP units when operating in an arctic environment. | 1 mechanic (63B10) | Provide mobility support to MP units. | Provide support as designated by the commander. | None |
| MP Detachment (C2) | Provide command and administrative personnel for MP law enforcement teams. | 19503LA has a platoon HQS; 3 individuals. 19503LD has a commander and support personnel; 8 individuals. | Provide L&O. 19503LA provides platoon-level C2. 19503LD provides company-level command, control, and support. | Commands and controls law enforcement teams, investigations teams, MWD teams, and physical-security teams. | None |

## Table 1-1. Unit Descriptions (continued)

| MP Unit | Mission | Action Elements | Support Role | Capabilities (Include But Are Not Limited to [See TOEs for Details]) | Augmentation Needed for Mission Accomplishment |
|---|---|---|---|---|---|
| MP Detachment (Old Guard) | Perform the 5 MP functions. | 4 squads. Each squad has 3 three-man teams.<br>Total: 12 teams | Support the HHC old guard battalion. | Provide support as directed by the commander. | None |
| **MP LAW AND ORDER (L&O) UNITS** | | | | | |
| MP HQ Team | Provide C2, personnel administration, and logistical support to attached L&O augmentation teams. | Commander and support personnel | Provide L&O. | Provide C2, personnel administration, and logistical support to attached L&O augmentation teams. | None |
| MP Operations Team | Provide mission supervision, staff planning, and technical supervision required to support the L&O mission. | L&O operations officer and operations staff | Provide L&O. | Provide overall mission supervision, planning, employment, and coordination of support. Provide technical supervision and administrative support of MPI, TAI, and force protection MP. Provide an evidence custodian/police intelligence NCO to be responsible for confiscated property used in criminal offenses and to receive and collect the initial analysis of criminal/operational information/intelligence. | None |
| MP Desk or Desk and Record Team | Serve as the primary control point for L&O-related incidents and operations initiated by MP patrols. | 2 three-man teams | Provide L&O. | Receive/record complaints, dispatch investigative persons, maintain control of offenders/detained persons, maintain the status of investigations, prepare reports, and provide criminal data to the police intelligence NCO for analysis. | None |
| MP Traffic Accident Investigations Team | Provide the technical capability to investigate traffic accidents. | 1 two- or three-man team:<br>19517AD00 has 1 two-man team.<br>19523LE00 has 1 three-man team. | Provide L&O. | Augment MP capabilities to enforce MSR regulations; man checkpoints, roadblocks, and dismount points; patrol traffic areas; and perform traffic escort duties. | The 19517AD00 accident investigation team is dependent on the supported unit for transportation. |
| MP Investigation Team | Provide the technical capability to investigate criminal incidents and conduct surveillance operations. | One, two, or three-man teams:<br>19517AE has 1 two-man team.<br>19533LA has 1 one-man team.<br>19533LB has 1 one-man team.<br>19533LC has 1 three-man team. | Provide L&O. | Investigate criminal incidents, conduct surveillance, work with HN military and civilian police, and collect criminal and operational intelligence.<br>19533LA00 provides supervision for the teams.<br>19533LB00 provides investigative support for populations of not less than 700 troops.<br>19533LC00 provides investigative support for populations of not less than 2,100 troops. | Dependent on the supported unit for transportation |

**Table 1-1. Unit Descriptions (continued)**

| MP Unit | Mission | Action Elements | Support Role | Capabilities (Include But Are Not Limited to [See TOEs for Details]) | Augmentation Needed for Mission Accomplishment |
|---|---|---|---|---|---|
| MP Force Protection and Physical-Security Teams | Provide the expertise and technical capability to assist units in safeguarding personnel, equipment, and facilities. | One, two, or three-man teams:<br>19517AF has 1 two-man team.<br>19593LA has 1 three-man team.<br>19593LB has 1 one-man team. | Provide L&O and area security. | Conduct vulnerability assessments and identify and prioritize critical facilities and key terrain within the AO. Identify mission-essential activities that are vulnerable to criminal acts or disruptive activities. Ensure that activities are inspected to determine if safeguards are adequate. Assist base/base cluster commanders with the development of internal defense plans. | The 19517AF is dependent on the 19517AB operations team for transportation. |
| MP Detachment (Patrol Supervision Team) | Provide supervision for 3 MP teams. | 1 MP NCO | Provide L&O. | Plan, direct, and supervise the employment of assigned or attached MP teams. | Dependent on the supported unit for transportation |
| MP Detachment (Motor and Dismounted Patrol Team) | Perform L&O operations. | 1 three-man team | Provide L&O. | Protect designated personnel or facilities. Perform route and area reconnaissance, enforce MSR regulations, provide refugee and straggler control, and disseminate information. Perform peacekeeping operations to maintain surveillance over an area, observe activities, and report findings. Preserve or establish L&O. Apprehend absentees or deserters (US military personnel) in conjunction with civil law enforcement agencies. | The dismounted patrol team is dependent on the supported unit for transportation, when required. |
| MP Detachment (Registration Team) | Perform vehicle and firearm registration. | 1 three-man team | Provide L&O. | Service up to 10,000 personnel for registration of individuals, vehicles, and firearms on a closed post or area. Issue credentials prescribing limits of circulation and privileges. Process up to 50 personnel daily for fingerprints and photography services. | Dependent on the supported unit for transportation |
| Senior Military Customs Inspector Supervisor | Perform technical supervision, staff planning, and coordination for customs inspection teams. | 1 MP NCO | Provide L&O. | Supervise two senior military customs teams to ensure that personnel, equipment, and material meet customs, immigration, Department of Agriculture, and other federal agency requirements for units and personnel redeploying to the US. | None |

**Table 1-1. Unit Descriptions (continued)**

| MP Unit | Mission | Action Elements | Support Role | Capabilities (Include But Are Not Limited to [See TOEs for Details]) | Augmentation Needed for Mission Accomplishment |
|---|---|---|---|---|---|
| Senior Military Customs Inspection Team | Conduct customs inspections and train and supervise redeploying unit personnel to augment US customs inspectors. | 1 five-man team | Provide L&O. | Conduct inspections and advise redeploying units. Train selected individuals from the redeploying unit to augment US customs efforts.<br><br>NOTE: One team supports one port of embarkation. | The redeploying unit undergoing customs inspection, to augment US Customs efforts. The number of personnel is proportional to the size of the unit. |
| **MILITARY WORKING DOG (MWD) UNITS** | | | | | |
| MWD Kennel Master Team | Supervise MWD teams. | One- or two-man team.<br>19537AA has 1 two-man team.<br>19583LG has 1 one-man team. | Provide L&O, area security, and I/R operations support. | Supervise, plan, and coordinate MWD operations and support requirements. Provide MWD handler/ canine proficiency certification. | None |
| MWD Explosive, Narcotics, and Patrol Team | Detect explosives and controlled substances and search for, detect, and control personnel in support of crime scene searches. Provide personal protection, MOUT, health and welfare inspections, and I/R and customs operations. | One-man/one-dog team.<br>The 19537A000-series TOEs have 3 teams each.<br>The 19583L000-series TOEs have 1 team each. | Provide L&O, area security, and I/R support. | The 19537A-series dog teams are capable of providing a 24-hour explosive, narcotics, and tracking handler/dog requirement or three concurrent short-duration missions each requiring 1 handler/dog team. The duration of the missions will vary based on the climate, the environment, and the individual dog's ability. The 19583L-series dog teams provide short- duration missions based on the dog's ability. Explosive and narcotics teams provide MWD patrol support when not employed with explosive and narcotics detection. | None |
| **MILITARY POLICE INTERNMENT/RESETTLEMENT (I/R) UNITS** | | | | | |
| MP Command (I/R) | Provide command, control, staff planning, and supervision of I/R operations performed by all assigned or attached elements. | Command and staff elements | Provide I/R operations support. | Provide command, control, and staff planning for 2 or more MP brigades. | MP IRIC<br>AND<br>MP EPW evacuation detachment |

### Table 1-1. Unit Descriptions (continued)

| MP Unit | Mission | Action Elements | Support Role | Capabilities (Include But Are Not Limited to [See TOEs for Details]) | Augmentation Needed for Mission Accomplishment |
|---|---|---|---|---|---|
| MP Brigade (I/R) | Provide command, control, and staff planning for I/R operations performed by all assigned or attached MP elements. | Command and staff elements | Provide I/R operations support. | Provide command, control, and staff planning for 2 to 7 MP I/R battalions or up to 21 MP I/R battalions when augmented by 7 brigade liaison detachments (BLDs). Plan and provide staff supervision of I/R collection and evacuation operations. Coordinate with HN military territorial organizations and civilian police authorities, nongovernment organizations, private volunteer organizations, and US federal agencies on I/R matters. Provide coordination and support for out-of-theater evacuation of EPWs, if required. Support posthostilities operations. | MP IRIC<br>AND<br>MP EPW evacuation detachment<br>AND<br>MP I/R BLD (TOE 19453LD00)<br><br>**NOTE: The BLD expands the staff planning and coordination capabilities on a ratio of 1 BLD to 3 MP I/R battalions.** |
| MP Battalion (I/R) | Provide command, staff planning, administration and logistical support for the operation of an I/R facility. | Command and staff elements | Provide I/R operations support. | Operate an internment facility for either EPWs/CIs or DCs or a confinement facility for US prisoners. (Never more that one category at the same time.) | MP detachment (I/R) (EPW/CI) for EPW/CI/DC internment missions<br>AND<br>MP detachment (I/R) (confinement) for US prisoner confinement missions<br>AND<br>MP company (EG) for security of the evacuation and/or movement of US prisoners or EPWs/CIs/DCs<br>AND<br>MP company (guard) for prisoner guard services<br>AND<br>Corps/EAC transportation assets for movement of EPWs/CIs/DCs |

**Table 1-1. Unit Descriptions (continued)**

| MP Unit | Mission | Action Elements | Support Role | Capabilities (Include But Are Not Limited to [See TOEs for Details]) | Augmentation Needed for Mission Accomplishment |
|---|---|---|---|---|---|
| MP Detachment (I/R) (EPW/CI) | Augment the MP battalion (I/R) to provide supervision, administration, combat health support, and logistical support for operating an internment facility interning EPWs/CIs or housing DCs. | 2 compound control sections (2 three-man teams), a work project section (1 two-man team), a personnel section, and a supply section. | Provide I/R operations support. | Provide command, control, and supervision of up to 1,000 EPWs/CIs or 2,000 DCs. Provide supply and subsistence support (to include supervision of food preparation) for up to 1,000 EPWs/CIs or 2,000 DCs. Direct the activities related to assigning and supervising work projects for EPWs/CIs. Augment the MP I/R battalion in processing and maintaining records for up to 1,000 EPWs/CIs or 2,000 DCs. | MP company (guard) for prisoner guard services |
| MP Company (Escort Guard) | Provide supervisory and security personnel for evacuating and/ or moving EPWs/CIs. | 4 platoons with 3 squads each. Each squad has a squad leader and 3 three-man teams.<br>Total: 36 teams | Provide I/R operations support. | Provide security for the movement of the following numbers of POWs or CIs by the methods indicated:<br>Marching - 1,000 to 1,500.<br>Vehicle - 1,500 to 2,000.<br>Rail - 2,000 to 3,000.<br>Air - 2,125 to 2,406. | Corps/EAC transportation assets for movement of EPWs/CIs |
| MP Company (Guard) | Provide guards for EPWs/CIs or US prisoners, installations, and facilities. | 3 platoons with 3 squads each. Each squad has a squad leader and 2 five-man teams<br>Total: 18 teams. | Provide I/R operations support. | Provide security for a confinement facility containing up to 500 US military prisoners. Provide guards for securing 1 EPW/CI compound containing up to 2,000 EPWs/CIs or 4,000 DCs. Provide security guards for 3 railway terminals, each having up to 8 tracks. (Includes EPW/CI rail movement operations and protection of sensitive material within the terminal.) Provide guards for 1 military installation or facility up to 240,000 square yards in size, containing sensitive material. (This unit can provide guards for material transit.) | None |
| MP I/R Processing Squad | Process EPWs/ CIs. | 1 squad leader and 1 eight-man processing squad | Provide I/R operations support. Augment the I/R battalion processing capability to meet operational requirements. | Provide processing capabilities of about 8 EPWs/CIs per hour. | None |

### Table 1-1. Unit Descriptions (continued)

| MP Unit | Mission | Action Elements | Support Role | Capabilities (Include But Are Not Limited to [See TOEs for Details]) | Augmentation Needed for Mission Accomplishment |
|---|---|---|---|---|---|
| I/R Brigade Liaison | Expand the MP brigade (I/R) or ASCC C2 capabilities. | Liaison officer and support staff | Provide I/R operations support. | When assigned to the MP I/R brigade, provide staff augmentation that expands the brigade's staff planning, coordination, and C2 capabilities for 3 MP I/R battalions. When assigned to the ASCC, provide I/R staff augmentation and a liaison link with allied/HN forces to ensure that the care and handling of US captured EPWs/CIs is in compliance with the Geneva Conventions. | None |
| I/R Camp Liaison Team | Provide continuous accountability of EPWs/CIs captured by US forces and transferred to an allied/HN for internment. | Liaison officer and support staff | Provide I/R operations support. | Provide advice, as requested, to commanders and staffs of allied/HN-operated internment facilities. Verify arrival, forward records, and provide continuous accountability for US captured EPWs/CIs interned in allied/HN facilities. Monitor prisoner treatment to ensure compliance with the Geneva Conventions. Receive/certify allied/HN requests for reimbursement of expenses associated with interning EPWs/CIs captured by US forces. | None |
| I/R Processing Liaison Team | Process and verify the transfer of EPWs/CIs captured by US forces to a allied/HN facility. | Liaison officer and support staff | Provide I/R operations support. | Provide processing and transfer capability for US-captured prisoners to an allied/HN EPW/CI facility. | None |
| MP Detachment (EPW Evacuation) | Provide evacuation support of EPWs. | EPW coordination officer and staff plus 10 one-man POD/POE teams and 3 two-man POE teams | Provide I/R operations support. | Coordinate administrative, logistical, and transportation support for up to 7 PODs and 6 POEs used for out-of-theater EPW evacuation and the escort guard personnel that support the movements. Coordinate for theater security at PODs/POEs and intermediate refuel points. | Responsible POD/POE security forces |

**Table 1-1. Unit Descriptions (continued)**

| MP Unit | Mission | Action Elements | Support Role | Capabilities (Include But Are Not Limited to [See TOEs for Details]) | Augmentation Needed for Mission Accomplishment |
|---|---|---|---|---|---|
| MP Detachment (Internment Resettlement Information Cener [IRIC]) | Provide a US central agency in each theater of war to receive, process, maintain, disseminate, and transmit the required information and data relating to EPWs, American POWs, CIs, and DCs within the theater. | Command and staff personnel | Provide I/R operations support. Serve as the single source for collection and storage of EPW/CI information in theater. Forward information to the National Prisoner of War Information Center at DA. | Collect, process, and disseminate (to authorized agencies) information regarding interned personnel detained in theater and those released to the custody of allied/HN authorities within theater. Receive, document, and disseminate to the theater commander information received regarding American POWs, CIs, and foreign nationals who are captured, missing, or otherwise detained. Receive, store, and dispose of personal property belonging to interned personnel who have died, escaped, or been repatriated and any property belonging to enemy soldiers killed in action that is not disposed of through grave registration channels. | None |
| MP Detachment (I/R) (Confinement) | Augment an MP battalion (I/R) in operating a confinement facility. Provide supervision, administration, combat health support, and logistical support of US military prisoners. | Corrections officer and confinement facility staff | Provide I/R operations support. | Provides a control team for the command, control, and supervision of US military prisoners and a disciplinary guard team to supervise custodial personnel for a maximum custody of 500 prisoners. Provide administration, health service, supply, and food service personnel to augment the battalion in providing these services for internees. (The food service personnel supervise and train US military prisoners working in the internee dining facility.) | MP battalion (I/R) for command, staff, planning, and operational support<br>AND<br>MP company (guard) for prisoner guard services |

**Table 1-1. Unit Descriptions (continued)**

| MP Unit | Mission | Action Elements | Support Role | Capabilities (Include But Are Not Limited to [See TOEs for Details]) | Augmentation Needed for Mission Accomplishment |
|---|---|---|---|---|---|
| MP Confinement Facility Detachment | Provide C2, staff planning, administration, and logistical support for a confinement facility for US military prisoners. | Commander and facility staff | Provide I/R operations support. | Provide command, control, and administrative support for a confinement facility. | MP guard company to provide exterior perimeter security and required guard support to the confinement facility<br>AND<br>MP correctional team supervisor (TOE19553LE) to provide supervision of the guard force<br>AND<br>MP correctional team work supervisor (TOE 19553LF) to provide supervision of prisoners |
| MP Correctional Team Supervisor | Provide correctional supervision of prisoner work guard forces within a confinement facility. | 2 two-man teams | Provide I/R operations support. | Provide correctional supervision to guard force personnel and US military prisoners within a modular-configured confinement facility. | None |
| MP Correctional Team Work Supervisor | Provide direct control and supervision of military prisoners within a confinement facility. | 1 two-man team | Provide I/R operation support. | Provide correctional control, supervisory functions, and escort duties for military prisoners within a modular-configured confinement facility. | None |
| **MILITARY POLICE CRIMINAL INVESTIGATION (CID) UNITS** | | | | | |
| MP Group (CID) | Provide command, control, staff planning, and supervision for all CID elements within the theater. | Command and staff elements | Provide L&O. | Provide command, control, staff planning, and supervision over all CID elements within the theater.<br>**NOTE: The Criminal Investigation Command is a stovepipe organization.** | None |

**Table 1-1. Unit Descriptions (continued)**

| MP Unit | Mission | Action Elements | Support Role | Capabilities (Include But Are Not Limited to [See TOEs for Details]) | Augmentation Needed for Mission Accomplishment |
|---|---|---|---|---|---|
| MP Battalion (CID) | Provide command, control, staff planning, and supervision for all assigned or attached CID elements. | Command and staff elements | Provide L&O. | Provides command, control, staff planning, and supervision over all assigned and attached CID elements. | None |
| MP Detachment (CID) HQ Cell | Provide C2, evidence custody control, and investigative administration support. Coordinate for personnel administration and logistic support. | Commander and support staff | Provide L&O. | Provide C2 of assigned CID SA sections and/or teams performing all CID operations in their area of responsibility. | None |
| MP Detach-ment (CID) | Provide criminal investigative support to Army commanders at all echelons. | 2-man teams. The number of teams varies by organization and is as follows:<br>DSE heavy - 4<br>DSE light - 3<br>Section A - 4<br>Section B - 4<br>Supervisory team - 1<br>Senior team - 1<br>SA team - 1 | Provide L&O. | Supervise and conduct criminal investigations. | When directed, MP CS, L&O, and MWD support |

## Chapter 2

# Battle Command

*Command is the authority a commander in military service lawfully exercises over subordinates by virtue of rank and assignment. Leaders possessing command authority strive to use it with firmness, care, and skill.*

*FM 101-5-1*

Battle command is the exercise of command in an operation against a hostile, thinking opponent. Battle command includes visualizing the current state and the desired end state, then formulating concepts of operations to get from one state to the other at the least cost. In addition to visualizing and formulating concepts, battle command encompasses assigning missions; prioritizing and allocating resources; selecting the critical time and place to act; and knowing how and when to make adjustments in the fight. Battle command enables MP commanders to lead, prioritize, and allocate assets required in support of the Army commander. MP commanders must observe, orient, decide, and act on their decisions quickly. Information is the key element in the battle-command process; therefore, the commander must have accurate and timely information upon which to base his decisions.

## OVERVIEW

2-1. The battle command of MP units is typically decentralized due to the nature of their CS functions, METT-TC, and the needs of the Army commander. This places the burden of sound, timely decision making to the lowest levels. MP leaders must develop a keen sense of situational awareness and visualization, and they must constantly track the actions of supported units.

## BATTLEFIELD VISUALIZATION

2-2. The ability to visualize the battlefield is a critical element of battle command. Battlefield visualization is an essential leadership attribute and is critical to accomplishing the mission. It is learned and attained through training, practice, experience, technical and tactical knowledge, and available battle-command technologies. It results when the MP commander understands the higher commander's intent, his assigned mission, the enemy, and the friendly force's capabilities and limitations. See Appendix D for further information on command technologies.

2-3. Battlefield visualization includes the MP commander's view of what his forces will do and the resources needed to do the mission. He envisions a sequence of actions that will cause his MP forces to perform at the desired end state. Ultimately, the MP commander's battlefield vision evolves into his intent and helps him develop his concept of operations.

## COMMANDER'S INTENT

2-4. The commander's intent is a key part of Army orders. It is a clear, concise statement of what the force must do to succeed with respect to the enemy, the terrain, and the desired end state. It provides the link between the mission and the concept of operations by stating key tasks. These tasks, along with the mission, are the basis for subordinates to exercise initiative when unanticipated opportunities arise or when the original concept of operations no longer applies. MP leaders at all echelons must ensure that the mission and the commander's intent are understood two echelons down (see FM 101-5).

## CONCEPT OF OPERATIONS

2-5. The commander's intent does not include the method by which the MP units will accomplish the mission. This method is called the commander's concept of operations. It must—

- Convey the commander's vision of how to accomplish the mission in a manner that allows his subordinates maximum initiative.
- Build around intelligence gathering and the precise employment of MP resources.
- Provide the basis for task organization, scheme of maneuver, terrain organization, tasks to subordinates, and synchronization.

## COMMAND AND CONTROL RELATIONSHIPS

2-6. MP units are assigned to, attached to, or placed under the operational control (OPCON) of MP or other units they support. OPCON is the authority to perform command functions over subordinate forces. This includes organizing and employing commands and forces, assigning tasks, designing objectives, and giving authoritative direction necessary to accomplish the mission. MP $C^2$ relationships may be changed briefly to provide better support for a specific operation or to meet the needs of the supported commander. MP units may be placed under the OPCON of another unit commander for short-term operations. The MP unit remains in this relationship only as long as it is needed for that operation.

> MP support to the Bosnian municipal elections consisted of one division and two corps MP companies. These MP assets, attached to Task Force (TF) Eagle, were task-organized from different sources. The division MP company and the PM cell were organic to TF Eagle's mechanized infantry division headquarters, but the two corps MP units were from US Army Forces Command (FORSCOM) MP battalions in CONUS.

## SUPPORT RELATIONSHIPS

2-7. MP units on the battlefield provide two types of support—general support (GS) and direct support (DS). Corps and EAC MP units provide GS to their respective corps/EAC subordinate commands. Light, airborne, and air-assault MP companies provide GS to their respective divisions. Heavy-division MP companies provide GS to the division rear and DS to the division's subordinate brigades.

## STAFF RELATIONSHIPS

2-8. The PM for each level of command is that command's advisor on MP combat, CS, and CSS operations. The PM—

- Advises the commander and staff about MP abilities/capabilities.
- Supervises the preparation of plans and dictates policies.
- Coordinates MP operations.
- Assists and supervises the interaction of supporting and supported units.
- Reviews current MP operations.
- Coordinates with allied forces and HN military and civil police.
- Ensures that MP plans and operations supporting the commander's tactical plan are carried out.
- Recommends when and where to concentrate the command's MP assets.
- Supervises or monitors MP support in the command's AO.

2-9. The PM works daily with the commander and staff officers who employ MP resources and whose AORs influence MP support. The PM works closely with the coordinating staff at the appropriate command level to coordinate MP support. He ensures that MP planning is practical and flexible, that plans are coordinated with staff sections and subordinate commands, and that plans reflect manpower and resources needed by MP. (This includes the need for $C^2$, fire support, equipment, and supplies. It also includes construction, communication, transportation, and aviation support.) As new information is received, the PM reviews, updates, and modifies the plans. He ensures that the echelon commander gets the necessary MP support.

2-10. In the absence of specific directions or orders, the PM plans the use of MP assets. He evaluates the current operations and projects the future courses of action (COAs). He bases his plans on assumptions consistent with the commander's intent and a thorough knowledge of the situation and mission. The PM considers—

- METT-TC.
- Current estimates developed by the intelligence preparation of the battlefield (IPB) and the police information assessment process (PIAP).
- The environment within the AO. This includes the climate, the terrain, and obstacles. It also includes the legal authority and status of the force; the width, depth, size, and location of built-up areas; and the attitudes and abilities of the local populace.

- The types of units operating in the area (to include joint, combined, multinational, and interagency units) and the missions and capabilities of these units. This knowledge is imperative to understand their capability to counter threats in their area.
- The specific missions of MP units in the area and the impact that rear-area security operations will have on the ability of these units to perform other functions.
- Personnel, vehicles, and equipment in the MP units.

2-11. Coordination and communication between the PM and Army commanders is essential. Such actions ensure timely and efficient MP support to all levels of command during any operation. The informal, technical chain of coordination is an open line of communication between PMs at different echelons. The informal chain of coordination fosters cooperation and help among the MP elements at each echelon. For instance, when the division PM needs more assets to accomplish added missions, he initiates coordination with the corps PM. If the corps PM can provide support, the division PM formalizes his request for assistance through the division Assistant Chief of Staff, G3 (Operations and Plans) (G3).

# Chapter 3

# The Threat

*In the 40-odd years of the Cold War, in many locations around the world, the Army performed a deterrent role as part of the containment strategy. In other places, at other times, the Army fulfilled the Nation's expectation in operations too small to be called "wars," although no less dangerous. To the soldier on the ground, Operations Urgent Fury in Grenada and Just Cause in Panama were indistinguishable from combat operations of their forefathers. Operations Provide Comfort in Iraq and Restore Hope in Somalia, although peace operations, also proved to be dangerous.*

*FM 100-1*

The end of the Cold War has reduced, but not eliminated, the most immediate threat to the security of the US and other western nations. However, the absence of a dominant, identifiable threat has produced a far more complex and confusing strategic environment than the one that was present during the Cold War. Forward-deployed and CONUS-based ARFOR and civilians are and will continue to be engaged in a range of military actions. These actions stem from deterring conflicts to conducting peacetime engagement operations to providing support to civil agencies at home and abroad.

## OVERVIEW

3-1. During the past decade, the US has deployed forces in multiple operations that have included crisis response in combat situations as well as participation in noncombat activities. The Army's presence in South Korea, Saudi Arabia, and Kuwait and its deployments to Somalia, Haiti, Bosnia, and Kosovo are clear indicators that the military must be prepared to face not only the traditional threat, but also a nontraditional, nonecheloned enemy. To support Army commanders successfully, MP leaders must understand the nature and complexity of these threats and how they can potentially affect the desired strategic, operational, and tactical end states.

## REAR-AREA AND SUSTAINMENT OPERATIONS

3-2. The rear area for any particular command is the area extending forward from its rear boundary to the rear of the area assigned to the next lower level of command. This area is provided primarily for the performance of support functions. Operations in the rear area assure freedom of action and continuity of operations, sustainment, and $C^2$. Sustainment operations are those that enable shaping and decisive operations by assuring freedom of action and continuity of operations, CSS, and $C^2$ (see FM 3-0). Sustainment operations include the following elements:

- CSS.
- Rear-area and base security.
- Movement control.
- Terrain management.
- Infrastructure development.

3-3. During the Cold War, the danger to rear areas included forces that would be deployed in support of major soviet-style operations. The adversaries using the soviet model could be expected to engage in intense combat activity in their enemy's rear area. Their forces were prepared to penetrate into the enemy's rear and to attack and destroy its reserve forces and rear-area installations. To protect the rear areas, the MP were among the first mobile fighting forces available to the battlefield commander and thus, a source of combat power. Today, the Army commander uses the MP's flexibility and their modular-force training, adaptability, and mobility to serve as a combat multiplier throughout his entire AO. During sustainment operations, the MP perform all functions to ensure freedom of maneuver in support of the overall operational effort.

3-4. Failure to protect our forces during sustainment operations normally results in failure of the entire operation. Sustainment operations determine how fast ARFOR reconstitute and how far they can exploit success. The likelihood of MP units encountering the enemy and engaging in direct combat (not only in the rear area, but also during sustainment operations) cannot be underestimated.

3-5. Threats to rear-area and sustainment operations exist throughout the full spectrum of military operations. These threats may be related or independently engaged, but their effects are frequently cumulative. Threats to rear-area and sustainment operations are usually theater-dependent and are not limited to those outlined in this manual. Joint Publication (JP) 3-10 further discusses the threat in the rear area. Although JP 3-10 defines the threat in the context of a JRA, MP leaders can expect the same level of activity anywhere that US forces are deployed.

## RECEPTION, STAGING, ONWARD MOVEMENT, AND INTEGRATION OPERATIONS

3-6. Reception, staging, onward movement, and integration (RSOI) operations consist of essential and interrelated processes in the AO that transform arriving personnel and materiel into forces capable of meeting operational requirements. During RSOI operations, the threat encountered will depend mostly on the type of entry, the nature of the operation, and the enemy. During major contingencies, forces deploy from power-projection platforms within the US or forward bases. The PM must plan MP support during the initial stages of the deployment to ensure the protection of follow-on forces and the detection of potential threats (see FM 100-17-3).

3-7. MP support to RSOI operations includes, but is not limited to—

- Conducting AS operations to counter or prevent enemy actions against marshalling and staging areas.
- Conducting convoy, airport, and rail security operations.
- Conducting populace- and resource-control operations.
- Conducting other physical-security and force-protection measures.
- Conducting other MP functions (as determined by the PM).

## THREAT LEVELS

3-8. The threat is divided into three levels. These levels provide a general description and categorization of threat activities, identify the defense requirements to counter them, and establish a common reference for planning guidelines. MP leaders must understand that this does not imply that threat activities will occur in a specific sequence or that there is a necessary interrelationship between each level.

### LEVEL I

3-9. Level I threats include the following types of individuals or activities:

- **Enemy-controlled agents.** Enemy-controlled agents are a potential threat throughout the rear area. Their primary missions include espionage, sabotage, subversion, and criminal activities. Their activities span the range of military operations and may increase during both war and military operations other than war (MOOTW). These activities may include assassinating or kidnapping key military or civilian personnel or guiding special-purpose individuals or teams to targets in the rear area.
- **Enemy sympathizers.** Civilians sympathetic to the enemy may become significant threats to US and multinational operations. They may be the most difficult to neutralize because they are normally not part of an established enemy-agent network, and their actions will be random and unpredictable. During war and MOOTW, indigenous groups sympathetic to the enemy or those simply opposed to the US can be expected to provide assistance, information, and shelter to guerrilla and enemy unconventional or special-purpose forces operating in the rear area.
- **Terrorism.** Terrorists are among the most difficult threats to neutralize and destroy. Their actions span the full spectrum of military operations.
- **Civil disturbances.** Civil disturbances, such as demonstrations and riots, may pose a direct or indirect threat to military operations. Although this threat may not be of great impact during war, it may significantly change and affect MOOTW.

### LEVEL II

3-10. Level II threats include the following types of forces:

- **Guerilla forces.** Irregular and predominantly indigenous forces conducting guerrilla warfare can pose a serious threat to military forces and civilians. They can cause significant disruptions to the orderly conduct of the local government and services.

- **Unconventional forces.** Special-operations forces (SOF) are highly trained in unconventional-warfare techniques. They are normally inserted surreptitiously into the rear area before the onset of an armed conflict. They establish and activate espionage networks, collect intelligence, carry out specific sabotage missions, develop target lists, and conduct damage assessments of targets struck.
- **Small tactical units.** Specially organized reconnaissance elements are capable of conducting raids and ambushes in addition to their primary reconnaissance and intelligence-gathering missions. Small (size or capability), bypassed conventional units, as well as other potential threat forces, are also capable of conducting raids and ambushes to disrupt operations.

### LEVEL III

3-11. Level III threats are made up of conventional forces. Potential threat forces are capable of projecting combat power rapidly by land, air, or sea deep into the rear area. Specific examples include airborne, heliborne, and amphibious operations; large, combined-arms, ground-forces operations; and bypassed units and infiltration operations involving large numbers of individuals or small groups infiltrated into the rear area, regrouped at predetermined times and locations, and committed against priority targets. Level III forces may use a combination of the following tactics as a precursor to a full-scale offensive operation:

- **Air or missile attack.** Threat forces may be capable of launching an air or missile attack throughout the rear area. It is often difficult to distinguish quickly between a limited or full-scale attack before impact; therefore, protective measures will normally be based on the maximum threat capability.
- **Nuclear, biological, and chemical (NBC) attack.** Commanders must be aware that NBC munitions may be used in conjunction with air, missile, or other conventional-force attacks. The NBC weapons could also be used at Level I or II by terrorists or unconventional forces in order to accomplish their political or military objectives.

### THREAT-LEVEL MATRIX

3-12. Table 3-1 lists the threat levels and their likely appropriate responses. The threat levels listed are based on the type of threat. The table should not be construed as restricting the response options to any particular threat.

## THREAT PRIORITIES

3-13. The threat will attempt to perform the following operations against targets in the rear area:

- Detect and identify targets.
- Destroy or neutralize operational weapons-system capabilities.
- Delay or disrupt the timely movement of forces and supplies.

**Table 3-1. Threat Levels**

| Threat Level | Example | Response |
|---|---|---|
| I | Agents, saboteurs, sympathizers, and terrorists | Unit, base, and base-cluster self-defense measures |
| II | Small tactical units, unconventional-warfare forces, guerrillas, and bypassed enemy forces | Self-defense measures and response forces with supporting fires |
| III | Large tactical-force operations (including airborne, heliborne, amphibious, infiltration, and bypassed enemy forces) | Timely commitment of a TCF |

- Weaken the friendly force's $C^2$ network.
- Disrupt support to combat forces.
- Set the stage for future enemy operations.
- Create panic and confusion throughout the rear area.

3-14. Typical examples of enemy priority targets include the following:

- NBC-weapons storage sites and delivery systems.
- Key command, control, communications, computers, and intelligence (C4I) facilities.
- Air-defense artillery (ADA) sites.
- Airfields and air bases.
- Port facilities.
- Main supply routes (MSRs) and MSR checkpoints.
- Key LOC.
- Reserve assembly areas (AAs).
- Troop barracks.
- Critical civilian and logistics facilities.

## THREAT LOCATION

3-15. The fact that the Cold War has ended does not imply that our traditional threat has ended. North Korea and Iraq are constant reminders of this fact. For the near future, Army commanders will fight units with Cold-War-era equipment and tactics. The Army trains and is prepared to fight an enemy capable of interfering with our freedom of maneuver throughout the battlefield. On an extended battlefield with asymmetric threats, the danger to high-value assets (HVAs) (including CSS, $C^2$, communication nodes, and MSRs) only increases. The idea that the danger to the rear area decreases as you travel farther away from the forward edge of the battle area (FEBA) is not true. Threat intensity does not depend on geographical location; it depends on what operations the enemy believes must be initiated (and to what degree) to achieve its objective in the rear area. Military commanders depend on the MP to delay and defeat threats in their AO with a mobile reaction force.

## COMMUNICATIONS ZONE

3-16. The nature of the COMMZ will encourage Level I and II threats to concentrate along the LOC and other areas of military significance. MP units will encounter an enemy that is capable of disrupting operations throughout the COMMZ while employing terrorist activities, enemy-controlled agent activities, enemy sympathizers, and saboteurs. If the enemy is Level III capable, MP leaders must expect infiltrations and air, missile, or NBC attacks as a precursor to a major Level III operation.

## CORPS REAR AREA

3-17. The activities in Levels I and II will be similar in composition and density as in the COMMZ, but they will target key corps units, key facilities, and corps sustainment capabilities. The threat activities, especially at smaller unit levels, may even precede hostilities. MP leaders must be alert and prepared to encounter unconventional forces conducting diversionary or sabotage operations and small combat units conducting raids, ambushes, or reconnaissance operations or collecting special warfare intelligence. With the fast tempo of offensive operations, MP leaders must also be alert and prepared to encounter bypassed forces that can disrupt operations in the corps rear area.

## DIVISION REAR AREA

3-18. The division rear area (DRA) contains many types of CS and CSS units and conducts many complex operations. As in the COMMZ and the corps rear area, the full spectrum of Level I, II, and III activities may occur in the DRA. The main target will be the division's HVA (including key $C^2$ facilities; airfields; artillery, aviation, and air-defense assets; LOC; and essential CSS units). The threat may conduct diversionary attacks, sabotages, raids, ambushes, and reconnaissance operations to affect the commander's freedom of maneuver and the continuity of operations. Unlike corps MP, the likelihood of division MP encountering bypassed enemy forces is expected. Failure to delay or defeat these forces will impact division operations.

## OTHER TYPES OF THREATS

3-19. As US forces are deployed throughout the world, they will have to face nontraditional, asymmetric threats (other than those listed in Table 3-1, page 3-5) that may be geographically specific. As part of situational awareness, and in coordination with military intelligence (MI) and CID personnel, MP leaders must evaluate and assess the impact of these threats in their AO. A TO is vulnerable to any or a combination of the following threats:

- National or international organized crime.
- Narcotics traffickers.
- Narcotics terrorists.
- Extremist groups.
- Paramilitary groups.
- Ethnic or religious disputes.
- Trade in illegal weapons or strategic materials.

3-20. MP leaders must be aware that other threats exist and that they have the same potential as the Level I and II threats to disrupt operations in rear-area or sustainment operations. In some instances, the above threats' capabilities or the massing of personnel may have the same potential threat as a Level III threat.

## COUNTERING THE THREAT

3-21. The disruption of rear-area and sustainment operations directly affects military efforts. Three types of forces may be used to counter the threat in these areas—a base/base-cluster self-defense force, a response force, or a tactical combat force (TCF).

### BASE/BASE-CLUSTER SELF-DEFENSE FORCE

3-22. A base cluster is established when the appropriate echelon rear-operations cell or command post (CP) places geographically contiguous or noncontiguous bases under the control of a headquarters. The base cluster becomes the next higher tactical $C^2$ headquarters of those bases. The rear-operations cell or the rear CP may also establish a base cluster for a corps support group (CSG), an area support group (ASG), or other CSS units operating in the corps or division rear areas.

3-23. US ARFOR have the inherent responsibility to contribute as many forces as possible for base defense and local security for themselves and their facilities, installations, and activities. Each base and base-cluster commander must develop a defense plan to detect, defeat, and minimize the effects of Level I and limited Level II threat attacks on his base or base cluster (including NBC attacks). To maximize the unit's mission accomplishment, defense plans must be flexible and allow for differing degrees of security based on the probability of threat activity. Defense plans are given to MP units operating near the base or base cluster. The base commander most often employs a series of defense measures providing internal and perimeter security. His internal reaction forces use organic weapons to neutralize and defeat most low-level threat activity. Although not fully equipped to engage major conventional or unconventional enemy forces that may confront him, a base commander must deploy his personnel to defend themselves until MP, HN, local police, or combat forces (if available) can respond.

### RESPONSE FORCE

3-24. A response force is summoned when the base or base cluster is faced with threat forces that are beyond their self-defense capability. If the MP are the designated response force, they must—

- Coordinate with the supported bases or base-cluster commanders to conduct a joint IPB.
- Review base and base-cluster self-defense plans.
- Exchange signal-operating-instructions (SOI) information.
- Identify MP contingency plans to counter likely enemy activities.
- Integrate ADA, engineer, chemical, field-artillery (FA), Army-aviation, and close-air-support (CAS) fire support into their plans (if available).

3-25. MP units help the base or base cluster return to its primary mission by defeating Level II threats. MP units closely watch likely avenues of enemy approach, possible landing zones (LZs), drop zones (DZs), $C^2$ facilities, and other key installations. They accomplish this through the MMS, AS, and PIO functions.

3-26. If MP units are not the designated response force, they may become the initial response force for units within their AO. When this occurs, they block, delay or, if possible, destroy enemy elements within their capability. If the attack is by a larger or more capable force, they will maintain contact and continue to develop the situation or delay until the appropriate response force appears or the battlefield commander commits the TCF.

## TACTICAL COMBAT FORCE

3-27. When the MP response force encounters or engages threats beyond its ability to defeat, it immediately notifies the higher headquarters. The battlefield commander will then evaluate the situation and commit the TCF to defeat the Level III threat. The TCF is normally a combined-arms organization tailored by the corps or division G3, based on METT-TC. The TCF normally receives fire, aviation, or other support needed to fight and defeat the threat. Once the TCF is identified and before it is committed to battle, it will conduct direct coordination with the MP or other response forces regarding the exchange of reconnaissance information, battle-handoff procedures, and contingency plans for TCF operations. Once the TCF is committed, the MP unit performing as the response force becomes OPCON to the TCF commander.

## MP AS THE TCF

3-28. The MP brigade or battalion is capable of conducting TCF missions only when properly augmented. MP augmentation may be in the form of fires, small combat units, aviation assets, or CAS. The specific type of augmentation is METT-TC dependent. Additionally, the MP $C^2$ headquarters must receive the respective liaison officers to ensure that augmentation forces are synchronized and employed according to their capabilities. The MP commander's situational awareness and battlefield visualization are key elements to TCF operations. Once designated as the TCF, the MP unit commander establishes liaison with the appropriate rear CP to obtain—

- The current rear-area IPB.
- The friendly unit disposition.
- Defense plans.
- Priorities for protection.
- The fire-support plan.

3-29. Based on the above information, the MP commander conducts his own IPB and develops a concept of operations. He then forwards it to the appropriate higher echelon for coordination and approval.

3-30. The MP's ability to employ organic MP assets as part of the TCF is limited by the following factors:

- **MP availability.** Normally, all MP assets available are committed at all times. The specific function and scope that MP units perform during the operation are determined by the Army commander's needs, the intensity of the conflict, and the availability of MP resources. The commander, with advice from the MP leader, must decide which MP operations must be scaled back, delayed, or shifted before the MP unit can be designated as part of the TCF.
- **MP dispersion.** MP units are normally displaced over a large geographical area. Technological capabilities and mobility allow them to operate over great distances. In today's battlefield, a typical MP company employment covers between 1,000 and 1,200 square kilometers and performs numerous missions in support of all five functions. The distance between elements, the reprioritization and movement of other MP units, the difficult terrain, poor roads, and bad weather may slow down the MP's commitment as a TCF.

# Chapter 4

# Military Police Functions

*Military police support the Army commander's mission to win the battle. They help the commander shape the battlefield so that he can conduct decisive operations to destroy enemy forces, large or small, wherever and whenever the Army is sent to war.*

*MAJ(P) Anthony Cruz*

The MP Corps supports shaping and sustainment operations while performing its five basic functions as a flexible, economy-of-force organization. Through these functions, MP units are able to provide the commander with an array of CS operations across the full spectrum of military operations (see Table 4-1).

**Table 4-1. MP Functions**

| | Functions | | | | |
|---|---|---|---|---|---|
| | **MMS** | **AS** | **PIO** | **L&O** | **I/R** |
| **Subtasks** | Support to river-crossing and breaching operations and passage of lines<br>Straggler and dislocated-civilian control<br>Route R&S<br>MSR regulation enforcement | Recon operations<br>ADC<br>Base/air-base defense<br>Response-force/ TCF operations<br>Critical site, asset, and HRP security<br>Force protection/ physical security<br>Antiterrorism | IPB support<br>PIAP<br>Active and passive roles<br>Information collection and dissemination<br>Joint, interagency, and multinational coordination | Law enforcement<br>Criminal investigations<br>US Customs operations<br>Related L&O training | EPW/CI handling<br>US military prisoner handling<br>Populace and resource control<br>Dislocated civilians |
| NOTE: Subtasks not all-inclusive. | | | | | |

## OVERVIEW

4-1. MP assets are limited. Specific functions are performed at any given time and are determined by the supported commander's need, the intensity of the conflict, and the availability of MP resources. The supported commander, through the command's PM, sets the priorities for MP operations.

4-2. The PM (based on METT-TC and the available assets) continuously evaluates the trade-off between the MP support that the commander requires and the MP support that can be provided. To meet the priorities set by the commander's tactical plan, the PM recommends the allocation and employment of MP assets for MP combat, CS, and CSS operations.

# MANEUVER AND MOBILITY SUPPORT

4-3. The MMS function involves numerous measures and actions necessary to support the commander's freedom of movement in his AOR. The MP expedite the forward and lateral movement of combat resources and ensure that commanders get forces, supplies, and equipment when and where they are needed. This is particularly important in the modern battlefield where there is a greater geographical dispersal of forces and lengthened LOC.

4-4. The MP maintain the security and viability of the strategic and tactical LOC to ensure that the commander can deploy and employ his forces. The MP support the commander and help expedite military traffic by operating traffic-control posts (TCPs), defilades, or mobile patrols; erecting route signs on MSRs or alternate supply routes (ASRs); or conducting a reconnaissance for bypassed or additional routes. The MP move all units quickly and smoothly with the least amount of interference possible.

4-5. As part of the MMS function, the MP support river-crossing operations, breaching operations, and a passage of lines. They also provide straggler control, dislocated-civilian control, route reconnaissance and surveillance (R&S), and MSR regulation enforcement.

## RIVER-CROSSING OPERATIONS

4-6. US forces conduct river-crossing operations to move a large force across a river obstacle with a minimum loss of momentum. The MP play a vital role by assisting the force commander in crossing the river as quickly and efficiently as possible. The crossing is usually planned and conducted by the headquarters directing the crossing. As such, a division crossing operation is conducted by a corps. Whether a brigade or division is crossing, the division MP company may also cross to provide uninterrupted support to the division. In these instances, there is a total reliance on the corps MP to support the crossing. The same is true for breaching operations and a passage of lines.

4-7. MP support for river-crossing operations reduces the crossing time and promotes the efficient movement of vehicles. It reduces congestion, speeds the crossing, and enables the maneuver commander to continue his momentum toward his primary objective. The MP establishes staging and holding areas and TCPs to control movement to and from these areas (according to the traffic-control plan). The MP may be called on to provide security for crossing forces at the crossing sites. In most cases, the MP TCPs and engineer regulation points (ERPs) are located on both sides of the river to improve communications and coordination between the units.

4-8. MP employment for river crossing is influenced by METT-TC. The number and placement of MP assets supporting a river-crossing operation varies with the size of the crossing force, the direction of the crossing (forward or retrograde), and the degree of enemy resistance expected or encountered.

4-9. The MP operating inside the crossing areas are OPCON to the crossing-area commander for the duration of the operation. The MP operating outside of the crossing area are under the command of their appropriate echelon commander.

4-10. The main thrust of MP support to river-crossing operations is within the immediate river-crossing site. The MP direct units to their proper locations (such as holding areas and staging areas) and ensure that units move through the area within the time listed on the movement schedule. This is a highly critical aspect of river crossing because the number of crossing sites is limited. MP assets are placed where they can stress MMS operations on MSRs leading into the crossing area.

4-11. The MP also provide AS to allow crossing forces to cross the river without losing momentum or forces. On both near and far sides, the MP are used to recon the crossing unit's flanks and rear to enhance security (see FM 19-4).

## BREACHING OPERATIONS

4-12. Breaching operations are conducted to allow forces to maneuver through obstacles. Obstacle breaching is the employment of a combination of tactics and techniques to advance an attacking force to the farside of an obstacle that may be covered by fire. It is perhaps the most difficult combat task a force can encounter. Breaching operations begin when friendly forces detect an obstacle, and they end when the battle handover has occurred between the follow-on forces and a unit conducting the breaching operation (see FM 90-13-1).

4-13. The MP support breaching operations in numerous ways. MP assets are employed based on METT-TC, the available resources, and the commander's priorities. As a minimum, MP support may include, but is not limited to—

- Establishing TCPs along routes leading to or departing from the breaching site.
- Establishing holding areas.
- Establishing TCPs at the breaching site.
- Assisting engineers with temporary route signs.
- Establishing straggler-control operations.
- Conducting AS operations.

4-14. The most critical MP support is provided at the breaching site. The MP provide the commander with a means to control traffic flow to appropriate lanes. When multiple lanes branch off of a single far-recognition marker, the MP assist in directing the formation through various lanes. They also assist in modifying the traffic flow when lanes have been closed for maintenance or expansion. The MP conduct close coordination with the crossing-force commander and the TF commander executing the breaching operation. The MP enable the commander to make last-minute changes in traffic flow, thereby giving him increased flexibility to react to the enemy situation.

## PASSAGE OF LINES

4-15. A passage of lines is an operation in which a force moves forward or rearward through another force's combat positions with the intention of moving into or out of contact with the enemy. The passage of lines is a high-risk military operation that requires close coordination between the passing unit, the stationary unit, and the MP providing the support.

4-16. The MP help reduce confusion and congestion during a passage of lines. They provide security in areas surrounding passage points and passage lanes to ensure that the passing unit has priority for using routes to and through the areas. The headquarters directing the operations sets the route's priority. The MP can support a forward, rearward, or lateral passage of lines. Before the actual operation, the MP in the AO conduct an area or zone reconnaissance to become familiar with the routes to, through, and beyond the area of passage. This enables the MP to extend the commander's $C^2$ by providing directions at passage points and by guiding the units through the passing lanes. Maintaining unit integrity and reducing incidents of stragglers is vital to maintaining the passing unit's momentum in a forward passage of lines. The MP perform aggressive straggler- and dislocated-civilian-control operations to prevent possible infiltration of the enemy.

4-17. A passage of lines is usually planned and coordinated by the headquarters directing the passage. A division's passage of lines is planned and coordinated by the corps headquarters. The detailed plans are made and coordinated between the divisions involved. Close coordination between the division and corps PMs is essential. An MP unit may be the unit involved in passing through the lines of another unit. When conducting a delay of a Level II threat, the MP are likely to conduct a passage of lines with the TCF. To avoid fratricide, close coordination between the MP response-force commander and the TCF is imperative (see FM 19-4).

## STRAGGLER CONTROL

4-18. Mobile patrols, TCPs, and checkpoint teams return stragglers to military control as part of their operations. Most stragglers are simply persons who become separated from their command by events in the CZ or while moving through the COMMZ. If a straggler is ill, wounded, or in shock, an MP must give him first aid and, if needed, call for medical evacuation (MEDEVAC). If a straggler is uninjured, an MP directs him to his parent unit or to a replacement unit (as command policies dictate). The MP ensure that stragglers attempting to avoid return to their units are escorted back to their command (as a minimum).

4-19. The MP set up special posts for straggler control following NBC attacks or major enemy breakthroughs that result in large numbers of lost, dazed, and confused military personnel. Mobile MP teams operate between posts, and they also direct or collect stragglers. Straggler collection points may be needed if many stragglers are present in a combat theater. If allied forces are present in the theater, each nation establishes a collection point for its own personnel. MP teams are aware of each allied location and are prepared to assist allied soldiers in returning to their respective command. The MP use available transportation assets to transfer stragglers from TCPs and checkpoints to a straggler collection point. At the collection point, they are screened and sorted for removal to a medical facility or returned to their units to reconstitute the tactical commander's combat force.

4-20. The MP report information about stragglers with whom they come in contact. This information is compiled by the MP headquarters and forwarded through appropriate channels to the higher command. Information given by stragglers that is of immediate tactical value is reported without delay.

## DISLOCATED-CIVILIAN CONTROL

4-21. The MP expediting traffic on MSRs may encounter dislocated civilians that could hinder military traffic. The MP assist and divert dislocated civilians from MSRs and other areas to I/R facilities. They may also deny the movement of civilians whose location, direction of movement, or actions may be a threat to themselves or to the combat operation. The HN government is responsible for identifying routes for the safe movement of dislocated civilians out of an AO. If needed, the MP assist the civil-affairs unit and the HN in redirecting dislocated civilians to alternate routes established by the HN government.

4-22. The US forces do not assume control of dislocated civilians unless requested to do so by the HN or unless operating in an environment with a hostile government. When the senior US commander assumes responsibility, the MP coordinate with civil affairs to set up TCPs at critical points along the route to direct dislocated civilians to secondary roadways and areas not used by military forces. As required, MWD teams may be used as a show of force or as a deterrent to assist with uncooperative personnel.

## ROUTE R&S

4-23. The MP conduct hasty and deliberate route reconnaissances to obtain information on a route and nearby terrain from which the enemy can influence troop movement. A route reconnaissance focuses on continually monitoring the condition of MSRs, ASRs, and other areas. MP patrols look for restricting terrain, effects of weather on the route, damage to the route, NBC contamination, and enemy presence or absence. When enemy activity is spotted, the MP report it, maintain surveillance, and develop the situation. To gather information for proposed traffic plans, they look at the type and number of available routes; and they check load classifications, route widths, obstructions, and restrictions.

## MSR REGULATION ENFORCEMENT

4-24. The MP undertake MSR regulation enforcement to keep the routes free for DSS operations. MP units support the command's MSR regulation measures as stated in the traffic-regulation plan (TRP). The TRP contains specific measures to ensure the smooth and efficient use of the road network. It assigns military route numbers and names, the direction of travel, highway regulation points, and preplanned MP TCPs. Most importantly to the MP, it gives the route's control classification. The MP ensure that classified routes are used only by authorized traffic. Vehicles traveling on roads too narrow for their passage or on roads unable to support their weight can obstruct the route.

4-25. To expedite traffic on MSRs, the MP operate special circulation control measures such as—

- Temporary route signing.
- Static posts such as TCPs, roadblocks, checkpoints, holding areas, and defilades at critical points.
- Mobile teams patrolling between static posts and monitoring traffic and road conditions.

4-26. They also gather information on friendly and enemy activities and help stranded vehicles and crews. The MP also place temporary route signs to warn of hazards or to guide drivers unfamiliar with the route. Using these measures, the MP exercise jurisdiction over the road network in the AO and coordinate with the HN (whenever possible) to expedite movement on MSRs.

## AREA SECURITY

4-27. The MP perform the AS function to protect the force and to enhance the freedom of units to conduct their assigned missions. The MP who provide AS play a key role in supporting forces in rear-area and sustainment operations. The MP are a response force that delays and defeats enemy attempts to disrupt or demoralize military operations in the AO. The MP's mobility makes it possible for them to detect the threat as they aggressively patrol the AO, MSRs, key terrain, and critical assets. The MP's organic communications enable them to advise the appropriate headquarters, bases, base clusters, and moving units of impending enemy activities. With organic firepower, the MP are capable of engaging in decisive operations against a Level II threat and delaying (shaping) a Level III threat until commitment of the TCF.

4-28. Throughout all aspects of the AS function, the MP perform counteractions to protect the force and to prevent or defeat a Level II threat operating within the MP's AO. MP countermeasures may include implementing vulnerability assessments, developing procedures to detect terrorist actions before they occur, hardening likely targets, and conducting offensive operations to destroy the enemy. The MP use checkpoints and roadblocks to control the movement of vehicles, personnel, and materiel and to prevent illegal actions that may aid the enemy. The use of these control measures serves as a deterrence to terrorist activities, saboteurs, and other threats. However, at the same time, checkpoints and roadblocks expose the MP to these potential threats. To counter this fact, the MP may upgrade or harden vehicles and defensive positions.

4-29. The MP provide combat power to protect the $C^2$ headquarters, equipment, and services essential for mission success. The MP provide the battlefield commander with a light, mobile fighting force that can move, shoot, and communicate against any threat. Major subtasks associated with the AS function include reconnaissance operations; area damage control (ADC); base/air-base defense; response-force operations; and critical site, asset, and high-risk personnel (HRP) security.

### RECONNAISSANCE OPERATIONS

4-30. As part of their AS mission, the MP serve as the eyes and ears of the battlefield commander by seeking out the enemy and reporting information obtained by recon patrols. The MP conduct area and zone reconnaissances, screening, surveillance, and countersurveillance to gain information to help guard against unexpected enemy attacks in the AO. The MP monitor likely avenues of approach and potential LZs and DZs. They become familiar with towns and other populated areas, ridgelines, woods, and other terrain features from which the enemy can influence movements along road networks. The MP pay close attention to areas near facilities designated critical by the commander. These areas include key MSR bridges and tunnels, depots, terminals, logistics-support bases, ammunition supply points (ASPs), communications centers/nodes, and $C^2$ headquarters. The MWD teams provide explosive detection and personnel detection/tracking capabilities that enhance reconnaissance operations (especially in urban terrain).

## AREA DAMAGE CONTROL

4-31. MP units take measures to support ADC before, during, and after hostile actions or natural and man-made disasters. The ADC actions integrate CS and CSS functions for many units. Engineers, medical personnel, and Army aviators work closely to ensure quick relief operations. The MP provide MSR regulation enforcement, refugee control, and some local security when required. As with reconnaissance operations, the MP may use MWD explosive-and personnel-detection capabilities to augment all MP missions in rear-area and sustainment operations.

## BASE/AIR-BASE DEFENSE

4-32. The MP are the base and base-cluster commanders' links for detection, early warning, and employment against enemy attacks. The information gathered is dispersed throughout the rear area to help apprise the commander of enemy activities near bases. Base defense is the cornerstone of rear-area security. When the threat exceeds the base/base-cluster capability, the base/ base-cluster commander requests MP assistance through the appropriate $C^2$ element.

4-33. Air-base defense requires special MP coordination with the US Air Force (USAF). The MP treat air bases like any other base or base cluster. A USAF air base may house the base-cluster commander, or it may be a cluster by itself. The MP are responsible for the air base's external defense. Its internal defense is primarily the responsibility of the Air Force's security forces. The security force provides in-depth defense for weapons, weapons systems, command centers, personnel, and other priority resources established by the base commander.

4-34. The security force is trained and equipped to detect, delay, and deny Level I and II threats. If a Level III threat is present, the security force is tasked with delaying actions; however, the HN, a sister service, or other support must be employed to defeat this threat. If the security force requires assistance to defeat a Level II threat, it may rely on MP response forces or another response force to assist in the defense. If available, the MP response force will react to the air-base defense, just as it would for any other base or base cluster within the MP's AO. However, the key to successful MP employment depends on the critical exchange of information before and during the MP employment. Good communications, an understanding of the defense plan, and liaison operations are vital in preventing responding forces from entering a situation that could result in fratricide.

## RESPONSE-FORCE OPERATIONS

4-35. The MP are the base and base-cluster commanders' response force against enemy attacks in rear-area or sustainment operations. The MP gather information about the enemy while performing missions throughout the AO. This information provides commanders with enemy activity near bases. When needed, the MP provide a mobile response force to respond to bases under attack and to destroy the enemy. A base commander's defense plan is the cornerstone for protecting rear-area and sustainment operations. The base commander is responsible for defeating all Level I threats. When this threat exceeds his capabilities, he requests MP support. The MP located near bases or patrolling or conducting AS operations consolidate their forces, respond as quickly as possible, and conduct combat operations to destroy the enemy. If needed, the MP conduct a battle handover to the TCF.

4-36. MP forces performing as a response force are capable of conducting the following offensive operations:

- A movement to contact.
- A hasty ambush.
- A hasty attack.
- A delay.
- A call for fire.
- A repel attack against critical sites.
- A defense of critical sites.

4-37. To conduct these missions, the MP consolidate into squads or platoons to delay, defeat, or defend against the threat. See FMs 71-3 and 71-100 for more information on battle-handover operations.

## CRITICAL SITE, ASSET, AND HRP SECURITY

4-38. The MP perform their AS function across the entire designated AO. When the MP provide security around a critical site or asset, they usually provide a mobile security screen, taking advantage of its weapons and communications platforms. This standoff protection detects and defeats the threat before it can move within direct-fire range of the facilities. The MP may be tasked to provide detail security to key facilities, assets, and personnel.

4-39. The MP provide security to major CPs and other facilities within the AO. Their employment maximizes mobility, lethality, and communications capabilities as a security screen. They may be required to establish local AS measures (such as checkpoints and listening/observation posts) to further protect these facilities. The MP provide internal access-control points to critical facilities, and they act as a response force. When the critical CP relocates, the MP provide in-transit security. Other types of critical site security include ASPs; deep-water ports; petroleum, oil, and lubricants (POL) terminals and pipelines; trains and railways; and air bases.

4-40. The MP may provide convoy security for top-priority units transporting especially critical supplies to combat forces. MP assets should be employed primarily on aggressive patrolling, route, area, and zone reconnaissance measures that would create a safe and secure environment for all types of vehicular and unit movement.

4-41. The MP and the CID provide protective services to designated key personnel by providing access control to restricted areas within CPs, providing in-transit security, or providing static security measures around the clock. The MP coordinate with the CID when close-in protection of key personnel is needed. The MP and the CID also provide training for personal-protection countermeasures. The MWD teams may be employed to enhance MP and CID detection and protection capabilities.

## INTERNMENT AND RESETTLEMENT

4-42. The Army is the Department of Defense's (DOD's) executive agent for all EPW/CI operations. Additionally, the Army is DOD's executive agent for long-term confinement of US military prisoners. Within the Army and through the combatant commander, the MP are tasked with coordinating shelter, protection, accountability, and sustainment for EPWs/CIs. The I/R function addresses MP roles when dealing with EPWs/CIs, dislocated civilians, and US military prisoners.

4-43. The I/R function is of humane as well as tactical importance. In any conflict involving US forces, safe and humane treatment of EPWs/CIs is required by international law. Military actions on the modern battlefield will result in many EPWs/CIs. Entire units of enemy forces, separated and disorganized by the shock of intensive combat, may be captured. This can place a tremendous challenge on tactical forces and can significantly reduce the capturing unit's combat effectiveness. The MP support the battlefield commander by relieving him of the problem of handling EPWs/CIs with combat forces. The MP perform their I/R function of collecting, evacuating, and securing EPWs throughout the AO. In this process, the MP coordinate with MI to collect information that may be used in current or future operations.

4-44. Although the CS MP unit initially handles EPWs/CIs, modular MP (I/R) battalions with assigned MP guard companies and supporting MWD teams are equipped and trained to handle this mission for the long term. A properly configured modular MP (I/R) battalion can support, safeguard, account for, guard, and provide humane treatment for up to 4,000 EPWs/CIs; 8,000 dislocated civilians; or 1,500 US military prisoners.

### EPW/CI HANDLING

4-45. The MP are tasked with collecting EPWs/CIs from combat units as far forward as possible. The MP operate collection points and holding areas to temporarily secure EPWs/CIs until they can be evacuated to the next higher echelon's holding area. The MP escort-guard company assigned to the MP brigade (I/R) evacuate the EPWs/CIs from the corps's holding area to the COMMZ's internment facilities. The MP safeguard and maintain accountability, protect, and provide humane treatment for all personnel under their care.

4-46. In a mature TO, MP (I/R) units process EPWs/CIs collected by MP teams and other units in the CZ. MP guard companies assigned to the MP (I/R) units guard EPWs/CIs at designated camps (see FM 19-40).

### POPULACE AND RESOURCE CONTROL

4-47. Populace and resource control (PRC) denies adversaries or insurgents access to the general population and resources. The MP supports civil-affairs personnel and the tactical commander in planning and conducting PRC programs employed during all military operations. These programs may consist of curfews, movement restrictions, resettling dislocated civilians, licensing, ration control, regulation enforcement, amnesty programs, inspecting facilities, and guarding humanitarian-assistance distributions. The MP also direct dislocated civilians to resettlement camps where they are cared for while NGOs work to coordinate their relocation.

4-48. The MP's security capability, acceptability, and interface with the populace make them suitable as the primary forces of choice in these operations. The MP I/R units are specifically designed to fill this need (see FM 41-10).

### US MILITARY PRISONERS CONFINEMENT

4-49. The MP detain, sustain, protect, and evacuate US military prisoners. Whenever possible, soldiers awaiting trial remain in their units, unless reasonable grounds exist to believe that they will not appear at the trial, the pretrial hearing, or the investigation or that they will engage in serious criminal misconduct. Under either of these two pretrial confinement instances, the commander must also reasonably believe that a less severe form of restraint (such as conditions of liberty, restriction in lieu of apprehension, or apprehension) are inadequate. When these circumstances exist and other legal requirements are met, US military personnel may be placed in pretrial confinement under the MP's direct control. Convicted military prisoners are moved as soon as possible to confinement facilities outside of the AO.

4-50. MP confinement operations parallel (but are separate from) the MP's EPW/CI operations. No member of the US armed forces may be placed in confinement in immediate association with an EPW or other foreign nationals who are not members of the US armed forces. A confinement facility is maintained within the TO only if distance or the lack of transportation requires such a facility. When military prisoners are retained in theater, temporary field detention facilities may be established in the CZ and a field confinement facility in the COMMZ (see FM 19-40 and AR 190-47).

## LAW AND ORDER

4-51. The L&O function consists of those measures necessary to enforce laws, directives, and punitive regulations. The MP's L&O function extends the battlefield commander's $C^2$. The MP, in close coordination with the CID, work to suppress the chance for criminal behavior throughout the AO. By coordinating and maintaining liaison with other DOD, HN, joint, and multinational agencies, the MP at all levels coordinate actions to remove conditions that may promote crime or that have the potential to affect the combat force. Crime-prevention measures and selective enforcement measures are also performed as part of other functions. For example, the MP investigate traffic accidents and regulate traffic as part of the MMS function. The L&O function includes major areas such as law enforcement, criminal investigations, and support to US Customs operations. The primary units conducting L&O are the L&O detachments, customs teams, and CID units. Both the MWD team and the MP company (CS) also support the L&O function.

## LAW ENFORCEMENT

4-52. Law-enforcement operations assist the battlefield commander in preserving his force. The MP dedicate assets to conduct law-enforcement operations based on the commander's needs. Since MP L&O and CID assets may be limited during the initial stages of any operation, the PM depends on the supported commander's development of an effective crime-prevention program and uses established investigative tools (such as inquiries and AR 15-6 investigations) to enforce rules. This will allow the PM to employ limited MP assets to perform other functions. Law-enforcement operations include responding to civil disturbances, conducting raids, investigating traffic accidents, conducting vehicle searches, supporting the commander's force-protection program, and providing support to HN and civil-enforcement agencies. Law enforcement also includes employing special-reaction teams (SRTs), marksman/observer (M/O) teams, and MWD teams.

4-53. The Army conducts counterdrug-support operations that generally fall within several DOD counterdrug-mission categories. The MP support the Army's role rather than directly participating in civil law-enforcement activities (such as searches, seizures, and arrests). When tasked, the MP provide training to law-enforcement agencies in common soldier skills, physical security, and tactical planning and operations. US Code (USC) Title 18, Posse Comitatus Act, Section 1385 prohibits the use of DOD assets to enforce civilian law (federal and state) except in cases and under circumstances expressly authorized by the Constitution or by an act of Congress.

4-54. In multinational operations, the MP may assist with the creation of multinational police units. Circumstances that may support the establishment of these police forces include existing or negotiated terms of international agreements or security-assistance programs, a multinational operational agreement, or appropriate military directives. The MP provide the capability to train foreign MP and/or reconstitute indigenous constabulary forces as part of stability and support operations. The MP can provide the initial mentoring to these forces and provide temporary, emergency law-enforcement capabilities until the foreign military or civilian police forces are functional (see FM 100-8).

## CRIMINAL INVESTIGATIONS

4-55. The MP investigate offenses against US forces or property committed by persons subject to military law. Investigations against minor crimes (such as low-value, personal-property thefts or simple assaults) are normally investigated by the MP's L&O detachment. Investigations against major incidents involving wrongful death, serious bodily injury, and war crimes are referred to special agents of the USACIDC. The USACIDC conducts death investigations in the absence of HN agreements or in conjunction with the HN. The USACIDC special agents are authorized to investigate any alleged criminal conduct in which there is an Army interest unless prohibited by law or higher authority.

4-56. The USACIDC's investigative authority and investigative responsibility outside of the US are determined by international treaty or agreement (including status of forces agreements [SOFAs]), the policies of the HN government, the US ambassador, and AR 195-2. In the absence of such provisions, the following guidelines apply:

- On Army-controlled installations, the USACIDC has the authority to investigate alleged crimes.
- Outside of an Army-controlled installation, the USACIDC may investigate after coordinating with HN authorities.

4-57. In all environments, the USACIDC has the responsibility to investigate all felony crimes involving Army personnel, DA civilians and agencies, and companies working for the Army. The USACIDC investigates war crimes and crimes involving personal and government property affecting the Army's mission (logistics security [LOGSEC]). Other investigations (such as those based on international treaties, SOFAs, and joint investigations with the HN) may be undertaken if requested by the supported commander in support of the overall Army mission. See Chapter 9 for a complete discussion of the USACIDC.

## US CUSTOMS OPERATIONS

4-58. The MP support the US Customs Service (USCS), the US Department of Agriculture (USDA), other federal agencies, joint staffs, and commanders who enforce the laws and regulations of the US concerning customs, agriculture, and immigration border clearances. Support to the USCS also includes assistance to federal agencies to eliminate the illegal traffic of controlled substances and other contraband through Army channels. MP support to customs operations are normally performed by specially trained MP customs teams. Although other MP units are not trained in all facets of customs operations, they may assist MP customs teams, the USCS or the USDA, and other federal agencies in the enforcement of applicable laws and regulations. When tasked, the MP/CID supports the investigation of violations of US Codes, DOD or DA regulations, and applicable provisions of SOFAs.

4-59. The MP report violations of customs laws, regulations, inspections, and investigative results to the installation's PM, the supported commander, and affected units. During redeployment from outside the continental US (OCONUS) to CONUS installations, the MP support the USCS or USDA efforts to ensure that personnel, equipment, and materiel meet customs, immigration, and agriculture requirements as stated by all applicable laws and regulations. As with other functions, MWD teams may be employed in support of customs operations for the detection of explosives or narcotics.

> During operation Just Cause, an MP platoon temporarily assumed the customs mission at the main terminal of the Torrijos International Airport located just outside of Panama City. Their mission supported the air evacuation of foreign-national civilians and the redeployment of some of the initial-entry US forces.

## POLICE INTELLIGENCE OPERATIONS

4-60. The PIO function supports, enhances, and contributes to the commander's protection program, situational awareness, and battlefield visualization by portraying relevant threat information that may affect his operational and tactical environments. This threat information—whether it is police, criminal, or combat information—is gathered while conducting MP functions. The PIO function—

- Demonstrates the MP's/CID's capability to collect relevant threat information actively or passively.
- Ensures that all information collected while conducting MMS, AS, I/R, and L&O functions continues to be reported through the proper channels so that it can be analyzed by the Intelligence Officer (US Army) (S2) or the Assistant Chief of Staff, G2 (Intelligence) (G2) with support from the appropriate MP echelon.
- Coordinates with USACIDC elements to employ data developed by the USACIDC's programs. These programs include—
  - The Combating Terrorism Program as outlined in AR 525-13 and CIDR 195-1.
  - The Criminal Intelligence Program (CIP).
  - Personal-security vulnerability assessments (PSVAs).
  - A crime threat analysis.
  - Logistics-security threat assessments (LSTAs).

**NOTE: The MP/CID must ensure that criminal information is released according to existing controls and restraints.**

- Maintains constant liaison and communication with the higher echelon S2/G2; psychological-operations (PSYOP) units; HN police and other law-enforcement agencies; joint, combined, interagency, and multinational forces; the staff judge advocate (SJA); the CMOC; civil-affairs teams; and the force-protection officer.

4-61. The MP brigade commander, the battalion commander, and the PM are responsible for the PIO function. As such, each one must determine the best way to employ the available staff resources to monitor the execution of the PIO function within his command.

### IPB PROCESS AND PIO FUNCTION

4-62. The IPB is a systematic, continuous process for analyzing the threat and environment in a specific geographic area. It is designed to support staff estimates and military decision making. Applying the IPB process helps the commander selectively apply and maximize his combat power at critical points in time and space on the battlefield by—

- Determining the threat's likely COA.
- Describing the environment the unit is operating within and the environmental effects on the unit.

4-63. The IPB process consists of the following four steps:

- **Define the battlefield environment.** The S2/G2 identifies the battlefield characteristics that will influence friendly and threat operations, establishes the limits of the area of interest (AOI), and identifies gaps in current intelligence holdings.
- **Describe the battlefield effects.** The S2/G2 evaluates the environmental effects with which both sides must contend. The S2/G2 identifies the limitations and opportunities that the environment offers on the potential operations of friendly and threat forces. This evaluation process focuses on the general capabilities of each force until COAs are developed later in the IPB process. This environmental assessment always includes an examination of terrain and weather, but it may also include discussions of characteristics of geography and infrastructure and their effects on friendly and threat operations.
- **Evaluate the threat.** The S2/G2 and his staff analyze the command's intelligence holdings to determine how the threat normally organizes for combat and conducts operations under similar circumstances. When facing a well-known threat, the S2/G2 can rely on historical databases and threat models. When operating against a new or less known threat, he may need to develop his intelligence databases and threat models concurrently.
- **Determine the threat's COA.** Given what the threat normally prefers to do and the effects of the specific environment in which he is operating, his likely objectives and the COAs available to him are determined. The S2/G2 develops enemy COA models that depict the threat's available COAs. He also prepares event templates and matrices that focus intelligence and identify which COA the enemy will execute (see FM 34-130).

4-64. Although the S2/G2 has the staff responsibility for the command's IPB, he is not the only one who conducts or needs to understand and use the IPB. Every Army commander and staff member must understand and apply the IPB process during the staff planning process. The MP on the battlefield are no exception. The MP employ the IPB process as their first step in developing and implementing the PIO function within their commands.

4-65. During the IPB process, the S2/G2 uses all available databases, intelligence sources/products, and related MI disciplines to analyze the threat and the environment. The PIO function supports this process by providing the S2/G2 with collected police, criminal, and combat information that can directly and significantly contribute to the success of the MI effort. In addition to the combat information, the PIO function provides additional information on possible criminal threats and COAs that may support the S2's/G2's IPB process and that can be used by the commander to upgrade the force-protection posture.

## POLICE INFORMATION ASSESSMENT PROCESS

4-66. Like the S2/G2 uses the IPB process to continuously analyze the threat and the environment in a specific geographical area, MP leaders use the PIAP as a tool to continuously collect, organize, interpret, and gain access to police/ criminal information in support of the IPB process. Criminal trends may have an impact on the tactical scenario, and the PIAP is a method used to consider this threat and its impact on friendly forces. MP leaders cannot use the PIAP as a substitute for the IPB process—the PIAP complements the IPB (see Appendix E).

## PASSIVE MODE PIO

4-67. Every MP conducts the PIO function in a passive mode during their normal day-to-day operations and across the full spectrum of military operations. In the passive mode, PIO are not a stand-alone function; as such, they cannot be separated from other MP functions.

4-68. During the performance of MMS, AS, I/R, and L&O functions, the MP develop and exchange information with other organizations in the AO. The MP obtain information through contact with civilians, NGOs, IHOs, local and HN police, multinational police, and other security forces. If the MP receive, observe, or encounter information (police, criminal, or combat) while performing these functions, they will immediately submit a report to relay information up the chain of command. This report may be in the form of a size, activity, location, unit, time, and equipment (SALUTE) report; a spot report (SPOTREP); or another appropriate report. When the higher echelon (brigade, battalion, or PM) receives this information, it is simultaneously integrated into the ongoing IPB/PIAP and forwarded to the higher echelon S2/G2 (see Figure 4-1, page 4-16).

4-69. If police/criminal information is obtained, the MP—

- Update the previous police/criminal estimates provided to the S2/G2.
- Identify new or potential criminal threats or trends in the AO.
- Consider recommending that the supported commander upgrade the force-protection level.
- Notify adjacent units of the potential criminal threat that may affect their forces.
- Consider reprioritizing MP support to the identified threat area.
- Share information with HN/local police and other agencies.

4-70. If combat information is obtained, the MP—

- Forward the information to the higher headquarters S2/G2.
- Forward the information to the MP chain of command, integrate it into the MP's IPB process and, if necessary, take appropriate action.
- Notify the adjacent unit of the potential threat that may affect their forces.

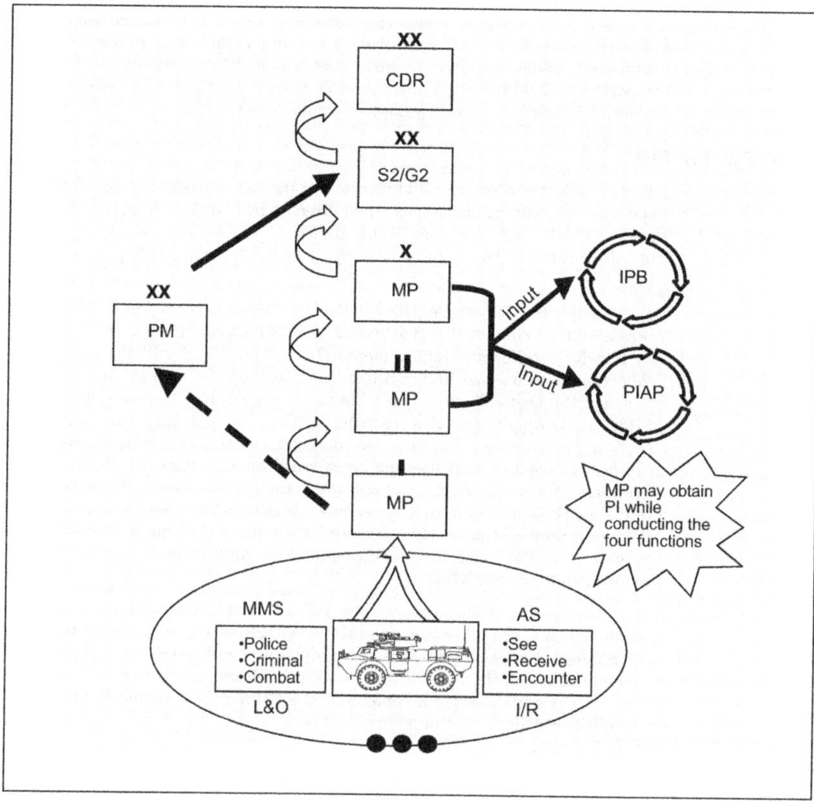


**Figure 4-1. PIO—Passive Mode**

> During Operation Uphold Democracy, an MP team was conducting a TCP as part of a cordon and search operation in Port-au-Prince, Haiti. While performing the task, two civilians approached the MP team informing them of criminal activity in the neighborhood. During the interview, the MP team prepared a sketch of a house and surrounding areas. The team also obtained information describing the criminals and their weapons. Recognizing that the criminal activity was in fact the actions of a political/mercenary group named FRAP, the MP team radioed the platoon leader and forwarded the field interview to higher headquarters. Two days later, a unit from the 10th Mountain Division raided the house, capturing weapons, ammunition, and equipment.

4-71. The preceding vignette demonstrates the MP performing the PIO function in the passive mode. The MP team received the information while conducting a TCP and submitted it through the appropriate chain of command, which resulted in an action taken. This example stresses the importance of submitting information up the chain of command regardless of whether it may be police, criminal, or combat information.

## ACTIVE MODE PIO

4-72. The MP perform the PIO function in the active mode and across the full spectrum of military operations when directed by higher headquarters. In this mode, the MP conduct specific MMS, AS, I/R, and L&O missions with the intent to collect information actively in support of the S2's/G2's IPB process or the PIAP.

4-73. When the S2/G2 identifies a gap in the command's knowledge of the threat and the current threat situation, it may be recommended to be included as priority intelligence requirements (PIR). The S2/G2 will then develop a collection plan to assist him in filling this gap. Part of his collection strategy is to select the best collectors available to cover each intelligence requirement. After a thorough analysis (which includes availability, capability, and performance history), the collection manager identifies which collection asset can best be used in response to a given requirement, and the Operations and Training Officer (US Army) (S3)/G3 tasks the asset. If the S2/G2 determines that the MP is the right force to serve as collectors, the S3/G3 will go through the appropriate request channels and task the PM. The S2/G2 will then provide the PM with a specific guideline and a prioritized collection requirement (see FM 34-2).

4-74. On receipt of the mission, the PM will conduct a mission analysis to decide which specific MP function (MMS, AS, I/R, or L&O) is needed to satisfy the S2's/G2's requirement. Once the analysis is completed and the appropriate function selected, the PM will then task subordinate units with the collection mission. Once the mission is completed, the PM may receive another collection tasking or continue with previous MP tasks.

4-75. When the MP commander or the PM conducts the PIAP, he may also encounter a police/criminal information gap. This gap may become the MP commander's police/criminal information requirements (PCIR). If the gap cannot be filled with available data from the S2/G2, the CID, the HN, and other agencies, the MP commander/PM may task subordinate MP units or request CID assistance to support the collection effort. On receipt of the mission, the MP will then conduct a mission analysis and decide which specific MP function (MMS, AS, I/R, or L&O) is needed to satisfy the PM requirement. Once the mission is completed, the MP may receive another collection tasking or continue with previous MP tasks (see Figure 4-2, page 4-18).

4-76. Since any soldier on the battlefield can report police or criminal information, the MP commander or the PM must constantly coordinate and communicate with the S2/G2, PSYOP, and other agencies to obtain information that could be of MP/CID interest. This constant coordination is a key factor for ensuring that the MP/CID has visibility over the police/criminal information that is reported through non-MP channels.

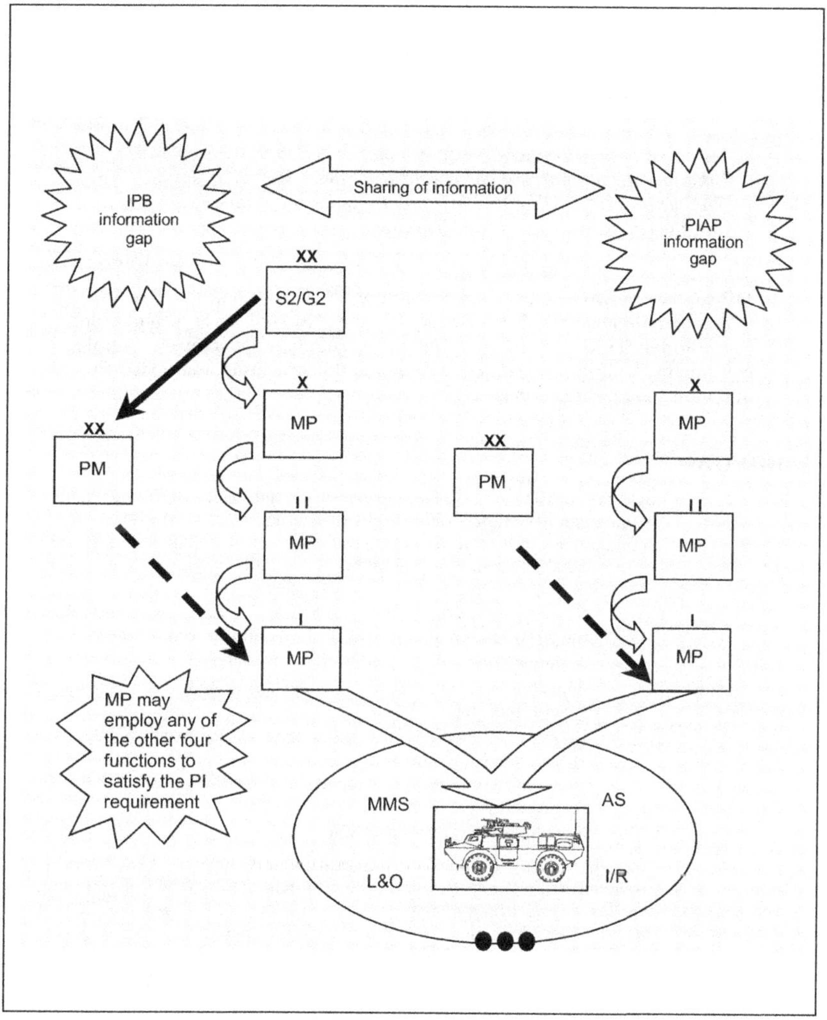


**Figure 4-2. PIO—Active Mode**

# Chapter 5

# Military Police Support to Echelons Above Corps

*MP units supporting EAC perform combat, CS, and CSS operations. Like the MP supporting corps and divisions, MP units supporting EAC units perform the five MP functions based on available assets and the supported commander's needs.*

*USAMPS*

MP support to EAC includes support to the ASCC and the TSC. The ASCC is responsible for Army Title 10 requirements in support of a combatant commander. This support includes recruitment, organization, supply, equipment, training, servicing, mobilizing, demobilizing, administration, and maintenance functions.

## OVERVIEW

5-1. The ASCC may also be responsible for significant DOD- and combatant-commander-designated Army support to other services. As the senior Army commander in the AOR, the ASCC commander tailors and designates ARFOR to accomplish operational-level tasks while conducting major land operations. The ASCC's operational responsibilities include—

- Recommending the proper employment of Army-component forces to the joint-force commander or to the subunified commander.
- Accomplishing operational missions as assigned.
- Selecting and nominating specific Army units for assignment to subordinate theater forces.
- Informing the combatant commander of the Army's CSS effects on operational capabilities.
- Providing data to the supporting operations plans (OPLANs) as requested.
- Ensuring signal interoperability.

5-2. The ASCC provides administrative and logistics (A/L) services to assigned ARFOR and to those of subordinate JFCs. When appropriate, the ASCC delegates the authority for support tasks to a single subordinate Army headquarters. In major operations, the TSC (along with other EAC support commands) would be the ASCC's lead organization for planning, coordinating, executing, or providing required support functions (see FM 100-10).

5-3. The TSC is the senior Army support organization in a theater. Its commander reports to the ASCC or ARFOR commander. The TSC normally operates at the operational level of CSS with links to the strategic and tactical levels. Unity of command is the critical element that the TSC brings to the fight. The TSC is a multifunctional organization that centralizes the command, control, and supervision of support functions at EAC as directed by the ASCC or ARFOR commander. The TSC's mission is to maximize throughput and follow-on sustainment, including all CSS functions, of ARFOR and other designated supported elements. The TSC is capable of synchronizing logistics and other support operations for the ASCC. It provides area support to EAC units in the COMMZ and sustainment support to tactical forces. This support may include supply, procurement, property disposal, maintenance, transportation, field services, health services, civil-military affairs, MP support, engineer support, religious support, finance support, and personnel and administrative services.

5-4. Units and commands requiring support coordinate with the TSC support-operations staff to secure their initial support, to reestablish support, or to resolve support problems. In a fully developed theater, the TSC coordinates with a corps support command (COSCOM) for support of combat forces, although direct coordination with a division support command (DISCOM) is sometimes necessary. The TSC, augmented by a rear operations center, is also responsible for security operations as directed by the ASCC/ARFOR commander (see FM 63-4).

# MILITARY POLICE SUPPORT

5-5. MP support to EAC units is provided through an array of multifunctional MP units. The nature of the operation, METT-TC, and the requirements of the supported commander will determine which type of MP unit is appropriate to augment, assign, attach, or place under OPCON to an EAC unit. The types of MP units that support EAC include CS, I/R, CID, and L&O teams (such as MWD or customs teams).

## THE MP BRIGADE (CS)

5-6. The MP brigade (CS) provides MMS and AS to extended LOC within the COMMZ. These supply corridors include ports, inland waterways, railways, pipelines, airfields, MSRs, and MSR critical points. The MP support the users of the COMMZ's LOC by aggressively patrolling the area along the LOC. They play an important role in securing rear areas by performing combat operations against the threat. When properly augmented, the MP brigade headquarters may serve as the TSC's/ASCC's TCF headquarters. The MP provide MMS on the COMMZ MSRs leading into the corps's rear area. The MP implement the plans of HN and US staff elements to control the forward movement of combat resources along the LOC.

5-7. If resources are available, the MP brigade (CS) provides escorts to move US noncombatants (if present) from AA points to theater embarkation terminals. Until the MP brigade (I/R) arrives in theater, the CS MP units also perform EPW, confinement, and other operations normally performed by the MP I/R units.

5-8. The organization of an MP brigade (CS) supporting EAC includes the following:

- A brigade headquarters and headquarters company (HHC).
- Up to six MP battalions (each with up to six companies).

- Numerous L&O detachments and MWD teams.

5-9. Additionally, the ASCC's PM or commander may attach or direct OPCON of customs teams to the MP brigade (CS). Battalion and company organization in the MP brigade is the same as that in the corps MP brigade; however, METT-TC determines the number of battalions and companies. The MP brigade (CS) has additional MP companies to provide security for EAC-specific units/activities such as—

- A unified command.
- An ASCC and TSC headquarters.
- LOC seaports, airfields, and railways.
- EAC ammunition storage areas.
- EAC petroleum terminals and pipelines.

5-10. While the corps MP brigade (CS) MWD teams are employed in a GS role, the MP brigade (CS) MWD teams are employed to augment seaport security and to conduct inspections of postal items to detect explosive materiel and narcotics.

5-11. The $C^2$ within the MP brigade (CS) is consistent with that in any Army brigade. The MP brigade commander works directly for the EAC commander, the battalion commander works for the brigade commander, and the company commander works for the battalion commander. However, this usual $C^2$ relationship may be altered briefly (based on METT-TC) to enhance the overall EAC combat capability for responding to a Level II threat. For example, MP units operating within an ASG's AO may be under the OPCON of the ASG's rear-area operations center (RAOC), which directly tasks MP units responding to Level II threats. The same is true for placing MP units under the OPCON of the EAC's TCF headquarters for responding to Level III threats. Any conflict in mission priorities is resolved through MP $C^2$ channels.

5-12. The MP brigade commander is both the MP brigade commander and the EAC's PM. He employs his assets according to METT-TC and the commander's concept of operations. Factors affecting his employment of MP assets include the—

- Nature of the operation (joint, combined, or multinational).
- HN's ability to provide MP-related support (such as port security).
- Custody and location of EPWs/CIs during internment operations until I/R units arrive in theater.
- Number of kilometers of the MSR in relationship to movement-control requirements.
- Number and kinds of critical facilities.
- Number of HRP requiring close-in security.
- HN's ability to control the civilian populace, refugees, and dislocated civilians.
- Supply distribution strategy.
- Risk acceptance and threat in the AO.
- Communications requirements (such as using teams as relays).

5-13. Whenever possible, the MP brigade's AO coincides with the territorial responsibility of the supported command. The MP brigade commander assigns the MP battalion's AO by the above factors as well as by METT-TC. For example, the AOR for an MP battalion may be a large population center of larger geographical areas in which CSS complexes and MSRs are located. But as employment factors and the commander's needs change, so will the MP's AOR. The MP brigade commander must move and tailor his forces to meet the current and projected mission requirements. Unlike many other EAC assets, MP units require 100 percent mobility to shift AOs frequently and rapidly. The following vignette depicts the required MP flexibility to support EAC operations:

> During Operation Desert Shield/Storm, MP units were supporting and moving units throughout their AOs at a such an extraordinary rate that many of them had to relocate their headquarters multiple times just to keep pace with their changing AO.

5-14. Most EAC MP assets are employed along LOC and around areas of high troop concentration. Few EAC MP are dedicated to support fixed commitments (such as ports, air bases, and headquarters [discussed in paragraph 5-8]). When supporting fixed commitments, MP units provide a mobile security screen, and they man static positions when directed or when based on METT-TC. The MP brigade (CS) may have to plan for and actually perform the evacuation and internment of EPWs/CIs and the confinement of US military prisoners until the MP brigade (I/R) arrives.

## THE MP BRIGADE (I/R)

5-15. US policy requires that all persons held in Army custody be accorded humane care and treatment from the moment of custody to their final release. The policy applies to detained or interned civilians as well as to EPWs and confined US military personnel. This policy is equally binding on all US troops (see FM 19-40).

5-16. The ASCC or the TSC supports US laws, regulations, policies, and international agreements by providing personnel, administrative, morale, internment, resettlement, and confinement services for the TO. The TSC's MP brigade (I/R) in the TO provides this support. However, since most I/R units are in the reserve components, the initial I/R operations (as mentioned above) may have to be conducted by the MP brigade (CS). Once the I/R unit arrives in the AO, it is responsible for—

- Providing firm but humane control of EPWs/CIs and dislocated civilians.
- Coordinating with HN personnel, military territorial organizations, civilian police authorities, NGOs, private volunteer organizations, and US federal agencies on matters pertaining to I/R operations.
- Performing $C^2$ operations for all I/R units.
- Controlling, employing, and releasing EPWs/CIs as set forth by the Geneva convention and other international laws and by the UN and other governmental bodies.
- Handling US military prisoners.

5-17. In a mature theater in which large numbers of EPWs are captured, the EPWs' requirements may exceed the capacity of the MP brigade (I/R). In this instance, an I/R command is established. An MP command (I/R) has two or more MP brigades (I/R) and will normally be assigned to the ASCC. When the MP command (I/R) assumes OPCON of the MP brigades (I/R) from the lower echelon, it assumes that echelon's I/R mission.

5-18. If the US decides to transfer captured EPWs/CIs to the HN or to another nation, the US must ensure that the nation is a party to the Geneva convention and is willing and able to comply with the convention. In this case, the number and type of I/R MP units required for processing and retaining EPWs/CIs before the transfer is based on agreements and on METT-TC. Additionally, the MP brigade (I/R) is assigned I/R teams that are located at the processing and transfer points and at the HN or third-country EPW camps. The MP brigade (I/R) liaison team will supervise these dispersed teams to ensure that the HN or the third country provides adequate care and security of US-captured EPWs/CIs and that accountability is maintained according to the Geneva convention.

5-19. The MP brigade (I/R) HHC is the $C^2$ element for the brigade's assets. It consists of the following elements:

- A brigade command section.
- A company headquarters.
- An Adjutant (US Army) (S1).
- An S2.
- An S3.
- A Supply Officer (US Army) (S4).
- A Civil-Affairs Officer (US Army) (S5).
- A Communications Officer (US Army) (S6).
- Finance and accounting.
- Medical operations.
- Public affairs.
- A unit ministry team.
- An SJA.
- An inspector-general (IG) section.

5-20. Other brigade assets may include the following:

- An MP I/R BLD.
- An I/R information center.
- An MP escort-guard company.
- An MP I/R battalion headquarters.
- MP I/R (EPW/CI) detachments.
- MP I/R (confinement) detachments.
- MP guard companies.
- MWD teams.
- Processing squads, processing liaison teams, camp liaison teams, and evacuation teams (all as required to support EPW transfer or to conduct an out-of-theater evacuation).

5-21. The MP brigade (I/R) subordinate units are employed most often in the COMMZ near CSS facilities and are placed near sea, air, and rail terminals. They receive, process, and intern EPWs/CIs; confine US military prisoners; or assist in the resettlement of refugees or dislocated civilians.

5-22. The MP brigade (I/R) escort-guard company supports the evacuation of EPWs/CIs from the CZ. The company has a company headquarters and four platoons. The escort-guard company requires sufficient mobility to transport MP personnel to the CZ and to operate mobile teams while escorting the EPWs/CIs. The MP go forward to the corps's holding area to take custody of the EPWs/CIs. They may go forward to division collecting points, if distances and resources permit. Using any means of available transportation, the MP ensure that the EPWs/CIs are quickly evacuated to MP battalions (I/R) in the COMMZ. Close coordination with the EAC and corps movement-control centers and the corps MP brigade is required to ensure that transportation assets returning to the COMMZ are employed to evacuate EPWs/CIs from the corps's holding area. Walking wounded EPWs/CIs are evacuated by the same means as other EPWs/CIs, while litter patients are evacuated through medical channels. Guarding EPWs/CIs while in the MEDEVAC channels and during their hospitalization is not an MP mission; therefore, there is not an MP force structure to support this mission. In most instances, the impact of having the MP perform this mission causes trade-offs in missions for which they are responsible.

5-23. The theater MP brigade (I/R) and out-of-theater MP brigade (I/R) subordinate units will evacuate EPWs to internment sites within CONUS (if directed). The theater brigade structure is based on the projected capture rate over time and available out-of-theater transportation assets (frequency and capacity). The out-of-theater brigade structure is based on the total EPW/CI population, the number of internment sites, transportation nodes, and escort requirements. Theater escort-guard MP move the EPWs/CIs to the seaport and aerial port of embarkation (SPOE/APOE). The escort-guard MP assigned to the out-of-theater brigade escort the EPWs/CIs from the theater ports of embarkation (POEs) to the out-of-theater internment sites. The out-of-theater brigade is assigned an I/R evacuation detachment, which is employed at and coordinates the evacuation from the theater POEs, through the out-of-theater ports of debarkation (PODs), to the out-of-theater facilities.

## THE MP BATTALION (I/R)

5-24. The MP battalion (I/R) is a modular organization and can be configured to operate internment facilities for EPWs/CIs, confine US military prisoners, or resettle dislocated civilians. When performing EPW/CI internment operations, the MP brigade (I/R) has up to 7 MP battalions (I/R); when augmented with the appropriate number of BLDs, it has up to 21 MP battalions (I/R). The ASCC, the TSC, and the MP brigade (I/R) must consider that the requirement to establish an MP battalion (I/R) internment facility is resource intensive. Therefore, MP I/R units, other supporting units, supplies, and equipment for the EPWs/CIs should arrive in theater ahead of the projected EPW/CI arrival at the internment facilities. Early arrival should be based on the time required to establish fully operational facilities (construct and man) and resupply operations before the EPWs/CIs arrive.

5-25. The MP battalion (I/R) has a command section, a company headquarters, and various staff sections. The staff sections provide the core battalion-level capabilities to conduct internment operations. The modulated design expands as the EPW/CI population increases. The battalion is assigned up to four detachments, two guard companies and, if needed, a processing squad. When fully operational, an MP battalion (I/R) operates an enclosure capable of interning 4,000 EPWs/CIs. The battalion mission centers on eight 500-man compounds. The battalion operates the compounds in close proximity to maximize its resources for the security and internment of the EPWs/CIs. This includes the resources needed to employ EPWs/CIs as a labor force according to the provisions of the Geneva convention.

5-26. Each MP detachment (I/R) (EPW/CI) operates two 500-man compounds and provides augmentation to the battalion staff sections to support 1,000 EPWs/CIs. Each guard company is capable of providing security for 2,000 EPWs/CIs. The guard company has a company headquarters and three platoons. The guard company requires sufficient mobility and communications to support routine battalion missions. While minimum mobility and communications is required to support EPW/CI internment operations, on-site guard personnel must often move considerable distances guarding labor groups performing work projects throughout the COMMZ.

5-27. The MP (EPW/CI) processing squad is capable of processing eight EPWs/CIs per hour and includes interpreters to support the processing. If processing squads are required to augment MP battalions (I/R), the operational requirements will be based on METT-TC.

### MP-Battalion Resettlement Operations

5-28. The basic organization used for EPW/CI internment is used for resettlement operations. The primary mission-focus change is from guarding EPWs/CIs to protecting and controlling dislocated civilians. As such, an MP battalion (I/R) with four detachments and two guard companies is capable of supporting 8,000 dislocated civilians. However, the MP battalion (I/R) may require augmentation to conduct L&O operations associated with the resettlement. Augmentation may include the full scope of PM functions (operations, investigations, physical security, MP-station operations, and patrols) and civil affairs.

### MP-Battalion Confinement Operations

5-29. When configured with the MP detachment (I/R) (confinement), the MP battalion (I/R) is capable of confining US military prisoners. The MP detachment (I/R) (confinement) provides trained corrections and support personnel required for confinement operations. As with the EPW/CI configuration, the modular confinement structure expands as the US prisoner population increases. With three confinement detachments and three guard companies, the MP battalion (I/R) has a maximum capacity of handling 1,500 US prisoners. Generally, only one MP brigade battalion (I/R) is configured for confinement, but the actual number will depend on the number of US prisoners requiring confinement within a theater. While theater policy for confinement operations remains with the ASCC commander, it is the MP brigade (I/R) that executes the mission.

5-30. When possible, soldiers awaiting trial remain in their units unless reasonable grounds exist to believe that they will not appear at the trial, the pretrial hearing, or the investigation or that they will engage in serious criminal misconduct. Under either of these two pretrial confinement instances, the commander must also reasonably believe that a less severe form of restraint (such as conditions of liberty, restriction in lieu of apprehension, or apprehension) is inadequate. When these circumstances exist and other legal requirements are met, US military personnel may be placed in pretrial confinement under the MP's direct control. Commanders may choose to establish field confinement facilities within their AO. However, corps and division MP companies have the expertise to operate only a field detention facility for a limited period of time. These units cannot operate a confinement facility and have neither the resources nor the capability to operate such a facility on an extended basis. Therefore, all confinement is consolidated in the COMMZ whenever possible.

5-31. All assets of the MP battalion (I/R) (confinement) are employed to detain, confine, sustain, and protect US prisoners. As with the battalions conducting EPW/CI and resettlement operations, the battalion conducting confinement operations is generally located in the rear of the COMMZ, near logistics and transportation support. This allows US prisoners to be moved as quickly as possible from the corps's area to the COMMZ's confinement facility. Movements of US prisoners from the COMMZ to CONUS will be according to DA policy.

# Chapter 6

# Military Police Support to Corps

*Corps MP support their command by performing the MP functions critical to the success of their tactical commander's concept of operations.*

*USAMPS*

Corps are the largest tactical units in the US Army. They are the instruments by which higher echelons conduct operations at the operational level. Higher headquarters tailor corps for the theater and the mission for which they are deployed. They contain organic combat, CS, and CSS capabilities to sustain operations for a considerable period of time. Corps are capable of operating in a joint and multinational environment, providing $C^2$ for up to five divisions and covering up to 35,000 square kilometers.

## OVERVIEW

6-1. MP support to a corps is provided by an MP brigade (CS) assigned to each corps. The MP brigade provides combat, CS, and CSS throughout the corps's AO. However, subordinate MP units are not assigned to subordinate corps units. Instead, the MP brigade commander gives them an AO based on the corps commander's concept of operations. When possible, MP battalion AOs coincide with those of the CSG RAOCs.

6-2. The corps MP provide combat power within the command's rear area. They perform combat operations to counter Level II forces and to support the defeat of Level III forces. When properly augmented, the MP brigade may serve as the corps's TCF. The corps MP also provide a critical link between MP operations in the division and in the COMMZ. The corps MP support division commanders by helping the division MP conduct sustainment operations. The corps MP coordinate with the division MP for mutual support.

## THE MILITARY POLICE BRIGADE (COMBAT SUPPORT)

6-3. The MP brigade (CS) supporting a corps contains a brigade headquarters, up to six MP battalions (CS), numerous L&O detachments, and MWD teams. Each MP battalion (CS) has up to six MP companies (CS). As with the EAC's MP brigade (CS), the number of battalions and companies is determined by METT-TC. The corps MP brigade (CS) has additional companies to support each division and to provide security for the corps and COSCOM headquarters and corps ammunition storage areas (see Figure 6-1, page 6-2).

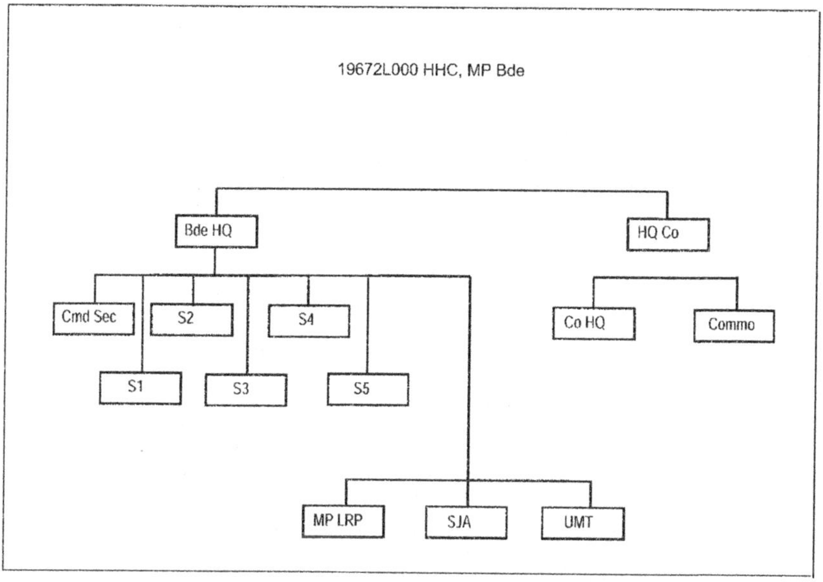


**Figure 6-1. Military Police and USACIDC Organizations**

6-4. The MP brigade HHC provides $C^2$ and A/L support to the brigade. The brigade HHC consists of a company headquarters and a brigade headquarters that contains the commander's immediate staff. The staff officers supervise the brigade's major organizational elements, including the—

- S1, S2, S3, S4, and S5.
- MP long-range plans (LRP) section.
- Communications section.
- SJA.
- IG.
- Public affairs.
- Unit ministry team (UMT).

6-5. The MP brigade (CS) command sergeant major (CSM) requires mobility and communications capabilities to execute his duties and responsibilities. The MP LRP section works with the corps G3 plans element, operating out of the corps main CP.

6-6. The MP battalion headquarters and headquarters detachment (HHD) provides $C^2$ for MP CS companies and any other assigned or attached MP elements. The battalion HHD consists of a detachment headquarters and a battalion headquarters that contains the battalion commander's staff. The staff officers supervise the major organizational elements, including the—

- S1, S2, S3, and S4.
- Communications and support sections.
- UMT.

6-7. As with the MP brigade (CS), the battalion CSM requires mobility and communications. The support section is vital to an MP battalion commander's ability to sustain his widely dispersed assigned or attached units during the performance of all five MP functions, primarily the MMS and AS functions.

6-8. The MP company (CS) provides support to an assigned AO. The company has a company headquarters, an MP operations center, a combat-medic section, and four platoons. The company headquarters provides maintenance, supply, communications, mess, and medical support to the unit. An MP operations center supports the unit's operation, conducting and planning for all five MP functions. The MP operations center includes three MP teams as the company-level response force.

6-9. The MP L&O detachments provide support to an assigned AO. The headquarters provides A/L support. The operation team plans and supervises desk operations, the traffic-accident and MP-investigation teams, and the force-protection teams. As with the EAC MP brigade (CS) detachments, requirements are based on the population supported and on METT-TC.

6-10. The number of assigned MWD teams is significantly less than those assigned to the EAC MP (CS) or (I/R) brigades. A kennel master, five explosive/patrol teams, and four narcotics/patrol teams are normally assigned to the corps MP brigade (CS). They are employed based on METT-TC.

## COMMAND AND CONTROL

6-11. The $C^2$ in a corps MP brigade (CS) is consistent with that of any Army brigade. The MP brigade commander commands the brigade and all attached personnel. Battalion commanders work for the brigade commander. The MP company commanders receive their orders and work for their respective battalion commanders and direct their platoon leaders according to mission requirements.

6-12. When corps MP assets are sent to augment the division PM or the MP company, they are placed under the OPCON of the division PM for as long as the unit is needed in the division area. This command relationship is applicable to company-size organizations and smaller. If an MP battalion (or larger MP unit) augments the division, then the division PM will not exercise OPCON of that MP unit. The MP battalion will be placed OPCON as designated by the division commander but not under the division PM. One L&O detachment is normally attached to the supported division and placed OPCON to the division PM. However, MP leaders at each level establish an MP $C^2$ relationship based on METT-TC and the supported commander's needs.

## EMPLOYMENT

6-13. Each corps MP brigade commander employs his assets according to METT-TC, the needs of the forces operating in his AO, and the priorities of the corps commander. Few MP assets in the corps area are employed to support fixed commitments. Instead, based on a broad consideration of the enemy and friendly situations, the corps MP are employed to support friendly forces engaged in combat, CS, and CSS operations. Since MP forces are dispersed throughout the corps area, the concentration of US forces, the location and vulnerability of critical sites, and the number of kilometers of the MSR to be controlled influence the designation of MP assets in the AO.

6-14. The MP brigade commander establishes his MP concept of operations based on the corps commander's concept of operations. His successful employment of MP assets depends on his foreseeing where the battle will be rather than where the battle is. Based on the rear-area IPB and PIAP, the MP brigade commander allocates and shifts resources to ensure the accomplishment of priority missions. This ensures the continuous support and forward sustainment of combat units and the safety of CSS units operating in the corps rear and sustainment areas.

6-15. To support the MP brigade commander's planning, the MP at brigade headquarters operate from several locations. Most of the staff locates in the vicinity of the corps rear CP where they can interface with the corps staff responsible for planning and executing rear operations. The control element of the brigade headquarters must be located where it can command and control its subordinate units. The brigade commander and his staff decide the best place to locate this element. The planning element of the headquarters locates near the corps main CP where it can interface with the corps commander's coordinating, special, and personal staffs. From there they monitor MP operations, integrate MP support with the corps plans cell for future operations, and learn the enemy situation through the G2's threat analysis almost immediately. The detailed information on rear-area activities and operations provided by the corps staff enhances the accuracy of the MP LRP.

6-16. The brigade S3 section provides the day-to-day planning and execution of an MP mission. The section provides a responsive CP that can relocate frequently throughout the AO. The S3 will normally provide liaison personnel to the corps rear CP, selected COSCOM units, interagencies, or other headquarters (based on mission requirements). When possible, battalion AOs coincide with the AOs of the CSG RAOCs. The MP brigade commander usually tailors battalions' AO boundaries to ensure responsive and flexible support across the corps's AO. He pays particular attention to the LOC behind the most heavily committed division and the critical bases and facilities in that area. He also ensures that the MP are available to respond quickly to combat operations throughout the entire corps rear area or during sustainment operations.

6-17. The MP brigade commander, coordinating with his battalion commanders, locates the MP companies where they can provide combat and CS power throughout the corps's AO. He bases his decision on the—

- Number and composition of urban areas.

- Location of CS and CSS units.
- Location of critical facilities (such as the headquarters, ammunition storage areas, and airfields).
- Expected threat.
- Level and frequency of support needed by the commander.
- Current and projected tactical situation.
- MSR network, including choke points and critical bridges and tunnels.
- Number of supported divisions and requirements.

6-18. The prioritization of MP missions is especially crucial during the early stages of the deployment when it is unlikely that an MP brigade commander will have a full complement of CS companies. Augmenting the division MP company with corps MP assets may not be possible at that time. Until it is, the MP brigade commander must locate corps MP assets to help meet the needs of the division while fulfilling the needs of the corps.

6-19. Like the brigade commander, battalion commanders plan the employment of their companies and platoons using METT-TC. Certain corps needs are constant. One MP company is allocated to provide security for the COSCOM, and one MP company is allocated to provide security for the corps's main CP. One platoon from that company may be used to secure the corps tactical CP or the jump CP. One or more platoons will also help secure the corps's rear CP. The number of MP assets allocated for a corps-level EPW/CI holding area and escort is based on METT-TC. However, a minimum of one platoon is dedicated to operate the corps's EPW/CI holding area and a minimum of one platoon per division is allocated for evacuating EPWs/CIs from division collection points. Additional MP assets may be allocated to provide security for the corps's ammunition storage area and ASPs supporting the divisions. The MP battalion commander places his companies where—

- MP assets support the brigade commander's concept of operations.
- The MP can support troop concentration, bases and base clusters, road networks, and critical areas.
- The MP can aggressively patrol critical terrain and monitor LZs and DZs to detect or deny enemy interference.
- The MP can respond to Level II threats.
- The MP can support the movement of combat resources throughout the AO.
- The MP can remove EPWs/CIs from division collection points.
- The MP can influence stragglers, refugees, and dislocated civilians.

6-20. Battalion commanders may choose to place a company behind the division rear boundary. This can help to ease the coordination between the corps MP and the division MP.

## Chapter 7

# Military Police Support to Division

*Division MP provide essential support to the forward tactical commander. The MP provide a flexible, mobile, and lethal force ready to be employed in combat or CS operations.*

*USAMPS*

The division is a large Army organization that trains and fights as a tactical team. Largely self-sustaining, it is capable of independent operations. The division is a unit of maneuver organized with varying numbers and types of combat, CS, and CSS units. Divisions fall mainly into two categories—heavy or light. The first category includes heavy divisions, their variants, and the Division XXI digitized heavy division. The second category includes the light infantry, airborne, and air-assault divisions. Each division can conduct operations over a wide range of environments. The success of Army operations depends on the success of its divisions.

## OVERVIEW

7-1. In corps operations, divisions are normally comprised of three maneuver brigades, each with up to nine maneuver battalions, artillery battalions, aviation battalions, and supporting CS and CSS units. Divisions perform a wide range of tactical missions and are self-sustaining for limited periods of time. The corps augments divisions as the mission requires. All divisions must be able to deploy and conduct offensive, defensive, stability, and support operations. Airborne and air-assault divisions must be able to conduct forced-entry operations. Divisions may be part of a JTF or serve as the ARFOR headquarters.

7-2. MP support to a division is provided by an organic MP company that includes the PM and his staff (assigned to each division) and is augmented by the corps MP brigade. The division MP company provides support in the DRA and in the maneuver brigade rear area. It is fully mobile in order to relocate frequently under short notice. While division MP companies are capable of performing all five MP functions, they must be augmented by corps MP/CID to fully perform all five MP functions throughout the division's battle space. The division PM must receive at least one corps MP company for GS missions in the DRA, one L&O detachment for L&O missions, and one CID detachment for criminal investigations and LOGSEC. In a mature theater (or based on METT-TC), the division PM can expect two corps MP companies. In addition, the division PM may request a corps MP battalion headquarters to provide $C^2$ to corps MP units operating in the DRA or to perform missions as a TCF. Because the need for MP support exceeds division organic assets (and many times exceeds augmenting corps MP assets), careful planning of MP employment is essential.

## DIVISION MILITARY POLICE COMPANY

7-3. The organization of the division MP company is set by specific tables of organization and equipment (TOE) and is designed to support a specific type of division. Regardless of the type of division, the assets in a division MP company enable a PM and his supporting staff to supervise the performance of all five MP functions. The PM coordinates the employment of the MP assets in the division's AO through his PM section. The division MP company contains a headquarters, a PM section, a medic section, and three to six platoons.

## COMMAND AND CONTROL

7-4. The $C^2$ of the division MP company extends downward from the tactical commander to the division PM. The division PM is the division commander's special staff officer. Under normal circumstances, he works for the chief of staff. The division PM has OPCON of the division MP company and MP assets that have been provided by the corps. The division PM plans and employs all of the organic, assigned, or attached MP according to METT-TC. He anticipates the support needs of the division commander and plans the employment to meet his needs. After learning the division commander's concept of operations and estimating the situation, the division PM decides which MP functions are critical to accomplish the commander's mission. He then recommends the allocation of resources, tasks his assets, and establishes the priority of support.

7-5. When the division MP company is augmented by corps MP platoons, the platoons are placed under the direction of the division MP company commander, who assigns an AO to each platoon. When corps MP assets augmenting the division make up a company, the company is attached to the division. The company then comes under the OPCON of the PM, who assigns it an AO. The augmenting corps MP company is usually given an AO from the division rear boundary forward. However, this assignment is flexible and based on the division commander's needs as set forth by the division PM. The four platoons of that corps MP company perform all five MP functions in that AO. Additionally, the corps MP brigade (CS) L&O detachment supporting a division is attached to the division and OPCON to the division PM.

7-6. The division MP company commander directs the employment of company assets. Through his company headquarters, he provides administrative, maintenance, and logistical support to the PM section and to the company's platoons that are dispersed throughout the division's AO. These sections are mobile to support the platoons without delay despite the extended distances that may be separating them. The company commander is also mobile. He travels the extended distances from one platoon area to another when checking on the status of his soldiers and resolving the problems affecting mission accomplishment. The division HHC provides mess support. Personnel administration is handled by the company headquarters and the DISCOM. The number and kind of assets in a division MP company and the configuration of its platoons and squads are determined by the type of division to which the company is assigned.

**NOTE: See the consolidated TOE update that is published yearly for complete details (base and objective) for all MP TOEs.**

7-7. At the direction of the division commander, the division band may be available to perform its secondary mission of supporting MP operations. When the intensity of the conflict prevents the band from performing its musical mission, the division commander may direct it to augment the MP company. The band, normally employed as a unit under the direction of the bandmaster, is under the OPCON of the MP company commander while augmenting the MP. When tasked to augment MP assets, band members may be employed to augment MP security at the division main CP or to augment perimeter security at the division EPW/CI collection point. The band is released to perform its primary mission as soon as the tactical situation permits.

## MILITARY POLICE EMPLOYMENT

7-8. The employment of division MP companies differs somewhat with the type of division to which they are assigned. In heavy divisions, the division MP companies provide GS to the DRA and DS to the division's maneuver brigades. When the platoon is providing DS to a maneuver brigade, the MP platoon leader coordinates all logistical support with the supported brigade. A platoon's DS relationship is seldom interrupted. However, METT-TC may require weighing MP support to the maneuver brigade tasked as the main effort or employing all platoons in a division MP company to function as a unit (such as during division-size river-crossing operations or deliberate breaching operations). In light divisions, the MP provide only GS to the DRA or during sustainment operations. Support exceptions are addressed under MP support to airborne and air-assault divisions, discussed later in this chapter.

7-9. In the division (where flexible support of an austere force is crucial), the division PM must have a clear understanding of situational awareness. To obtain current information for projecting MP needs in the division area, he must be mobile and be able to conduct split-cell operations. The assets available to the PM include the division MP company and at least one corps MP company. Corps augmentation is required for sustained operations and for special operations such as river crossings, dealing with dislocated civilians, and refugee internment or resettlement. The division PM coordinates with the corps PM and the MP brigade or CID commanders for—

- Evacuating and guarding EPWs/CIs from division to corps.
- Providing law-enforcement assistance to HN forces in the division's AO.
- Providing corps augmentation for the division's AO, convoy security, LOC security, AS, R&S, L&O, and other missions. These other missions may include augmentation for security of the division main CP, ASPs, and other critical facilities.
- Integrating, sharing, and exchanging police intelligence between corps and division MP elements.
- Providing CID support.

## SUPPORT TO THE HEAVY DIVISION

7-10. The Army's armored and mechanized infantry divisions (normally referred to as heavy divisions) provide mobile, armor-protected firepower that is normally employed for their mobility, survivability, lethality, and psychological effect (shock) on the enemy. These divisions destroy enemy armored and mechanized forces, and they can seize land areas and secure key terrain. Because of their strategic lift requirements, heavy divisions are slow to deploy from home staging bases into an AO. They have high consumption rates of supplies and have limited use in restrictive terrain. These capabilities and limitations are key factors in planning effective and efficient MP support throughout the division's AO.

7-11. In the heavy divisions, the PM section is organized to support split-cell operations at the main and rear CPs. The PM must be mobile to ensure that he is fully aware of the current status of critical MP operations. Therefore, the deputy PM (DPM) locates in one of the PM cells and—

- Handles the section's routine operations.
- Monitors ongoing division operations and MP support.
- Helps provide long-range planning and interface with the primary division staff.
- Forwards PM taskings to the division MP company commander.

7-12. The DPM and the operations sergeant normally set up operations at the division main CP where they can coordinate requirements with the division staff personnel. The company headquarters is initially located near the division rear CP in the division support area (DSA). Once augmentation arrives from the corps, the headquarters relocates with one of its GS platoons operating behind the brigade rear boundary. However, this location could be changed based on METT-TC, the supported commander's needs, and the company commander's idea of where he can exercise better $C^2$ for his unit.

7-13. The Army of Excellence (AOE) heavy division MP company has six platoons. Three platoons provide support to each maneuver brigade and are designated as DS. The other three platoons are designated as GS platoons. One MP platoon provides security for the division main CP; one provides security for the division's EPW central collection point; and one performs other MP operations within the division rear.

7-14. The GS MP platoons' AOs are configured based on METT-TC and the availability of MP augmentation from the corps. The DS MP platoons' AOs coincide with the supported maneuver brigade's boundary. Each platoon headquarters locates within its brigade's support area or any other area where it can best provide and receive support. To accomplish its mission, each DS platoon requires a minimum of two squads, each with three teams. One squad operates the EPW/CI collection point. The other squads perform MMS and AS operations. All MP platoons are capable of performing all five MP functions. However, performance of these functions is prioritized based on METT-TC and the division commander's concept of operations. The division PM, the company commander, and METT-TC dictate how these platoons should be tasked-organized to accomplish the mission.

## SUPPORT TO THE LIGHT INFANTRY DIVISION

7-15. The AOE light infantry division (LID) is one of the most rapidly and strategically deployable divisions. It fights as part of a larger force in war or conducts missions as part of a joint force in MOOTW. Its $C^2$ structure readily accepts any augmentation forces, permitting task-organizing for any situation. The augmentation required for the division is largely determined by METT-TC. The division's capabilities allow it to exploit the advantages of restricted terrain and limited visibility. It achieves mass through the combined effects of synchronized small-unit operations and fires rather than through the physical concentration of forces on the battlefield. These characteristics are key factors in planning and employing MP assets in support of the LID.

7-16. The LID MP companies are capable of performing all five MP functions. However, their performance of these functions is prioritized based on METT-TC and the division commander's concept of operations. Contrary to the heavy division MP company, the LID MP companies are much smaller. The constrained size of the LID MP companies makes corps MP augmentation crucial to the sustainment of MP operations. Additionally, the LID MP company is the only MP unit with the capability of antiaircraft support through the use of shoulder-fired air-defense weapons.

7-17. The company has three GS platoons to support the division. No platoons are provided to the maneuver brigade. One platoon is normally located in the vicinity of the division main CP so that its resources can help support CP security. Another platoon locates in the DSA and operates the division EPW/CI collection point. The last platoon has an AO configured according to METT-TC and the commander's priority of MP missions. Each GS MP platoon has a headquarters and three squads, each with two teams. The PM section is located in the vicinity of the division main CP. The exact location is based on the current operational status and on METT-TC.

## SUPPORT TO THE AIRBORNE DIVISION

7-18. The AOE airborne division can rapidly deploy anywhere in the world to seize and secure vital objectives. It conducts parachute assaults to capture initial lodgments, executes large-scale tactical raids, secures intermediate staging bases (ISBs) or forward operating bases (FOBs) for ground and air operations, and rescues US nationals besieged overseas. It can also serve as a strategic or theater reserve as well as a reinforcement for forward-presence forces. The airborne division can assault deep into the enemy's rear area to secure terrain or to interdict enemy supply and withdrawal routes. It can seize and repair airfields to provide a FOB and airheads for follow-on air-landed forces. It is capable of all other missions assigned to LIDs. The airborne division does not have sufficient armored protection to defeat heavier armored formations at close range. Therefore, engagements with enemy armored formations require special considerations. Antitank weapons in the division compensate for, but do not completely offset, this deficit.

7-19. Airborne divisions conduct operations in two phases—assault and defense. The division accomplishes the assault phase in three stages. First, they deploy; second, they establish the airhead; and third, they extend the logistics base and build their force.

7-20. Like other division MP companies, those assigned to the airborne division are employed to support their division commander's concept of operations. The airborne MP company has four platoons, each having three squads with two teams. Normally, the company headquarters and one of the platoon headquarters colocates with the PM section. The division PM's exact location depends on METT-TC. During the assault phase, the airborne division MP company provides DS to the assault brigade. The remaining platoons provide MMS and AS in the vicinity of the division main CP.

7-21. The nature of airborne operations makes the capture of EPWs likely. Therefore, during the first stage of the assault phase, the priority of MP support is given to EPW operations. After assembling the DZ or LZ, the MP collect EPWs captured during the assault. Combat elements are relieved of EPWs as far forward as possible. In airborne operations, EPWs are held for later movement to a central collection point. During the first stage of the assault, the MP perform limited straggler and refugee control and undertake AS operations, when possible.

7-22. MP support is reevaluated after the airhead is established in the second stage of the assault. The PM takes in consideration personnel and equipment flow, roadways, and security requirements to shift MP support priorities. As the entire operation matures, MP support may expand to include all five MP functions.

## SUPPORT TO THE AIR-ASSAULT DIVISION

7-23. The air-assault division executes tactical missions at operational depths to achieve strategic results. It is capable of launching brigade-size air assaults of nearly 4,000 soldiers from either an ISB or a tactical assembly area (TAA). Within 6 hours, this air-assault task force (AATF) can attack an opposing force, occupy and defend key terrain, or establish a FOB (out to 150 kilometers) from which even deeper operations can be executed. Air-assault operations are terrain independent, but they rely on suitable weather and a detailed attrition/assessment of enemy capability—particularly air-defense assets along air corridors and in the objective area. The air assault is preceded by detailed, lethal and nonlethal condition settings, culminating in a comprehensive condition check before execution.

7-24. As the corps AATF unit, the brigade combat team (BCT) task-organizes and habitually trains with both aviation lift and attack (Apache) battalions. An air-assault division MP company has four platoons, each having three squads with two teams. To facilitate operations, air-assault division MP platoons are habitually aligned with each of the three AOE BCTs; however, as with aviation assets, the MP are task-organized to support the main effort's BCT/AATF. When conducting these operations, two MP platoons may be tasked to provide support; one is available to posture and participate in the air assault, and the other conducts AS and MMS operations at the pickup zone (PZ). This second MP platoon may further provide MMS and accompany a ground-assault convoy (GAC) to the objective, or it may remain at the PZ to receive EPWs returning on aircraft from the objective area. The mission of conducting MMS along this vulnerable ground LOC is particularly critical if the AATF objective is a FOB from which further division operations will expand.

7-25. MP participation in the actual air assault competes for heavy lift with FA and air-defense systems, as well as with vital Class V resupply. However, the lethality and versatility of the MP bode well for their employment, and two MP platoons are available to support the brigade air assault as the division's center of gravity.

7-26. When possible, habitually aligned platoons remain with their brigades, and corps assets perform GS missions. However, when no corps assets are available and two division platoons are employed as stated above, the two remaining platoons conduct division EPW collection-point operations and other MP functions based on METT-TC. Normally, the EPW platoon and the MP company headquarters colocate in the DSA. As required (and based on METT-TC), airflow planning includes EPW/CI evacuation from the AATF/ FOB collection point back to the DSA. The PM section operates from the division rear CP to facilitate I/R operations and to coordinate MMS and AS with key logistical staff. Due to potentially extreme distances on the air-assault battlefield, the DPM normally locates with the division main CP to serve as a key G3 battle-staff member and to coordinate PIO with the G2.

## SUPPORT TO DIVISION XXI (HEAVY DIVISION)

7-27. Division XXI represents a significant change in the manner in which division operations are conducted. These changes are brought on by information-age capabilities; an increased integration of service components into an effective battle team; more lethal, survivable, and agile systems; and more capable soldiers and leaders. The Division XXI operates in a larger battle space and at a higher tempo than the AOE division. The division is evolutionary in design, but revolutionary in its use of information technology. It improves the Army's deployability while enhancing its ability to dominate in decisive fights. The following are characteristics of the Division XXI operational environment:

- **Multidimensional.** The division will operate in an extended battle space that goes beyond the traditional physical dimension of width, depth, and height. It includes portions of the electromagnetic spectrum; and it extends beyond the physical boundaries of the division through its communication and digital connectivity to other Army, joint, and coalition elements, even reaching back to CONUS.
- **Precise.** Precision operations go beyond a precision strike to include every aspect of military operations from deployment through combat and through redeployment or transition to other operations. Precision in decisive operations is enabled by three emerging capabilities. First, digitization provides soldiers and leaders at each echelon the information required for making decisions. Second, a full suite of strategic, operational, and tactical sensors linked to analytical teams fuses combat information into situational awareness across the battle space. Third, simulations enable Army elements to be tailored and operations to be planned, war-gamed, and rehearsed—yielding precision execution.

- **Nonlinear.** Nonlinear operations do not seek a battle-space grid of close, deep, and rear operations. Instead, the battle space is fluid, changing as METT-TC changes through the duration of mission preparation and execution. Another dimension of this characteristic is the synchronization of near-simultaneous operations to achieve nonlinear effects across the battle space.
- **Distributed.** Division operations are distributed or executed where and when required to achieve decisive effects concentrated at a decisive point. Dispersion empowers subordinates to operate independently within the commander's intent, leading to synergistic effects that exceed the effects of a centralized headquarters.
- **Simultaneous.** The concept of decentralization operations that are multidimensional, precise, distributed, and nonlinear yields the capability to conduct simultaneous operations across the battle space. Simultaneous operations seize the initiative and present the enemy leadership with multiple crises and no effective responses. Rather than a single, concentrated attack, the division executes a series of attacks (lethal and nonlethal) as simultaneously as possible.
- **Integrated.** Division operations are fully integrated with joint, multinational, and nongovernmental partners. Integrated operations enable the Army to leverage the full suite of capabilities that the services bring to the battle space.

7-28. The Division XXI heavy division is one of total integration—not rounding up or down. Reserve-component soldiers assigned to the division are an integral part of the division mission and task accomplishment. The division cannot operate without them. The division is focused on massing effects on the enemy—enabled by a new foundation of information technology and distributed logistics. The division is built around the pattern of "conduct decisive operations," accomplished by either fires or maneuver. The design performs complementary battle-space shaping in concert with the overall corps plan. Its long-range assets are not typically held in reserve nor just employed in the close fight; they are used to best support the corps and division commanders' plans of operations.

7-29. Although the basic structure of the division has not changed—three maneuver brigades, a division artillery, and a DISCOM—the internal structure and operational construction have changed in a near revolutionary way. The $C^2$ structure and the basic organizational structure are very similar to the divisions discussed above. However, by integrating the capabilities of digital technology and by using enhanced situational awareness, it is better able to execute combat operations.

7-30. The maneuver brigades have their own scouts—a brigade reconnaissance troop (BRT); therefore, they no longer rely exclusively on the battalion scouts. The maneuver battalions are reduced to three companies with a total number of 45 combat platforms. The division artillery has three Multiple-Launch Rocket System (MLRS) batteries per battalion. The cavalry squadron is under the aviation brigade's $C^2$ but retains and performs the division's traditional cavalry mission. Figure 7-1 depicts the organizational structure of Division XXI.

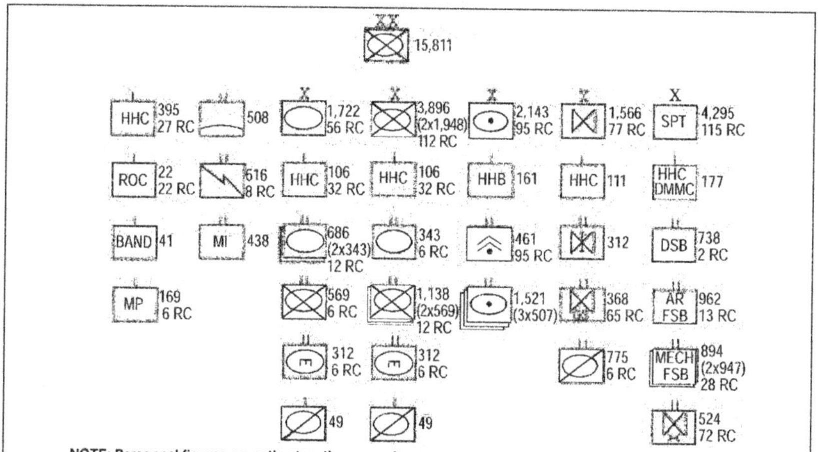


**NOTE: Personnel figures are estimates; they may change.**

**Figure 7-1. Division XXI Structure**

7-31. The division's CSS is centralized. Instead of each unit having its own organic CSS personnel, they are now centralized in CSS units. The CSS assets are reorganized and reassigned to the DISCOM. Forward-support companies (FSCs) are habitually associated with each maneuver battalion and forward-support battalions (FSBs) will support each maneuver brigade. This operational structure enables distribution-based rather than supply-based logistics concepts to support the division. The overall logistics footprint for the division is now reduced and the division has organic self-sufficiency to operate unsupported for up to 30 hours.

7-32. A major difference between the AOE fix-forward doctrine and the Division XXI maintenance support is the change in the actual repair site. Division XXI maintenance units directly supporting a given customer will diagnose faults and will replace components forward in the battlefield and will repair rearward. Units will be manned with maintenance personnel capable of performing DS and limited GS maintenance.

7-33. The division operates in a battle space two and a half times larger than an AOE division, and it will usually fight as part of a corps or a JTF. The division conducts offensive, defensive, stability, and support operations in cooperation with other elements of the corps/JTF to gain tactical or operational results significant to the joint-force commander's campaign goals.

7-34. Although the division is a heavy division, MP support is somewhat different from that of an AOE heavy division. The PM (based on METT-TC) establishes operations where MP assets can best be commanded. At the same time, overall MP operations can be synchronized to support operations within the division rear and the division commander's intent can be effectively supported. To this end, the PM conducts split-cell operations within the division main CP. Normally, the PM section is distributed between the security-operations cell (SOC) and the mobility/survivability (M/S) cell. The DPM (who works out of the M/S cell) is also part of the division planning team. He ensures that MP operations are integrated into division plans according to the PM's intent. The DPM attends coordination meetings and assumes coordination responsibility with the rest of the division staff when the PM relocates away from the division main CP.

7-35. The PM, as a staff officer, is involved with every aspect of planning and with the military decision-making process (MDMP). He uses the Force XXI Battle Command—Brigade and Below (FBCB2) $C^2$ system to understand the battlefield. He understands the mission and the commander's intent and priorities. The PM has complete situational awareness, establishes an effective staff relationship with the G3, and recommends the best way to employ MP resources.

7-36. The Division XXI heavy division MP company has five platoons. Although tasked-organized based on METT-TC and the commander's needs, three MP platoons provide DS for each maneuver brigade. The DS platoon requires a minimum of three squads (each having three teams) to accomplish the mission. The other MP platoons are normally employed forward in sector, providing AS and force protection for the division main and tactical (TAC) CPs and for other HVAs in the vicinity of the maneuver brigade rear boundary. As with the AOE heavy division, the Division XXI heavy division requires augmentation with at least one corps MP company for CS missions in the DRA and can expect two companies in a mature theater (based on METT-TC) as the main effort. In addition, the division may request an MP battalion headquarters to command and control CS MP operations in the rear or to conduct TCF operations.

7-37. While the division MP company is capable of performing all five MP functions, it must be augmented by the corps MP to fully perform all of these functions throughout a mature division's battle space. The PM prioritizes MP functions (based on the supported commander's needs), conducts a mission analysis, and requests needed corps assets through the division G3. Corps MP companies, L&O detachments, and CID assets are needed for sustainment operations. When corps MP companies and L&O detachments are attached to the division, they are placed under OPCON of the PM. The PM will then assign an AO and missions based on METT-TC.

7-38. The PM supports the division's distribution-based logistics concept by conducting aggressive R&S operations that secure the LOC. Since the main effort is to secure the LOC, traditional MMS missions (such as convoy security and escorts) are performed by exception. Through the allocation of increased MP to conduct LOC R&S operations aggressively, the division is able to detect and destroy threat forces before they interrupt operations in rear-area or sustainment operations.

7-39. Digitization gives the division the situational awareness that allows it to cover much larger distances and to move much faster. The systems that provide this digitization are generally unprotected HVAs on the battlefield, which require aggressive AS and force-protection operations.

7-40. The EPW operations are much more fluid, sometimes requiring the division GS MP to move forward and assume brigade collection points, thus allowing DS platoons to move forward with their brigades. Similarly, corps MP may be required to move forward and assume the division's central collection points as the division's rear boundary is pushed forward to shorten the LOC.

7-41. As with all AOE divisions, until corps L&O and CID detachments augment the division, L&O operations are on an exception basis. The MP collect intelligence during every facet of mission execution. The PM develops police intelligence in the division main CP through coordination with the SOC, the G2, the Assistant Chief of Staff, G5 (Civil Affairs) (G5), and the Assistant Chief of Staff, G6 (Communications) (G6).

# Chapter 8

# Separate Brigades and Initial/Interim Brigade Combat Teams

*The MP support separate brigades and IBCTs designed to provide the brigade commander with MP assets that can serve as a force multiplier and that can be employed as an economy of force. The brigade MP platoon is as lethal, flexible, and capable as any other platoon in the MP Corps.*

*USAMPS*

The Army uses separate brigades to inject a small but powerful force where it is needed. It must be able to fight and win while operating on its own for a sustained period of time. It must be able to defend itself on a 360-degree front in war or MOOTW. While there are some variations of separate brigades, it is the heavy separate brigade that is most commonly found within the force.

## OVERVIEW

8-1. The Army's IBCT is a full-spectrum, wheeled combat force. It is employed in all operational environments against all projected future threats. However, it is designed and optimized primarily for employment in small-scale contingency operations in complex and urban terrain, confronting low-end and midrange threats that may employ both conventional and asymmetric capabilities. The IBCT deploys very rapidly, executes early entry, and conducts effective combat operations immediately on arrival to prevent, contain, stabilize, or resolve a conflict through shaping and decisive operations. The IBCT participates in war (with augmentation) as a subordinate maneuver component within a division or a corps and in a variety of possible roles. The IBCT also participates (with appropriate augmentation) in stability and support operations as an initial-entry force or as a guarantor to provide security for stability forces by means of its extensive capabilities.

## SUPPORT TO SEPARATE BRIGADES

8-2. MP support to separate brigades is provided by a four-squad MP platoon organic to the brigade HHC. A separate PM cell within the brigade HHC serves as the $C^2$ element for the platoon (see Figure 8-1, page 8-2). Support to the platoon and the PM section for maintenance, supply, mess, and communications is provided by the brigade HHC. Since the platoon and PM section have no organic support, the MP leadership must perform close coordination for this support. However, the MP platoon must compete with other brigade HHC assets for priority of repair for weapons, vehicles, and communications equipment.

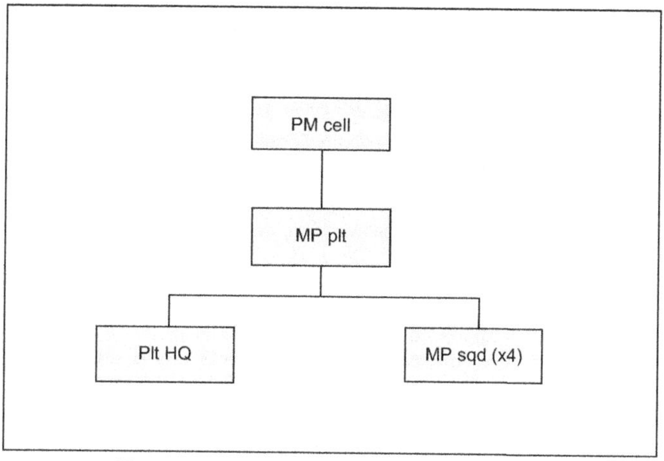


**Figure 8-1. MP Support to a Separate Brigade**

8-3. The $C^2$ of an MP unit supporting separate brigades also extends down from the tactical commander. The separate brigade PM has OPCON of brigade MP assets the same way the division PM has OPCON of division MP assets. This includes any MP assets that may have been provided from the corps. The PM advises the commander of a separate brigade on matters pertaining to MP operations. The platoon leader directs the execution of his platoon's mission based on the priorities set forth by the PM and the supported commander. Since the separate brigade's organic MP platoon is more robust than an MP platoon supporting a division maneuver brigade, METT-TC will determine the requirement for augmentation. However, corps L&O and CID augmentation is required.

8-4. The separate brigade's MP platoon is capable of performing all five MP functions. However, its resources are quite limited. Although the MP squads are employed according to METT-TC, the platoon supporting the separate brigade may have—

- One squad operating in the EPW collection point.
- One squad providing a mobile security screen and providing AS around the brigade's main CP.
- Two squads conducting MMS and AS throughout the brigade's rear area.

## SUPPORT TO THE INITIAL/INTERIM BRIGADE COMBAT TEAM

8-5. The IBCT is a divisional brigade. It is designed to optimize its organizational effectiveness and to balance the traditional domains of lethality, mobility, and survivability with the domains required for responsiveness, deployability, sustainability, and a reduced in-theater footprint. Its two core qualities are its high mobility (strategic, operational, and tactical) and its ability to achieve decisive action through a dismounted infantry assault. The major fighting components are its motorized infantry battalions. The IBCT has a unique reconnaissance, surveillance, and target acquisition (RSTA) squadron to enhance situational understanding.

8-6. To achieve a rapid deployment threshold, the brigade's design capitalizes on the widespread use of common vehicular platforms—particularly a highly mobile, medium-weight, combat/CS platform coupled with the minimization of the personnel and logistical footprint on theater. Encompassing a personnel strength of about 3,500 and preconfigured in ready-to-fight combined-arms packages, the entire IBCT can deploy within 96 hours of "first aircraft wheels up" and begin operations immediately upon arrival at the APOD. In essence, the APOD is the TAA. The IBCT cannot conduct forced-entry operations, but it provides the JFC with an improved capability to arrive immediately behind forced-entry forces and begin operations to shape the battle space and to execute decisive action to expedite decisions. Once committed, the IBCT can sustain operations for up to 180 days without relief.

8-7. The IBCT's organization is expandable either through augmentation or scalability (according to METT-TC) in any given contingency. The IBCT is scalable in terms of its ability to accept like-type additional forces to expand core tasks and functions already resident within the IBCT (for example, adding additional infantry or RSTA organizations). The IBCT is also capable of accepting augmentation consisting of units or elements that execute tasks or functions not resident within the IBCT (for example, adding armor, air-defense, MP, or aviation assets). In both cases, added units execute their normal mission-essential task list (METL) tasks and, therefore, will not require extensive training to deploy and operate with the IBCT. In both cases, the IBCT includes the command, control, and communications ($C^3$) capabilities necessary to permit the rapid integration of additional enabling capabilities, particularly for operations outside the scope of small-scale contingency operations (such as stability operations, support operations, and war).

8-8. The IBCT organization excludes other unit-based capabilities often provided in a division slice. However, for each missing capability, the brigade headquarters includes staff cells capable of conducting rudimentary planning and analyses to ensure that all functional-area considerations are incorporated in route planning and preparation for operations. The first MP elements in support of the IBCT are at the brigade headquarters level.

8-9. The MP planning cell is composed of a two-person (a major [MAJ] and a sergeant first class [SFC]) planning cell located within the HHC's maneuver-support cell and under the direct staff supervision of the brigade S3. The role of the MP planning cell is significantly different from that of a traditional division PM or a separate brigade PM. The main difference is in the IBCT's lack of organic or habitual MP assets. However, the absence of organic MP assets makes the job of the MP planning cell that much more critical. The MP planning cell must—

- Understand the organization, capabilities, and limitations of the IBCT.

- Conduct effective liaison with higher headquarters PM elements.
- Become an effective planner and anticipator of MP requirements.
- Prepare MP staff estimates and employment recommendations.
- Plan for MP deployment via air, sea, rail, or land.
- Task-organize MP units effectively and efficiently.
- Assume $C^2$ of incoming MP forces or, if operating under a division or corps, relinquish $C^2$ to the division/corps PM or battalion commander (if appropriate for effective employment of MP forces).

8-10. Depending on METT-TC, the brigade may be augmented by MP elements ranging from a platoon to a battalion. In any case, the MP planning cell then becomes a staff planner and coordinator of functional matters pertaining to—

- Force protection, physical security, and vulnerability countermeasures.
- Operations for collecting, processing, and evacuating (possibly) EPWs, CIs, and detainees.
- US military prisoner operations.
- Operations for processing and controlling dislocated civilians and refugees.
- Customs and counterdrug operations.
- Criminal investigations and CID-/MP-investigator support.
- PIO and establishing links with HN police agencies and other international/interagency law-enforcement agencies.
- MMS, AS, and L&O operations.
- Coordination of MP or CID division/echelons above division (EAD) augmentation forces, MWD teams (explosives, narcotics, and patrol), L&O detachments, and MP I/R units through the division, corps, and major Army command (MACOM) PMs.
- Training guidance to other US forces required to perform MP functions in the absence of MP forces.

8-11. Since the IBCT is a divisional brigade, the division PM and the IBCT MP planning cell play an important role in developing an optimum MP force package to support the brigade commander's concept of operations. Despite the brigade's early-entry timelines, the MP planning cell must consider and plan for MP augmentation forces as early as possible to free valuable combat resources. Small-scale contingency operations that result in numerous EPWs, civilian detainees, and refugees will hamper momentum and freedom of maneuver.

8-12. Once the initial brigade receives MP augmentation (see Figure 8-2), the MP priority of effort during the offense will focus on ensuring that routes remain unencumbered and secure for movement of ground combat, CS, and CSS forces. The MP's priority of effort during the defense will focus on conducting AS and counterreconnaissance along the LOC, $C^2$ centers, and CSS bases. The MP may conduct response-force operations or become part of the TCF.

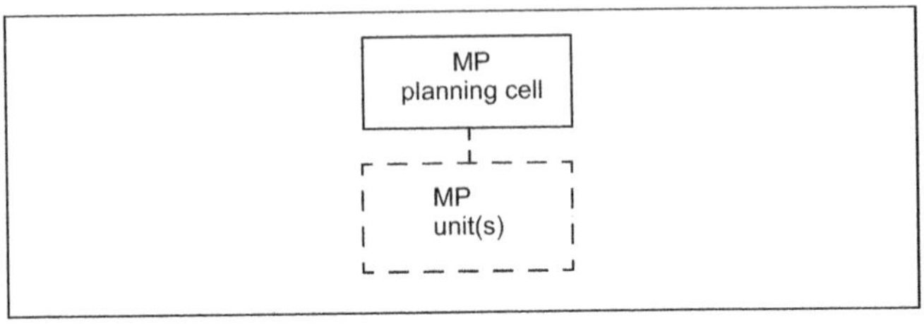


Figure 8-2. MP Support to the IBCT

8-13. Stability and support operations present some unique challenges. The IBCT may be deployed to a geographical area that is politically unstable, that lacks civil control, or that is in complete turmoil. The MP planning cell plans for and coordinates MP support according to available resources and the supported commander's needs. In this scenario (and based on METT-TC), an MP battalion TF could be required to deal with the challenges of—

- A significant number of refugees or dislocated civilians.
- AS or force-protection issues.
- MMS operations.
- Black-market and criminal investigations.
- Restoration of order.
- Civil unrest.
- Intervention of private and nongovernmental organizations.

# Chapter 9

# The United States Army Criminal Investigation Command

*The history of the USACIDC goes back to World War I, when General John J. Pershing organized the CID in France. Today, the USACIDC is the Army's sole agent responsible for investigating felony crimes on and off the battlefield. The USACIDC provides investigative support to commanders at all echelons.*

*USACIDC*

The USACIDC investigates offenses committed against US forces or property, or those committed by military personnel or civilians serving with US forces or where there is a military interest. The USACIDC agents investigate violations of international agreements on land warfare. They conduct special investigations at the direction of the USACIDC's commanding general (CG) or a higher authority. In general, the USACIDC's missions include—

- Investigating and deterring serious crimes.
- Conducting sensitive/serious investigations.
- Collecting, analyzing, processing, and disseminating criminal intelligence (CRIMINTEL).
- Conducting protective-service operations for designated personnel.
- Providing forensic-laboratory support.
- Maintaining Army criminal records.
- Enhancing the commander's crime-prevention and force-protection programs.
- Performing LOGSEC operations.

## OVERVIEW

9-1. The USACIDC's operations help the commander maintain discipline and order by preventing or investigating crimes that reduce a unit's ability to fight. During the investigation of serious crimes, the USACIDC concentrates its efforts on investigating serious crimes such as wrongful deaths, controlled-substance offenses, theft, fraud, sex crimes, and assaults. The USACIDC also conducts sensitive and special investigations involving matters pertaining to senior Army officials and those associated with classified programs.

9-2. The USACIDC agents collect, analyze, process, and disseminate criminal intelligence/information relating to crime within or directed toward the Army. Specific information relating to modus operandi, crime techniques, investigative leads, gang violence, and terrorism is shared with the appropriate intelligence and law-enforcement agencies. Conversely, USACIDC agents solicit and receive crime-related information from the MP and from local, national, and foreign law-enforcement and intelligence agencies. Special agents identify and evaluate crime-conducive conditions and indicators of potential attacks against Army property, facilities, or personnel. They then provide reports to the appropriate commander.

## COMMAND AND CONTROL

9-3. The USACIDC is a centralized (stovepipe) MACOM whose special agents in the field report through the USACIDC's chain of command (detachment to battalion to group) to the CG, who in turn reports directly to the Army Chief of Staff and the Secretary of the Army (see Figure 9-1).

9-4. As an independent MACOM, the USACIDC's tactical units are not under the $C^2$ of supported organizations. The USACIDC's elements may be temporarily attached to a supported commander when required to accomplish a nonstandard mission. Reassignment, promotion, accreditation, and disciplinary actions are retained by the USACIDC. Attachments will be coordinated with the appropriate USACIDC headquarters planning the specific mission and approved by the CG, USACIDC. Additionally, although there is no formal staff relationship, USACIDC commanders advise their supported commanders on criminal-investigation matters. This enhances the quality, reliability of information, support, and trust between USACIDC elements and their supported commanders.

## WARTIME SUPPORT

9-5. The USACIDC supports each echelon of command from the division to the ASCC. The theater USACIDC structure is comprised of a $C^2$ headquarters

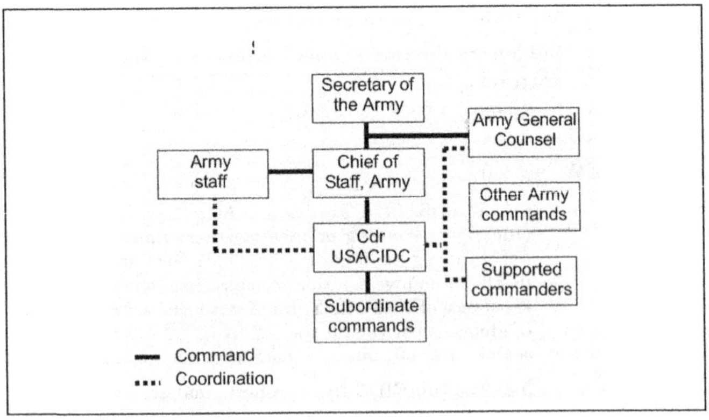


**Figure 9-1. USACIDC Chain of Command**

and mobile, modular, and tailorable investigative detachments. The USACIDC supports combatant commanders with the following functions:

- **LOGSEC.** Tracking and protecting materials and equipment from the manufacturer to the soldier on the battlefield.
- **CRIMINTEL.** Collecting, consolidating, analyzing, and disseminating intelligence associated with criminal and terrorists activities targeted at Army interests.
- **Criminal investigations.** Investigating war crimes and, in some cases, crimes against coalition forces and HN personnel.
- **Protective-service operations.** As in peacetime, protecting key personnel anywhere on the battlefield.

9-6. The USACIDC's LOGSEC function protects the Army's supply pipeline against criminal activities from the manufacturer, through logistics channels, all the way to the frontline soldier. It involves preventing, detecting, and investigating criminal and terrorist activities such as supply diversion, destruction, and sabotage or product substitution. The USACIDC's LOGSEC operations assess LOGSEC, identify weaknesses, and provide a prioritization of threats so that commanders can implement preventive measures to reduce the vulnerability of the logistics pipeline. Whenever possible, the USACIDC will initiate actions to recover logistical losses and return them to Army control.

The Army let a $22 million contract for the manufacture of its mobile kitchen trailer (MKT)-90 mobile field kitchens in 1990. These kitchens were distributed to Army units worldwide, including those units serving in Saudi Arabia during Operation Desert Shield/Desert Storm. After distribution of the MKT was well underway, a US government quality-assurance representative received information from the manufacturer's employees indicating that the kitchens were not built to contract specifications. This information led to an investigation by the USACIDC's major fraud procurement unit (MFPU). The investigation revealed that inferior material and manufacturing processes were substituted, which allowed the contractor to realize an additional $228,000 in profit. The US Army Environmental Hygiene Agency determined that the MKT-90 had serious sanitation defects and posed a safety hazard to troops using and being fed from these kitchens. The problem with the MKT was identified and expeditiously resolved to ensure the safety of soldiers in the field. The primary subject was indicted by a grand jury and subsequently pled guilty to one count of major fraud. He was found guilty, sentenced to 10 months confinement, and ordered to pay restitution.

9-7. The USACIDC's CRIMINTEL collection efforts focus on the identification and prevention of terrorist and nonterrorist crimes against US and allied military personnel, facilities, and other interests. Upon collecting this information, agents recommend countermeasures to combat subversive activities through coordination with MI, the PM, rear-area operations officers, and HN military and civil intelligence agencies. Additionally, when directed, the USACIDC becomes the lead US military investigative agency at theater level. When this happens, it is tasked with leading the prevention-of-terrorism effort from all services, not just the Army component.

> The MI analysts, the MP, and US Army criminal investigators deployed in Kosovo formed a combined all-source information center. This center provided threat assessments to units assigned to TF Falcon. Fifty-five percent of these assessments referenced some form of criminal activity. These intelligence products provided commanders with reliable information concerning potential incidents or criminal activity. With nearly 7,000 US soldiers deployed within the region, these assessments were absolutely crucial for the force-protection planning effort.
>
> The fusion cell within the information center developed intelligence products from national, theater, and operational sources. Due to the significant threat posed by criminal elements in the region, two CID military agents and two CID civilian analysts were attached to the fusion cell to facilitate the police-intelligence function. The CID personnel, in cooperation with MP soldiers, played a key role by linking CRIMINTEL to specific groups and events. The CRIMINTEL collection effort was specifically targeted on weapons, drugs, organized crime, and the identification of smuggling routes. The identification of smuggling routes resulted in a significant increase in the number of weapons being confiscated. The timely transfer of CRIMINTEL products to tactical units enabled a rapid response to serious confrontations, increased confiscation of arms and ammunition, and improved stability in TF Falcon's AO.

9-8. During war and MOOTW, the criminal-investigation effort includes—

- War crimes and, when directed, crimes against coalition forces and HN personnel.
- Major felony crimes committed by US personnel or EPWs/CIs and, when directed, crimes committed by foreign nationals in which there is an Army interest.
- Criminal acts by indigenous personnel, factions, and ad hoc groups.
- Special/sensitive investigations, hostage negotiations, polygraph operations, and force-protection operations.
- CRIMINTEL operations.

9-9. The USACIDC detachments provide enhanced coverage and protection for designated key and essential leaders during war and MOOTW. The protection requirement for senior JTFs and Army commanders may be significantly greater during MOOTW than during peacetime or war as the propensity of asymmetrical threats (such as criminal and terrorist groups) operating in the AO also increases.

## ORGANIZATION

9-10. The USACIDC is organized with an array of multifunctional units. As with many other Army organizations, its force is in a transition process. During peacetime, the USACIDC groups provide brigade-level area support for worldwide Army operations. The group provides $C^2$, staff planning, and administrative support for USACIDC elements assigned to an AOR or deployed to a contingency operation. During war, a group provides the same capabilities in support of an ASCC and provides $C^2$ for all USACIDC units in the theater. Regardless of the operational environment, the group ensures the connectivity between all USACIDC units. It establishes and maintains links with supported units and interagency, joint, multinational, allied, and HN authorities on matters pertaining to Army and USACIDC operations. The group headquarters has a command section, a detachment headquarters, an S1, an S2, an S3, an S4, an SJA, and a communications section. It provides $C^2$ for up to six subordinate battalions. Major differences in the AOE and Force XXI groups are in an increase of support and mission personnel and the movement of all polygraph support down to battalion level.

9-11. During peacetime, the USACIDC battalions provide area support for worldwide Army operations. They provide $C^2$, staff planning, forensic-science support, CRIMINTEL, polygraph support, technical guidance and supervision of USACIDC operations, and administrative support for USACIDC elements assigned to an AOR or deployed to a contingency operation. During war, the USACIDC battalion provides the same capabilities in support of each corps and TSC. Under AOE designs, one battalion supports each corps and TSC. Under the Force XXI design, up to two battalions support each TSC. Like the groups, each USACIDC battalion ensures connectivity and establishes and maintains links with other units/agencies within its AOR. The battalion headquarters consists of a command section, a detachment headquarters, an S1, an S2, an S3, an S4, and a communications section. It provides $C^2$ to five USACIDC detachments on a sustained basis. Its span of control can accommodate up to seven detachments for a short period of time. Major differences in the new battalion are the addition of a CSM, an S6 officer, a detachment commander and support personnel, and CRIMINTEL and forensic personnel.

9-12. The USACIDC investigative team is the smallest operational investigative element. Each team consists of two special agents (a warrant officer and a noncommissioned officer [NCO]). METT-TC requirements dictate that these teams have the capability to operate independently from the detachment headquarters. The investigative teams are the building block for both AOE and Force XXI USACIDC detachments. Both designs provide levels of flexibility to task-organize without splitting units apart. The detachment (with its complement of teams) performs the full range of criminal-investigative functions in all operational environments, but it has limited CRIMINTEL and LOGSEC management capabilities and no forensic or polygraph capability. Each detachment includes a headquarters section and up to eight investigative teams. METT-TC requirements dictate that these teams have the capability to operate independently from the detachment headquarters. The major differences between the AOE and Force XXI detachments' design begin with replacing the heavy and light division support elements (DSEs), the corps-area support element (CASE), the TSC-area support element, (TASE), and the port-area support element (PASE) designs with a modular design consisting of a standardized headquarters element and two special-agent sections. Each section consists of four investigative teams. Personnel administration, supply, and the bulk of administrative support found in the AOE organizations have been consolidated at battalion.

Additionally, the Force XXI detachment is commanded by a special-agent warrant officer.

9-13. Both the AOE and the Force XXI units retain the Army's standard dependencies upon other units for support requirements (religious, health services, finance, photographic processing, and so forth). The new USACIDC TOEs have enhanced A/L support capabilities. However, the new units still retain the AOE support requirement for food services and emergency Class III/V resupply from CS MP brigades, battalions, and companies.

## THE FIELD INVESTIGATIVE UNIT

9-14. Sensitive investigations are normally conducted by the field investigative unit (FIU). The FIU is a one-of-a-kind organization within the DOD that enables the Secretary of the Army to conduct sensitive investigations requiring access to special information or programs that are highly classified. The FIU works closely with the Army IG, the Judge Advocate General of the Army, and the Army General Counsel to support commanders in the special-operations and intelligence communities and in the area of classified acquisition programs. The FIU may also be assigned investigations involving senior Army personnel or those of special interest to the Army leadership.

## THE COMPUTER-CRIME INVESTIGATION UNIT

9-15. The computer-crime investigative unit (CCIU) deals with intrusions involving classified networks or multijurisdictional offenses. The CCIU works closely with MI and federal law-enforcement agencies to coordinate military actions, nonmilitary-affiliated offenders, and foreign-intelligence services. In addition, it provides technical assistance to CID elements that are conducting computer-related investigations.

## THE PROTECTIVE-SERVICE UNIT

9-16. The protective-service unit (PSU) provides worldwide protective services to designated personnel to protect them from assassination, kidnapping, injury, or embarrassment. The PSU plans, coordinates, and executes executive protection for—

- The Secretary of Defense.
- The Deputy Secretary of Defense.
- The Chairman of the Joint Chiefs of Staff.
- The Vice Chairman of the Joint Chiefs of Staff.
- The Secretary of the Army.
- The Army Chief of Staff.
- The Army Vice Chief of Staff.
- Others (as directed).

9-17. During war and MOOTW, the USACIDC routinely provides protective-service support to senior Army ground commanders and, in some cases, to JTF commanders. The USACIDC agents also provide training to the MP details providing protective services to senior field commanders.

The President of the US hosted the North Atlantic Treaty Organization's (NATO's) 50th summit anniversary celebration in Washington, DC during the period of 22 to 26 April 1999. The President invited 28 heads of state or prime ministers, 48 foreign ministers, 88 ministers and chiefs of defense, and the chairman of the military committee to the summit. During the conference, the principal guests resided in 23 hotels in Washington, DC. The responsibility for providing protection for the ministers, chiefs of defense, and the chairman for the military committee was given to the USACIDC's PSU. In preparation for this mission, additional training on protective-service operations and evasive driving techniques was given to over 200 reserve agents. These individual mobilization augmentees, troop program units, and other USACIDC agents were critical to the successful execution of this mission.

## US ARMY CRIMINAL-INVESTIGATIONS LABORATORY

9-18. Forensic-laboratory support is provided by the US Army Criminal-Investigations Laboratory (USACIL) that is currently located at Fort Gillem, Georgia. The USACIL is a nationally accredited forensic laboratory that provides all three services with dedicated, highly responsive, deployable, state-of-the-art scientific and forensic investigative support. The six forensic divisions of the USACIL include the following:

- Chemistry (serology, drugs, trace, and deoxyribonucleic-acid [DNA] testing).
- Questioned documents (alterations and comparisons of handwriting, typing, and printing).
- Firearms and tool marks (weapons and obliterated numbers).
- Photography (evidence, charts and photos for court).
- Latent prints.
- Computer forensics.

9-19. The USACIL collects, receives, and examines evidence and prepares reports of findings. It also provides expert-witness testimony at court-martials.

## US CRIME RECORDS CENTER

9-20. The US Crime Records Center (CRC) maintains the Army's criminal records. It maintains more than two million MP reports and USACIDC reports of investigations. These records are retained for at least 40 years before being destroyed. During their retention, the records are actively maintained and frequently researched.

During the course of an investigative search, civilian police authorities seized 19 hand grenades. A routine search of stolen property and criminal index files failed to identify the grenades as having been stolen from a nearby military installation. However, a search of the CRC files identified a USACIDC investigation relating to the theft of a case of grenades that occurred 20 years earlier. Further investigation revealed that the lot number of the recovered grenades was identical to that of the stolen grenades. The investigative leads resulting from the CRC file search contributed to the resolution of both cases.

9-21. The CRC conducts more than 10,000 criminal-history name checks each month to identify victims and perpetrators of criminal offenses. The checks are requested not only by USACIDC agents, but also by other military and civilian law-enforcement officials. The CRC is also the Army's agent for Freedom of Information Act (FOIA) and Privacy Act requests relating to criminal investigations and MP reports. The CRC also manages the Army's polygraph programs and their support to Army installations around the world.

## EMPLOYMENT

9-22. Whenever possible, USACIDC units are colocated with the MP. This provides unified MP support across the battlefield, and it facilitates logistical support (primarily mess) provided by MP units. However, the USACIDC maintains its vertical $C^2$. Based on METT-TC, the MP group (CID) will colocate with the supported ASCC and the MP battalions (CID) supporting the corps and EAC will colocate with the MP brigade (CS). When feasible, MP detachments (CID) will colocate with MP battalions/companies and division MP companies. Figure 9-2 illustrates a typical battlefield array for AOE organizations, and Figure 9-3 illustrates a Force XXI array.

9-23. Since USACIDC detachments are austere organizations with little organic A/L capability, they rely on the battalion for unit-level maintenance, supply, and personnel support. However, detachments must often operate independently when performing criminal-investigation functions throughout the AOR. During these times of long-distance separation, detachments must seek A/L support from the supported unit.

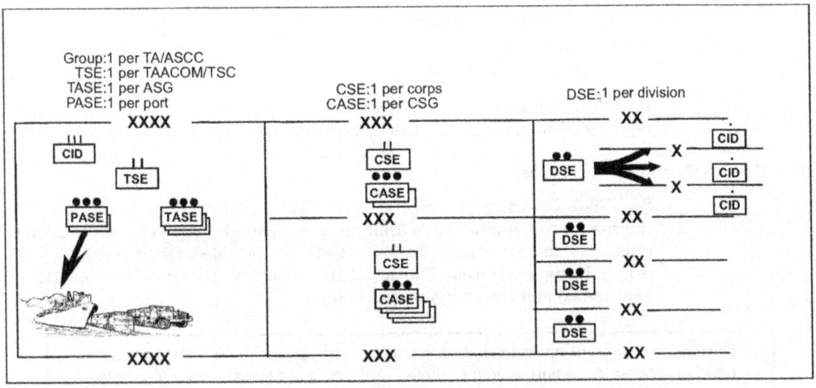


Figure 9-2. AOE Battlefield Array

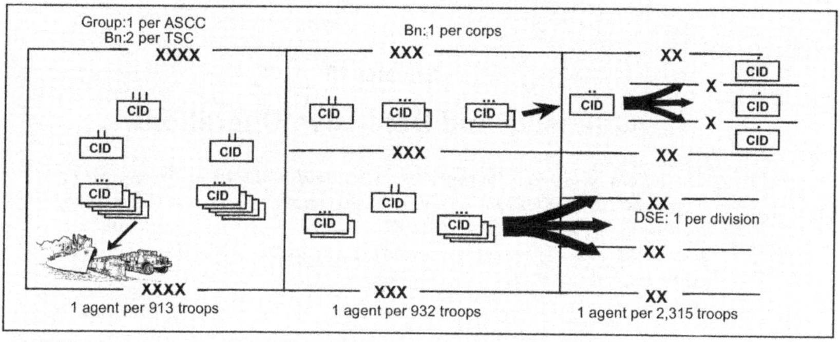


**Figure 9-3. Force XXI Battlefield Array**

## Chapter 10

# Offensive and Defensive Operations

*The whole of military activity must relate directly or indirectly to the engagement. The end for which a soldier is recruited, clothed, armed, and trained; the whole object of his sleeping, eating, drinking, and marching is simply that he should fight at the right place and the right time.*

*Clausewitz*

Army doctrine considers the full range of operations from war to MOOTW. An operation is a military action or the carrying out of a military mission needed to gain the objectives of any battle or campaign. Offensive and defensive operations normally dominate military operations in war and in some smaller-scale contingencies. An offensive operation is aimed at destroying or defeating the enemy and imposing our will on him for a decisive victory. Defensive operations defeat an enemy attack, buy time, economize forces, or develop conditions favorable for a counteroffensive that regains the initiative and defeats the enemy.

## OVERVIEW

10-1. The offense is the decisive form of battle. Offensive campaigns and major operations are designed to achieve operational and strategic objectives quickly and decisively at the least cost. Offensive operations combine the terrain and the force to achieve the objective. The main purpose of the offense is to defeat, destroy, or neutralize the enemy force. Offensive operations are characterized by a rapid shift in the main effort to take advantage of opportunities by momentum and simultaneous destruction of enemy defenses.

10-2. The immediate purpose of defensive operations is to defeat an attack. Military forces defend only until they can gain sufficient strength to attack. Commanders choose to defend when they need to buy time, to hold a piece of terrain, to facilitate other operations, or to erode enemy resources at a rapid rate while reinforcing friendly operations. The ARFOR conduct defensive operations as part of major operations and campaigns and in combination with offensive operations (see FM 3-0).

## SUPPORT TO OFFENSIVE OPERATIONS

10-3. MP support to offensive operations varies according to the type of operation being conducted. The MP functions support each echelon commander based on METT-TC. In the offense, MP priorities are often placed on MMS, I/R, and AS operations. Regardless of the offensive effort, MP leaders supporting an offensive operation must—

- Understand the commander's intent and his concept of operations.
- Anticipate that selective elements of the offensive force may need to pause, defend, resupply, or reconstitute while other forces attack.
- Anticipate changes in the operational tempo and prepare to weight the MP effort to meet the maneuver commander's intent.
- Provide MP support within the context of linear and nonlinear operations. They must understand how both types of operations affect the security of rear areas and LOC, the protection of C4I, the means of sustainment, and the land force's mobility.
- Support the maneuver commander's intelligence efforts by conducting the PIO function. The MP must observe, identify, and report information.
- Assist maneuver forces to mass. The MP protect $C^2$ nodes such as the CP that orchestrates the mass efforts and resources that sustain the offensive move. The MP must deny incursions by enemy reconnaissance forces into the AO. They also maintain surveillance, provide early warning, and impede and harass the enemy with supporting and organic fires.
- Speed the commitment of follow-on forces and reserves. They must also conduct MMS operations to assist the commander in a smooth and quick shift of forces to support the main effort.
- Anticipate the pursuit and exploitation by positioning MP forces that can support continuous operations. The MP support as far forward as possible while protecting resources needed to seize the opportunity. The EPWs/CIs are collected as far forward as possible. The MP coordinate the treatment of dislocated civilians and refugees with the HN or foreign forces.
- Tailor the MP support based on the unit being supported (corps, divisions, and brigades).
- Consider the location and composition of response forces and the TCF and the strength of rear-area threats.
- Integrate a force into the offensive plan that is capable of conducting aggressive combat operations against designated targets.

> During Operation Just Cause, an MP platoon received the mission to disable an armed Panamanian Defense Force (PDF) patrol boat docked near a key objective. After receiving the mission, the platoon leader maneuvered his force into position and attacked after receiving enemy fire. The platoon successfully disabled the patrol boat and captured the objective while suffering minimum casualties.

10-4. During offensive operations, MP support provided by teams, squads, and platoons provide the commander with an agile, flexible, versatile, and capable force ready to contribute to the overall mission success.

## SUPPORT TO DEFENSIVE OPERATIONS

10-5. The MP support defensive operations consistent with the commander's intent and priorities. They support a defending force by assisting subordinate elements in the execution of different, yet complementary actions. For example, the MP may support the maneuver of an attacking unit by conducting MMS operations while simultaneously providing AS for a defending unit in a base cluster. The MP support reflects the maneuver commander's focus—destroying an attacking force, retaining or denying key terrain, moving away from an enemy force, or a combination of the aforementioned. However, despite the intent of the defensive operation being conducted, MP leaders must—

- Conduct MMS operations to help a force maneuver and mass. The MP must anticipate the shift from the defense to the offense and assist the forward movement of reserves.
- Conduct AS operations to deny information to enemy reconnaissance elements seeking out the exact location of the defending force. The MP are positioned where they can control key terrain or improve the defensive capability of bases and base clusters. The MP conduct aggressive R&S to locate and deny enemy access to critical logistical facilities.
- Protect sustainment resources while supporting the lateral, forward, and rearward movement of forces.
- Consider the type and size of the AO, the LOC security, the rear threat, and the number of EPWs/CIs and dislocated civilians to determine how they may affect the movement of forces.

> During Operation Desert Shield, EAC, corps, and division MP units conducted a myriad of missions to support both the XVIII Corps's (Airborne) and the VII Corps's movement to their TAA. The MP's main focus was on MMS and AS throughout the AO to ensure freedom of movement and to protect US and coalition forces against terrorists and other threats in the rear area.

10-6. The success of MP operations in the defense depends on the MP leader's situational awareness, the commander's intent, and the precise employment of MP resources when and where they are needed. In the defense, the MP perform their five functions to assist the defending commander by providing a lethal mobile force that allows him to concentrate his efforts and resources in fighting the next battle.

# Chapter 11

# Stability and Support Operations

*ARFOR conduct stability operations in a complex, dynamic, and often asymmetric environment. Stability operations are usually nonlinear and noncontiguous. They are often time and manpower intensive. The purpose of support operations is to meet the immediate needs of designated groups for a limited time until civil authorities can accomplish these tasks without Army assistance.*

*FM 3-0*

The Army's mission is to prepare for war and, when deterrence fails, to achieve a quick, decisive victory. The DOD projects that many of its future missions will be stability and support operations. However, these operations are not new to the Army. The Army has participated in stability and support operations in support of national interests throughout history. The Army has protected its citizens, assisted nations abroad, and served America in a variety of other missions. The pace, frequency, and variety of stability and support operations have quickened in the last three decades. As demonstrated in Operations Provide Comfort, Restore Democracy, and JTF Los Angeles, these types of operations present a challenge to all services.

## OVERVIEW

11-1. Stability operations are conducted outside of the US and its territories to promote and protect US national interests by influencing political, civil, and military environments and by disrupting specific illegal activities. Stability operations may include both developmental and coercive actions. Developmental actions enhance a HN government's willingness and ability to care for its people. Coercive actions apply carefully prescribed limited force or the threat of forces to change the AO's environment.

11-2. Support operations provide essential supplies and services to relieve suffering and to help civil authorities prepare or respond to crises. In most cases, the ARFOR's efforts are focused on overcoming conditions created by man-made or natural disasters. The ultimate goal of support operations is to meet the immediate needs of designated groups (for a limited time) until civil authorities can accomplish these tasks without Army assistance.

11-3. Stability and support operations are distinctly different from offensive and defensive operations and must be analyzed differently. For instance, the application of the rules of engagement (ROE) will be unique to the operation. The restrictions on combat operations and the use of force must be understood and obeyed at all levels. MP leaders must clearly take the initiative and ensure that the applicable guidance is implemented and followed by all subordinates.

11-4. Stability and support operations have more diverse political considerations than are usually encountered in war. There are also complexities that must be addressed in joint, combined, and interagency stability and support operations. The Army's challenge is to execute these operations successfully in a joint or combined arena without degrading its capability to wage war successfully.

11-5. If the US is to fight and win future wars, it must develop, train, and employ its force structure in a manner that ensures success in every operation across the spectrum of military operations. Under these auspices, MP forces can provide commanders with especially valuable assets for stability and support operations. The MP have the unique capability of serving as a combat, CS, and CSS force. The domestic and international acceptability of the MP operational image frequently makes the MP the most appropriate force for stability and support operations. Additionally, the MP are flexible and capable of rapidly transitioning from one end of the spectrum to the other if the environment changes.

## FORCE SUITABILITY

11-6. The MP's capabilities and their operational and supporting tasks are as integral to stability and support operations as they are to offensive and defensive operations. MP branch-/core-related skills are highly compatible with the capabilities required for stability and support operations. The MP provide a highly capable, politically acceptable force that is suitable for a variety of missions. They possess robust moving, shooting, and communicating capabilities; and they project an assist, protect, and defend image that is particularly important when tailoring a force that requires significant capabilities but a low political profile. The MP provide a flexible, but limited, economy-of-force organization. Mission requirements will always exceed available MP resources, especially in stability and support operations. For this reason, MP assets must be prioritized to deliver the greatest mission capability. The specific operations that MP units perform at a given time are determined by the echelon commander's needs, the mission requirements, and the availability of MP resources. As discussed in previous chapters, the echelon commander, through the command's PM, sets the priorities for MP operations. As in offensive and defensive operations, the PM knows that while the MP force can perform all MP functions, they may not be robust enough to perform all assigned MP missions all of the time. The PM must continuously make trade-offs between the numbers and kinds of MP missions that can be supported.

11-7. The MP's capabilities are further enhanced by their training, mind-set, and experience in dealing with people in highly stressful and confusing situations. Although capable of conducting combat operations when needed, the MP are highly practiced in de-escalation and in employing the minimum-essential force to contain potentially violent situations. This mind-set serves as the framework for MP law-enforcement training and is especially applicable in MP support for stability and support operations. Furthermore, this mind-set is exercised and reinforced daily in peacetime law-enforcement operations. This is a significant capability. MP soldiers learn and receive constant reinforcement training in controlling a situation on the spot before it escalates and cannot be controlled. The MP have a unique understanding of human nature and are adept at handling emotion-laden situations.

> During operations in Kosovo, as both Serbs and Albanians attempted to create unfavorable conditions and responses by US forces, the MP kept their cool under fire in the streets of Gnjilane and Kamenica. The MP's reaction led to a quick understanding by all sides that the US Army MP were fair and just and understood the tense situations.

11-8. The MP stress four essential training competencies—skill proficiency, human dimension and attitude, camaraderie and teamwork, and leadership. These competencies are the foundation for all MP operational performances. They equip the MP to exercise discretion in dealing with others, to protect and assist those in need, and to accomplish the varying demands of MP missions in stability and support operations and, when needed, in war.

## STABILITY AND SUPPORT

11-9. Stability and support operations are performed in support of US authorities or in support of international bodies and foreign governments when so required by US authorities. The MP have long provided the essential support for stability and support operations; however, the frequency of these operations has increased significantly in recent years. Consequently, the MP have been deployed in greater numbers with greater frequency. MP units have a substantial history of successful, simultaneous, and continuous deployments in both a lead and a support role.

### SUPPORT TO DOMESTIC CIVIL AUTHORITY

11-10. The purpose of support to domestic civil authorities is to—

- Render humanitarian assistance.
- Provide disaster relief.
- Restore order.
- Combat terrorism.
- Support counterdrug operations.

11-11. Military operations supporting domestic civil authorities are governed by the provisions of the Stafford Act, the Posse Comitatus Act, and other laws and regulations. Peacetime experience in civil-military affairs equips the MP to oversee, assist, and train other units and agencies for law enforcement, sensitive security, and operations to control dislocated civilians (within the legal limits of a particular stability or support operation). For this reason, the MP are often needed after other forces have redeployed, and they stay until government order is restored. Since active-component military missions are constrained by the Posse Comitatus Act, nonfederalized national guard (NG) units are often employed to conduct these operations (particularly counterdrug operations). The NG units can operate under gubernatorial authority within state jurisdiction without Posse Comitatus restrictions. The NG MP units are the only NG units trained, skilled, and experienced in providing law enforcement. They offer civilian authorities a unique, specialized capability not available from other NG units.

## SUPPORT TO HUMANITARIAN ASSISTANCE AND DISASTER RELIEF

11-12. Humanitarian assistance and disaster-relief operations provide emergency assistance to victims of natural and man-made disasters and other endemic conditions. Natural disasters include earthquakes, floods, and hurricanes. Man-made disasters include riots and civil conflicts. Humanitarian assistance and disaster relief conducted by US forces alleviate urgent nonmilitary needs of a populace until the appropriate civilian agencies can provide the needed services without military support. These types of operations include, but are not limited to—

- Conducting searches and rescues.
- Evacuating, sheltering, sustaining, and protecting dislocated civilians.
- Making food, medical care and treatment, and other civilian-aid programs available.
- Reducing opportunities for criminal activity and restoring L&O.
- Maintaining other operations needed to ensure the well being of the affected population.
- Providing relief operations for man-made disasters. These operations are characterized by civil conflict, and they entail all of the requirements for other disasters but focus on the restoration of L&O.
- Providing force protection and security.

11-13. MP support can be an important asset in all disaster-relief operations. The MP are trained and equipped for decentralized operations. MP teams operate in highly mobile vehicles equipped with radios, which are invaluable in disaster-relief operations. Each MP platoon, company, and battalion headquarters is equipped with high-frequency radios. These capabilities allow the MP (even when widely dispersed) to maintain centralized communications; to provide L&O support; and to assist in search, rescue, and evacuation efforts. Because of their decentralized operations and communications equipment, the MP can also play a vital role in notification, ADC, and access and egress control. They can deploy with disaster-assessment teams to assist in advising and developing a disaster-assistance plan. The MP can also deploy to an affected area to—

- Assist in restoring order.
- Search for, rescue, and evacuate victims.
- Provide personnel and vehicular circulation controls.
- Provide ADC.
- Conduct detention operations.
- Provide physical and area security to protect life and property immediately following a disaster.
- Protect fire and emergency-services personnel.
- Gather, record, and report information.
- Provide force protection and security.

11-14. MP (I/R) battalions can support civil-affairs units by establishing I/R facilities to shelter, sustain, protect, and account for disaster-relief victims. These units can also train and assist government agencies (US and foreign) or police in the operation of facilities for dislocated civilians and detainees.

11-15. The MP possess capabilities critical to relief in man-made disasters characterized by civil conflict. The MP have the essential training and skills needed for relief operations that focus on restoring civil order. Restoration of civil order demands—

- **Objectivity and neutrality.** The objective of civil-disturbance control is to restore L&O, not to impose punishment or suppress peaceful dissent. Government forces quelling civil disorders must be emotionally objective and politically neutral. MP law-enforcement training stresses objectivity and neutrality, an important attribute for forces employed in stability or support operations.
- **Minimal use of force.** The application of more than minimum force may be detrimental to mission success. The MP are particularly suited for these situations, having been trained on the technique for the use and measured application of minimum forces to accomplish mission goals.
- **Public acceptance.** Government forces must avoid actions that will alienate the populace. Any action offensive to community sensitivities must be avoided. The MP receive training in crowd-control operations, interpersonal communications skills, and minimum use-of-force techniques—skills that are critical in gaining public acceptance.
- **Positive image.** In many disturbance operations, demonstrators and government forces compete for the approval and sympathy of the general population. The MP's assist-and-protect image provides the government with a distinct advantage in such situations.
- **Threat awareness.** Government forces must pay attention to and try to learn the organization, intentions, and activities of demonstrators. Equal attention must be paid to how the community will respond to the demonstrator's actions and the government's response. The MP interact daily with the populace, placing them in a position to develop criminal information and to determine the collective attitude of the populace.
- **Effective coordination.** Disturbance operations often involve many agencies from various levels of government. Thorough, timely, and coordinated preparation, planning, training, and execution are essential for mission success. MP civil-disturbance training (including planning and coordination measures) and their communications assets enhance coordination, particularly during mission execution.
- **Operational unity.** Multiple control forces representing different agencies with varying jurisdictional authority can pose $C^2$ problems. When unity of command cannot be achieved, operational unity must be achieved by colocating operational centers, integrating communications, and delineating organizational responsibilities. The MP are well trained in police and military operations, possessing a sound understanding of both civilian police and military measures, legal constraints, and effective coordination.

11-16. The MP operational tasks supporting man-made disasters vary according to the nature of the disaster and the operational environment. For example, in CONUS where the commitment of military forces to civil disturbances is considered a last resort and military involvement is limited by law, nonfederalized NG MP units have long been employed by civil authorities for their unique capabilities. MP forces can—

- Isolate and contain an area with barricades, roadblocks, and perimeter patrols. The MP can also assist civil-affairs units by enforcing pass and identification systems and public-utility controls.
- Provide mobile patrols to enhance the security of high-priority targets (such as buildings, utilities, and services that are critical to the community's economic and physical well being).
- Monitor, disperse, contain, or limit crowd movement through—
  - Observation points and patrols.
  - The communication of interest and intent to the crowd.
  - The channelization or diversion of the crowd.
  - Gaining the cooperation of the dissident leadership.
  - The proclamation of the illegal nature of the crowd's actions.
  - Show of force.
  - The use of crowd-control formations.
  - The apprehension of crowd members and leaders.
  - The application of minimum-force measures.
- Establish area control to prevent looting, to protect businesses and other likely targets, and to prevent arson. The MP accomplish this mission with saturation patrolling, including vehicle and foot patrols. With augmentation, the MP can perform air and water patrols. They also enforce populace control measures (such as ordinances to prevent gathering, permits to gather, restrictions on circulation, restrictions on the interference with government and public functions and personnel, restrictions on possessing weapons, and other measures instituted by civil authorities).
- Neutralize special threats (such as snipers and bombings) that are highly dangerous to both government forces and the community. Snipers pose a particularly grave danger in disaster-relief operations. The MP can take immediate protective actions, secure an area, isolate the threat and, when authorized, use SRTs to assist civilian or HN authorities in apprehending or neutralizing the sniper. The MP are trained to perform the necessary actions when an explosive device is discovered. In such an event, they can secure and evacuate an area, organize search teams, and isolate the site.

11-17. The MP can also support in rescue and recovery operations, which may include rescuing US or foreign nationals. Such operations are also employed to locate, identify, and recover sensitive materials deemed critical to national security. The operations may be conducted in benign or hostile environments. The MP provide OPSEC, and they process and account for civilians affected by rescue and recovery operations. The MWD teams can also help search for lost or hidden personnel. In stability and support operations, the MP can perform their police functions for rescue and recovery operations while keeping the perceived military signature low.

## SUPPORT TO COUNTERDRUG OPERATIONS

11-18. Counterdrug operations are measures taken to disrupt, interdict, and destroy illicit drug activities. In a counterdrug role, the DOD may offer certain forms of support to HN counterdrug personnel, to US civilian law-enforcement agencies operating in a HN, to Department of State (DOS) counterdrug personnel, and to the Bureau of International Narcotics Matters. Without direction from the National Command Authority acting under constitutional or statutory authority, US forces engaged in counterdrug activities may not engage in direct law-enforcement activities (such as apprehension, search, seizure, and other similar activities). As discussed earlier, the Posse Comitatus Act authorizes specific DOD assistance in counterdrug activities, but US military forces usually may not participate directly in arrests, searches, seizures, and similar activities. Conversely, as discussed earlier, USC Title 32 authorizes state governors to use NG MP units with full authority to engage directly in law-enforcement activities. When authorized, US advisors may accompany counterdrug forces on operations.

11-19. MP operational contributions can include the use of force, drug identification, crime-scene protection, customs operations, surveillance of criminal activity, serious-incident reporting, and other related activities. Employing the MP for these operations ensures force suitability and avoids a mismatch in terms of unit capability and mission effectiveness.

11-20. Border-screening operations consist of three separate but related measures—mobile patrolling, the use of listening/observation posts, and airfield surveillance. These measures are part of the MP's AS and MMS functions used in tactical rear-area operations. The MP's operational effectiveness may be further increased through augmentation by a civilian law-enforcement officer who would be responsible for search, seizure, and arrest actions. Active-component units are permitted to assist in marijuana eradication in support of domestic law-enforcement agencies (DLEAs). Marijuana-eradication operations are an application of the MP AS, mobility, and L&O functions, thus providing realistic training while supporting an operation.

11-21. As discussed in previous chapters, the MP are well suited to augment customs operations. They routinely perform customs inspections for US forces in Korea and Germany. The NG MP regularly conduct vehicle, cargo, and container searches under the supervision of US Customs agents and at US ports of entry.

11-22. The USACIDC agents can provide support to counterdrug operations. The CID units identify the sources of illegal-drug distribution systems. When directed, the MP and the CID complement interagency counterdrug efforts to stop the production, flow, and distribution of drugs and to provide planning, training, equipment, and facilities to support domestic counterdrug operations. The CID also supports CRIMINTEL programs, internal security, and site surveys at borders where Army units will deploy. In addition to the aforementioned support, the DOD General Counsel has ruled that MWD teams can be loaned to DLEAs to assist in detecting smuggled contraband.

## SUPPORT IN COMBATING TERRORISM

11-23. The Department of Justice (DOJ) is the lead federal agency for dealing with acts of terrorism within the US, its territories, and its possessions. Within the DOJ, the Federal Bureau of Investigation (FBI) has the lead. The FBI can train the police of friendly nations in antiterrorism and counterterrorism operations. Usually, US military forces act in a technical-advisory but not a law-enforcement capacity in combating acts of terrorism. It is often difficult to distinguish the acts of politically motivated terrorists from violent acts of criminals or individuals in society at large. In fact, all terrorist acts are criminal; however, they also may be part of an insurgency.

11-24. The MP are key players in the defensive planning, implementation, and control of combating-terrorism operations, whether in garrison or deployed to war or MOOTW. In whatever countries US forces operate, MP units conduct antiterrorism operations daily as part of their L&O and AS functions. The MP also respond to terrorist incidents. They help secure the area and search for evidence, and they help the lead agency. The MP provide advice and assistance in developing antiterrorism and counterterrorism programs for deployed US forces and DOS agencies and for police, paramilitary, and military forces of assisted nations (when so authorized). MP training, liaison, and joint patrols help strengthen HN-police organizations, improving their effectiveness and efficiency in combating terrorism. The MP also assist HN police in performing PRC operations.

11-25. The training of police (domestic and foreign) requires DA approval, and restrictions exist on some forms of training. Legal advice and coordination are a must before training can occur. The MP and the CID provide personnel and equipment for planning, advising, equipping, and training agencies to prevent, deter, and respond to terrorism. The training focuses on deterring terrorist acts such as bombings, kidnappings, assassinations, taking hostages, and hijackings. Related assistance may include training and equipment for explosive detection, management of hostage situations, physical security, protection of key personnel, and defensive/evasive driving methods. For counterterrorism operations on US military installations, the MP provide highly trained, specially equipped SRTs for situations requiring the selective use of force. Additionally, CID agents are highly trained in hostage negotiation, which increases the commander's options before resorting to force.

11-26. The MP and the CID assist in counterterrorism information-gathering activities through liaison and joint patrols with HN police. These activities are part of the PIO function. The MP have extensive contacts with civilian police, who in turn have contacts with the populace. These contacts can help surface criminal information concerning terrorist plans, recruitment, and support structures sooner than military sources. The CID can provide commanders with CRIMINTEL on a daily basis. At the same time, MP antiterrorism support includes measures to reduce the vulnerability of people and property. Such measures include—

- Personal awareness.
- Personal-protection operations and techniques.
- Crime prevention.
- Physical-security programs.

11-27. The MP and CID crime-prevention programs employ proactive measures to protect people and their property and US and HN property. Crime prevention is a continuous process of planning, implementing measures, evaluating effectiveness, modifying measures, and developing further information. The primary crime-prevention tool is the crime-prevention survey. It is used to examine the physical features of various critical sites. The MP physical-security programs help identify, reduce, eliminate, or neutralize conditions favorable to criminal, terrorist, and insurgent activities. The key measures used to accomplish this are physical-security inspections and surveys.

## SUPPORT TO NONCOMBATANT EVACUATION OPERATIONS

11-28. Noncombatant evacuation operations (NEOs) relocate civilians who are endangered in a foreign nation. The NEOs are classified as benign (unopposed) or hostile (opposed). MP roles in support to NEOs differ accordingly. In a benign environment, NEOs are usually conducted with the help and full cooperation of the affected nation, with little or no opposition to evacuation. The use of force is limited to self-defense (as outlined by the ROE) and protection of the evacuees. In a hostile environment, NEOs are opposed by forces with specific intentions of preventing or destroying the operation. In this scenario, the MP will follow the ROE as established by higher headquarters.

11-29. MP operations in NEOs closely parallel EPW/CI evacuation operations. The objective of both is to safely collect, secure, protect, transport, process, and account for people, whether they are EPWs or noncombatants. However, due to the dissimilarities of these groups, the techniques employed differ.

11-30. The MP can secure selected facilities, reception and transfer points, or other locations required to collect, process, and evacuate people. They can provide route reconnaissance and mobile security teams to escort HRP, DOS personnel, evacuees, and sensitive material. If needed, MP units assist civil-affairs personnel with the movement and control of dislocated civilians. The MP may support civil affairs by setting up and operating evacuation sites, holding areas for foreign nationals denied evacuation, and reception and processing stations. This support can be provided by MP CS units for a short duration and by MP I/R units for a prolonged duration. Additionally, the MP will establish and maintain liaison with HN police to ensure a mutual understanding of jurisdiction, to coordinate efforts, and to provide criminal and tactical intelligence. In hostile environments requiring the employment of combat forces, the MP would be employed in their traditional CS role.

## SUPPORT TO SECURITY-ASSISTANCE OPERATIONS

11-31. Security-assistance operations are one of the main tools of US foreign policy. A critical concern in these operations is the HN's ability to plan and manage its own defense resources. In security-assistance operations, the US provides defense materiel, military training, and defense-related services through grants, loans, credit, or cash sales. When authorized, the MP support security assistance with military and police training through the International Military Education and Training Program (IMETP). This program provides instruction and training to foreign military and civilian personnel on a grant/aid basis. Many foreign civil law-enforcement agencies request training in the following areas:

- Basic police procedures (such as patrolling and crime-scene protection).
- Physical security.
- Corrections.
- Civil-disturbance operations.
- Customs operations.
- Traffic control.
- Use of force.

11-32. The MP possess the expertise and experience to advise, train, and assist a HN's military and police forces. The MP's participation in the IMETP includes—

- Formal and informal instruction of foreign students.
- Technical education and tactical training with applicable training aids.
- Assistance to foreign police and military elements by MP mobile training teams (MTTs).
- Providing MP units or small, modular, functionally specific teams to support security-assistance needs. These teams can train HN police in both field and institutional facilities.
- Support to nation-assistance operations.

11-33. Nation-assistance operations support a nation's effort to promote development, ideally by using its national resources. This is achieved through education and the transfer of essential skills to the HN. The goals of such operations are to—

- Promote long-term regional stability.
- Develop sound and responsive democratic institutions.
- Develop supportive infrastructures.
- Promote an environment that allows for orderly political change and economic progress.

11-34. The MP support nation-assistance operations with training and joint MP and HN-police operations. The MP know and understand police operations, military operations, and $C^2$ procedures. Consequently, the MP are ideal for training the law-enforcement personnel in both field and institutional settings. The MP's nation-assistance goals are to—

- Enhance L&O in democratic societies.
- Improve efficiency and effectiveness.
- Promote the proper usage and maintenance of equipment.
- Establish a sound training base for police.

- Standardize procedures that enhance combined police operations.
- Promote friendship and goodwill toward the US.
- Preclude the need for US military interventions to counter acts of violence.

11-35. Combined police operations reinforce training and provide HN police with mobility, security, and communications to operate more effectively while conserving critical personnel resources. The MP provide support through training and the following:

- Law-enforcement, security, and criminal-information support to the HN police force (to include L&O operations and administration in a democratic society).
- Patrol and desk operations.
- Circulation-control operations.
- PRC and civil-disturbance operations.
- MWDs.
- Physical-security operations.
- Personnel-security operations.
- Area- and route-security operations.
- Counterdrug operations.
- Antiterrorism operations.
- Mass immigration operations.
- Customs and border operations.
- Confinement operations.
- Crime prevention.
- Community and police relations.
- $C^2$ of police operations.

11-36. When conducting nation-assistance operations, as well as other operations, the MP will often be tasked to support civil-affairs operations. Close coordination with the G5/S5 and liaison with the appropriate HN representatives have added emphasis to these types of operations. The scope of civil-affairs operations is influenced by the economic, social, psychological, and political background of the country and its people. Civil-affairs personnel will coordinate the MP's effort to restore stability, contribute to national development, and promote support for the host government. Civil-affairs personnel assist in preventing civilian interference with military operations and coordinate all other military and civil affairs (such as community relations, PRC, civil defense, and community- and area-security programs).

11-37. The MP can conduct ADC measures to assist the HN in the aftermath of a natural or man-made disaster. They help rescue and evacuate the injured for medical care, and they issue food, water, and essential supplies (See FM 41-10).

## SUPPORT TO PEACEKEEPING OPERATIONS

11-38. Peacekeeping operations (PKOs) are military operations conducted with the consent of the belligerents in a conflict. The PKOs are intended to maintain negotiated truces and to facilitate diplomatic solutions. The US may participate in a PKO under the auspices of an international organization, in cooperation with other countries, or unilaterally. A PKO usually occurs in an ambiguous situation in which there is extreme tension and the possibility of violence. This violence may range from terrorism and sabotage to minor military conflicts involving known and unknown belligerents. MP forces may serve as an element of a UN peacekeeping force or as part of a multinational observer group. The overall operational control of a peacekeeping force is exercised by a multinationally staffed military peacekeeping command. However, the commander of each assigned national force retains command integrity of his unit.

11-39. The basic MP force structure and appropriate augmentation are situation dependent. The size and composition of the MP forces are determined by diplomatic negotiations. Personnel spaces are allocated based on the US contingents' respective missions. The principle of consent affects the composition and function of the peacekeeping force. Consent applies to the degree of interest the disputing countries have when contributing their forces to participate in the peacekeeping effort.

11-40. Each PKO is unique but is generally characterized by constraints in the use of force. In a PKO, force is normally limited to self-defense, the defense of others, and the protection of national security. Therefore, the soldiers in a PKO must possess the skills required for warfare, but they must also be trained so that they are equally able to act as intermediaries. In this aspect, the MP are a distinct asset. They are experienced in a case-by-case approach to the use of force. They are experienced in using discretion and in the de-escalation of potential violence. The MP training in EPW operations are especially useful in handling EPWs if the peacekeeping force supervises the exchange of prisoners of war.

11-41. The peacekeeping force employs a combination of the following techniques to control the potential for violence inherent to PKOs:

- Observation is the peacekeeping force's primary responsibility and basic requirement. It provides timely and accurate reports of any suspicious situation, incident, or occurrence.
- Information gathering is a continuous requirement. Peacekeepers must be constantly alert to what takes place around them and to any changes or inconsistencies in the behavior, attitude, and activities of military and civilian personnel.
- Surveillance and supervision operations help oversee the implementation of agreements. The operations require restraint, tact, and patience.
- Complaint investigations must be thorough and objective, and they must result in a fair assessment.
- Negotiation and mediation skills are often required of peacekeepers. If peacekeepers can resolve minor problems at the lowest level, they can prevent major issues from rising, which may prevent increased tensions or the resumption of fighting.
- Patrolling (mounted and dismounted) among the population publicizes presence. It also serves as a means of gathering information, informally enhancing supervision, and investigating complaints.

11-42. The above techniques are compatible with MP battlefield functions. Operationally, the MP conduct route and area R&S, operate mounted or dismounted patrols, operate listening/observation posts, provide humanitarian assistance, investigate possible terrorist and criminal acts and cease-fire and sanction violations, and gather information. The MP provide a flexible, wide-ranging means of information dissemination. Additionally, they provide crucial support for force protection. The MP and the CID employ active and passive measures to deter and defeat threats directed at service members, their families, DOD civilians, facilities, equipment, and very important persons (VIPs). The MP use emerging technologies in force protection to conduct security, detection, and identification operations and to warn of intruders.

11-43. As part of a PKO, the MP conduct crowd- and riot-control operations, including the extraction of mob leaders. The MP are an ideal force for controlling antagonistic masses engaged in rioting, looting, and demonstrating. The MP understand how to make the transition from the lower end of the use-of-force spectrum to the use of lethal force, if so required or directed. Finally, the MP deploying with peacekeeping forces can provide early-on force protection, including headquarters and initial aerial port/ seaport of debarkation (APOD/SPOD) security. The MP's mobility, firepower, and communications provide critical reconnaissance, information-collection, and response-force capabilities and acceptability not found in combat forces.

11-44. When conducting crowd- and riot-control operations in a PKO or peace-enforcement operation (PEO), nonlethal weapons (NLWs) are an additional means of employing force for the purpose of limiting the probability of death or serious injury to noncombatants or belligerents. However, the use of lethal force must always remain an inherent right of individuals in instances when they, their fellow soldiers, or personnel in their charge are threatened with death or serious bodily harm. The NLWs add flexibility to the control of disturbances by providing an environment where military forces can permissively engage threatening targets with limited risk of noncombatant casualties and collateral damage (see FM 90-40).

11-45. The use of lethal force (employed under the standing ROE) will never be denied. At no time will forces be deployed without the ability to defend themselves against a lethal threat nor will they forego normal training, arming, and equipping for combat. Nonlethal options are a complement to, not a replacement for lethal force. They seek to expand a proactive response across the range of military operations.

11-46. The decision to use NLWs against an adversary during a confrontation should be delegated to the lowest possible level, preferably to the platoon or squad. However, this requires that all personnel (not just leaders) have a clear understanding of the ROE and the commander's intent.

11-47. Commanders and public-affairs officers must be prepared to address media questions and concerns regarding the use and role of NLWs. They must be prepared to address the role of NLWs, and they must make it clear that the presence of NLWs in no way indicates abandoning the option to employ deadly force in appropriate circumstances.

11-48. The PEOs are in the high end of the PKO's spectrum. These operations are just short of combat and require a different operational approach. MP operations to support PEOs correspond to MP battlefield functions supporting combat forces. The PEOs often pose special operational considerations for dealing with dislocated civilians. The MP provide PRC operations and I/R, when required.

## ORGANIZATIONS AND CAPABILITIES

11-49. MP units are well suited for stability and support operations given their unique capabilities and low force signature. The MP brigade, battalion, company, and I/R units may be tailored for stability and support operations based on their unique METT-TC.

### MP BRIGADE TF

11-50. In essence, the organization and capabilities of the MP organization in stability and support operations is the same as those for other operations. However, its augmentation is tailored based on METT-TC unique to stability and support operations and the commander's needs. The MP brigade has been the centerpiece of successful MP deployment during previous stability and support operations. The MP brigade provides comprehensive planning, $C^2$, and A/L support to assigned and attached subordinate units. The brigade also has the command and staff experience to coordinate and supervise HN security and law-enforcement activities.

11-51. The MP brigade headquarters provides senior MP leadership and the communications capabilities required to direct a variety of functional elements—CSS, PSYOP, public affairs, and combat-arms augmentation—in support of joint or unilateral TF operations. At authorized level of organization (ALO) 1, the MP brigade headquarters has an SJA section to coordinate the legal and jurisdictional issues inherent to stability and support operations. The brigade headquarters also has an S5 section and a liaison capability to coordinate support and assistance to HN agencies, friendly forces, other US forces, DOS personnel, and nongovernmental or private volunteer organizations.

11-52. The MP brigade TF (MPBTF), with its robust capabilities and low force signature, is well suited for the unique mission requirements and sensitivities of stability and support operations. The MPBTF can be employed as part of a combat TF in high-threat, opposed-entry operations or as the lead TF for a low-threat, unopposed-entry operation. The following are examples of previous MPBTFs and the operations in which they participated:

- **Operation Provide Comfort.** The 18th MP Brigade deployed to northern Iraq to provide humanitarian assistance and protection to the Kurds.
- **JTF Los Angeles.** The 49th MP Brigade (NG) provided civil-disturbance assistance to Los Angeles in the aftermath of the Rodney King trial.
- **Operation Hawkeye.** The 16th MP Brigade and units of the 89th MP Brigade provided security, protection, and humanitarian

assistance to the population of Saint Croix in the aftermath of Hurricane Hugo.

- **JTF Andrew.** The 16th MP Brigade provided humanitarian assistance in the aftermath of Hurricane Andrew.
- **Operation Sea Signal.** The 89th MP Brigade deployed to Guantanamo Bay, Cuba, to provide humanitarian assistance to Haitian migrants.
- **Operation Restore Democracy.** The 16th MP Brigade deployed to Haiti to provide humanitarian assistance.

11-53. Early deployments of the MPBTF headquarters or a mission-tailored portion of the headquarters positions a senior, effective MP $C^2$ headquarters to respond to all MP requirements. It can assist in the coordination with the media, governmental and nongovernmental agencies, and other forces, and it can coordinate civil-military aspects of the operation. The MP brigade headquarters can assimilate and employ additional forces without degrading operations. The augmentation of brigade headquarters is driven by METT-TC. With the brigade headquarters in place, subordinate or supporting units can be significantly expanded without the loss of $C^2$. In essence, little or no loss in operational momentum occurs when the $C^2$ element deploys first.

11-54. The MP LRP section of the MP brigade headquarters colocates and works with the higher headquarters' G3/Operations Directorate (J3) plans element. This allows the MP LRP to incorporate MP operations effectively into higher-level planning, and it enables an efficient transition to field operations. Also, because in peacetime this section functions as a PM cell, it provides the PM with efficient and effective technical support during stability and support operations. As the operations approach termination or enter sustainability and fewer forces are required, the brigade headquarters can redeploy, leaving a battalion TF in charge of MP forces.

## MP BATTALION TF

11-55. The MP battalion (CS) provides command, control, and support for assigned MP companies (CS). MP battalions (CS) are best employed as part of a TF—either as part of a combat TF (high threat, opposed) or a support TF (low threat, unopposed). With sufficient augmentation, the battalion can provide the logistical, administrative, personnel, and operational support needed to lead a diverse TF. During stability and support operations, an MP battalion TF may be placed under the OPCON of a brigade-size TF (MP or non-MP) or a division TF. If placed under the OPCON of a division TF, both the division PM and the MP battalion commander will synchronize their efforts to maximize the employment of available divisional and nondivisional MP assets in support of the division commander's concept of operations.

11-56. As with the MPBTF, the MP battalion's contributions to stability and support operations are not a new concept. The following are examples of previous battalion TFs and the operations in which they were involved:

- **Cuban Refugee Movement.** The 519th, 716th, 720th, and 759th MP Battalions deployed for humanitarian-assistance operations.
- **Operations Restore Hope and Continued Hope.** The 720th MP Battalion provided humanitarian assistance in Somalia.

- **Operation Sea Signal.** The 720th MP Battalion deployed to Guantanamo Bay, Cuba, to conduct humanitarian-assistance operations for Haitian migrants. Later, the 716th and 759th MP Battalions also deployed to Guantanamo Bay to perform like functions for the Cubans.

## MP COMPANY IN STABILITY AND SUPPORT OPERATIONS

11-57. The MP company is the MP Corps's primary mission performer. It executes a wide variety of missions. In these operations, MP companies—

- Provide mobility support for both vehicles and personnel.
- Provide AS, including security for critical personnel, sites, cargoes, and railways.
- Secure, safeguard, shelter, and control detainees and dislocated civilians.
- Restore order in civil disturbances and conduct PRC operations.
- Carry out limited L&O operations.
- Provide liaison, coordination, and training for all aspects of law enforcement to other agencies.
- Conduct police-intelligence collections and disseminate information to the populace, friendly forces, and other HN agencies.
- Support force protection and security.

11-58. The MP company possesses robust mobility, lethality, and communications. Each of the MP companies' three-man teams are equipped with either a high-mobility, multipurpose, wheeled vehicle (HMMWV), an up-armored HMMWV, or an armored security vehicle (ASV); an AN/VRC 91 radio; a Light-Vehicle Obscuration Smoke System (LVOSS); an MK-19 grenade machine gun; a squad automatic weapon (SAW); two M-16 or M-4 rifles; an M203 grenade launcher; and a 9-millimeter pistol as a side arm for each team member. In addition to conducting combat operations, the MP company commander has the capability to conduct limited L&O operations (desk and investigative operations). With a tremendous operational reach, the MP company will displace its platoons as far as possible to accomplish all of the MP functions.

## MP I/R UNITS IN STABILITY AND SUPPORT OPERATIONS

11-59. The MP I/R units are specifically equipped and trained to support, safeguard, account for, and provide proper and humane treatment for EPWs/CIs, US military prisoners, or dislocated civilians. When conducting dislocated-civilian operations, the MP I/R units support and interact with civil-affairs and PSYOP units who are responsible for monitoring and handling dislocated civilians. Should more than one category of personnel have to be supported, the MP I/R units will establish separate compounds for each to preclude contact with other populations. When operation of the facility is going to be more than a short-term endeavor, the intent is to involve (as appropriate) NGOs, IHOs, HN authorities, and dislocated civilians in the administration of the facility. The tasks performed by MP I/R units for stability and support operations are basically the same as those for war, but the operations and procedures are modified according to the operational environment and the categories of people to be housed in the facility (see FM 19-40).

11-60. Criminal detainees (who are detained against their will) are subject to controls and procedures that closely parallel those used for EPWs/CIs. The primary emphasis is on guarding and preventing escape until their disposition is determined by recognized civilian judicial/law-enforcement authorities. When the operation is conducted outside of the US and its territories, criminal detainees are treated according to established legal procedures.

11-61. Dislocated civilians (who are being assisted rather than detained) are provided aid, shelter, and protection. The emphasis is on protecting them from harm by natural forces or hostile personnel. A special category of personnel arises when the MP I/R unit is required to house personnel that are dislocated civilians but who must be detained against their will. Such is the case of mass migrants who flee their countries and find themselves under US custody while formal proceeding guidance is being developed. In this case, MP I/R units must be sensitive to the situation and attempt to strike a balance between security, shelter, protection, and detention procedures.

11-62. Military detainees (who are detained against their will) are subject to controls and procedures that closely parallel those used for EPWs/CIs. In stability and support operations, US forces detain opposing-force military personnel. Without a formal declaration of war, opposing military personnel who are captured cannot be categorized as EPWs. However, they are afforded many of the EPW's rights and privileges under the Geneva convention. Such a situation occurred during Operation Just Cause when PDF personnel were detained by US forces.

11-63. In an operational environment in which hostile groups are engaged against one another, an I/R facility may be set up to protect one group from another. In this case, the facility's purpose is to shelter, sustain, account for, and protect its dislocated civilians from the hostile group. The MP I/R units and other supporting units concentrate on providing AS to protect the facility from direct fire. Other MP or combat forces provide protections beyond the direct-fire zone. Accountability for the dislocated civilians is coordinated with the SJA and civil affairs. They focus on maintaining a record of the people in the facility and their physical conditions. In operations in which no hostile groups are engaged (such as natural disasters), the I/R facility may be set up to provide shelter, sustain, and account for personnel. There may not be a need for external security personnel.

11-64. The MP I/R $C^2$ structure for stability and support operations is based on METT-TC. The nature and complexity of the mission, the number and types of detainees, and the operational duration should be considered. At one end of the operations spectrum, an MP brigade (I/R) may be required; while at the other end, an MP battalion (I/R) may be assigned to an MP brigade (CS).

# Chapter 12

# Force Protection

*There is no more important responsibility than force protection. It is an inherent command responsibility and must be fully integrated into every unit's mission. A commander must continually review his unit's force-protection posture and crosswalk it with current and changing policy and threat levels. Force protection demands the personal involvement of commanders to ensure the best possible security consistent with the threat to personnel and mission-essential resources.*

*AR 525-13*

Force protection consists of those actions that prevent or mitigate hostile actions against DOD personnel (to include family members), resources, facilities, and critical information. It coordinates and synchronizes offensive and defensive measures to enable the joint force to perform while degrading opportunities for the enemy. It includes air, space, and missile defense; NBC defense; antiterrorism; defensive-information operations; and security to operational forces and means.

## OVERVIEW

12-1. The MP contribute significantly to the Army's overall force-protection effort. They are directly responsible for executing certain security-related measures (law enforcement and physical security) and for providing support to others (HRP security and antiterrorism), all of which contribute to force protection. Commanders use the Army's risk-management methodology and established security programs and capabilities within US forces to help prioritize and implement force-protection measures.

## SUPPORT TO FORCE PROTECTION

12-2. MP support to force protection is accomplished under the leadership of the respective command's PM. It is accomplished mainly through an active role in physical security, HRP security, law enforcement, and antiterrorism.

### PHYSICAL SECURITY

12-3. Physical security is an integral part of security of forces and means. Physical-security measures deter, detect, and defend against threats from terrorists, criminals, and unconventional forces. These measures include—

- Fencing and perimeter standoff space.
- Lighting and sensors.

- Vehicle barriers.
- Blast protection.
- Intrusion-detection systems and electronic surveillance.
- Procedural measures such as security checks, training and awareness, property accountability/inventory requirements, physical-security inspections of mission-essential or vulnerable areas (MEVAs), and physical-security surveys of installations.

12-4. These physical-security measures are not exclusively peacetime measures. They are designed to protect US personnel and equipment regardless of the mission or the geographical location.

12-5. The MP support the commander's security of forces and means programs by—

- Controlling or monitoring installation, base, or base-cluster access or entrance points.
- Monitoring intrusion-detection systems and providing a response force.
- Conducting physical-security inspections.
- Conducting perimeter security or site surveillance.
- Recommending the placement of walls, berms, gates, or barriers around designated MEVAs, high-value areas, or perimeters.
- Supporting the commander's risk-analysis effort.
- Conducting roving patrols, checkpoints, or roadblocks.
- Performing other physical-security measures as required by the commander.

## HRP SECURITY

12-6. The HRP security program provides additional security to designated individuals (and their families) whom by virtue of their rank, assignment, symbolic value, vulnerabilities, or location or a specific threat are at a greater risk than the general population. This security program is applicable across the full spectrum of military operations.

12-7. The MP support the commander's HRP security measures by—

- Conducting protective-service details for HRP (normally conducted by USACIDC special agents).
- Responding to duress alarms and intrusion-detection systems.
- Conducting a PSVA (performed by the USACIDC).
- Employing MWDs.
- Coordinating with local, state, and federal agencies as well as joint, multinational, and HN law-enforcement agencies.
- Performing other HRP security measures as required by the commander.

## LAW ENFORCEMENT

12-8. The Army's law-enforcement program supports the security of forces and means through the prevention, detection, response, and investigation of crimes. The MP, the USACIDC, and DOD police play a crucial role in the law-enforcement program. They provide a visible deterrent against a broad portion of the threat spectrum, investigate crimes, and provide an initial response force to threat incidents. As with the previous programs, the law-enforcement program is applicable anywhere across the spectrum of military operations.

12-9. The MP support law-enforcement operations by—

- Providing liaison teams with local, state, and federal agencies; HN police; and joint and multinational agencies.
- Employing SRTs and hostage-negotiation teams (normally provided by the USACIDC).
- Providing traffic enforcement, MSR regulation enforcement, and other route-control measures.
- Employing MWDs.
- Conducting MP investigations (criminal and noncriminal).
- Conducting patrolling, AS, and surveillance measures.
- Implementing applicable threat-condition (THREATCON) measures.
- Conducting and implementing other law-enforcement measures as required by the commander.

## ANTITERRORISM

12-10. Terrorism is characterized as the unlawful use of violence or the threat of violence to coerce or intimidate a government or a society. Protection against a terrorist threat requires both an offensive counterterrorism capability and a defensive antiterrorism program. The primary focus of antiterrorism is on training and awareness and thorough planning. Antiterrorism is not a stand-alone program. It leverages other force-protection-related programs (such as physical security and law enforcement) to provide much of the physical defense against terrorism.

12-11. The MP support the commander's antiterrorist program by—

- Collecting CRIMINTEL (within the limits of AR 190-45).
- Responding to terrorist threats and incidents.
- Performing aggressive patrols and R&S operations.
- Detecting suspicious activities.
- Enforcing joint and cooperative agreements with the FBI, the USACIDC, MI, and other pertinent agencies (home or abroad).
- Conducting information collection and dissemination.
- Protecting critical assets, facilities, and personnel.
- Performing other antiterrorism measures as required by the commander.

## Appendix A

# Metric Conversion Chart

This appendix complies with current Army directives which state that the metric system will be incorporated into all new publications. Table A-1 is a conversion chart.

**Table A-1. Metric Conversion Chart**

| Metric to English | | | English to Metric | | |
|---|---|---|---|---|---|
| **Multiply** | **By** | **To Obtain** | **Multiply** | **By** | **To Obtain** |
| **Length** | | | | | |
| Centimeters | 0.0394 | Inches | Inches | 2.54 | Centimeters |
| Meters | 3.28 | Feet | Feet | 0.0305 | Meters |
| Meters | 1.094 | Yards | Yards | 0.9144 | Meters |
| Kilometers | 0.621 | Miles (stat) | Miles (stat) | 1.5609 | Kilometers |
| Kilometers | 0.540 | Miles (naut) | Miles (naut) | 1.853 | Kilometers |
| Millimeters | 0.039 | Inches | Inches | 25.40 | Millimeters |
| **Area** | | | | | |
| Square centimeters | 0.1550 | Square inches | Square inches | 6.45 | Square centimeters |
| Square meters | 10.76 | Square feet | Square feet | 0.0929 | Square meters |
| Square meters | 1.196 | Square yards | Square yards | 0.836 | Square meters |
| **Volume** | | | | | |
| Cubic centimeters | 0.610 | Cubic inches | Cubic inches | 16.39 | Cubic centimeters |
| Cubic meters | 35.3 | Cubic feet | Cubic feet | 0.0283 | Cubic meters |
| Cubic meters | 1.308 | Cubic yards | Cubic yards | 0.765 | Cubic meters |
| Milliliters | 0.0338 | US liq ounces | US liq ounces | 29.6 | Milliliters |
| Liters | 1.057 | US liq quarts | US liq quarts | 0.946 | Liters |
| Liters | 0.264 | US liq gallons | US liq gallons | 3.79 | Liters |
| **Weight** | | | | | |
| Grams | 0.0353 | Ounces | Ounces | 28.4 | Grams |
| Kilograms | 2.20 | Pounds | Pounds | 0.454 | Kilograms |
| Metric tons | 1.102 | Short tons | Short tons | 0.907 | Metric tons |
| Metric tons | 0.984 | Long tons | Long tons | 1.016 | Metric tons |

## Appendix B

# Doctrine and Training Literature

Users of this manual must understand the elements of doctrinal literature and their relationship to each other. The commonly used terms doctrine and tactics, techniques, and procedures (TTP) are interrelated and mutually supportive; yet, each term has its own usage, level of detail, and place in the hierarchy of doctrinal publications. Because there is a close interrelationship, publications overlap in what they provide to the user. Figure B-1 illustrates the relationship of these doctrinal terms and publications.

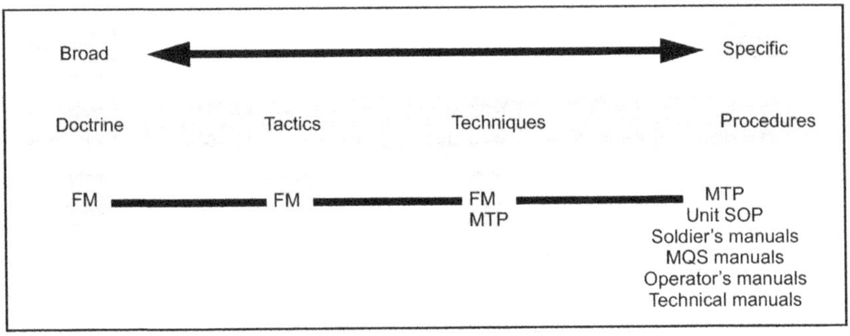


**Figure B-1. Doctrinal and Training Literature**

## DOCTRINE

B-1. Doctrine is the body of fundamental principles by which the military guide their actions in support of national objectives. It is authoritative, but requires judgment in application. Doctrine provides the Army's fighting philosophy, establishes a common language, and structures the manner within which units plan and conduct combat operations.

B-2. Doctrine is not absolute; it is applied by the leader to meet the circumstances of the situation being faced. Doctrine provides for this flexibility with broad fundamentals and principles for conducting operations. Leaders well grounded in doctrine understand how to synchronize the elements of combat power and how to apply TTP using the assets available to accomplish the mission.

B-3. Doctrine is described in FMs. FM 3-0 defines the systems of war fighting practiced by the US Army. FMs 100-15 and 71-100 expand on this foundation. This manual describes the CS doctrine of the MP; it is the capstone manual for MP operations. It defines the role, operational requirements, functions, and $C^2$ relationships of MP units. This manual is not oriented on a specific organizational structure. Leaders at all levels apply this doctrine to their unique organizational and operational situations and use various tactics and techniques to accomplish the doctrine.

## TACTICS

B-4. Tactics describe how the leader carries out doctrine (relating to the arrangement of forces for battle). Tactics include the—

- Employment of units in combat (to accomplish a mission assigned by the commander).
- Ordered arrangement and maneuver of units in relation to each other and to the enemy in order to use their full potential (a description of how the commander should arrange his forces and maneuver to accomplish a mission).

B-5. This is the part of doctrinal literature used for training in preparation for combat. Like doctrine, tactics presented in manuals are applied with judgment by the leader. The FMs prescribe "how to fight" and the mission training plans (MTPs) describe "what to train."

## TECHNIQUES

B-6. Techniques are the methods of performing any act, especially the detailed methods used by troops or commanders in performing assigned tasks. Techniques describe the basic methods of using equipment and personnel. Techniques give details on how commanders actually carry out assignments. They improve the force's efficiency by ensuring the uniformity of action or by ensuring that the actions of various individuals and elements complement those of other individuals or elements.

B-7. The FMs and MTPs provide a description of techniques using the personnel and equipment available in actual organizations. These techniques show (at a more detailed level) how to use available assets to carry out tactics.

## PROCEDURES

B-8. Procedures are the lowest level of detail. They address "how to" at the task level. Procedures are a particular course or mode of action that describe how to perform a certain task. Procedures include the—

- Standing methods used by units to accomplish tasks.
- Operating procedures for weapons and equipment.
- Crew drills.
- Staff actions and coordination requirements.
- Methods of target engagement by direct- and indirect-fire weapons.

B-9. Procedures are building blocks of individual- and collective-task accomplishment that serve as the foundation of tactics and techniques. Procedures are explained in unit standing operating procedures (SOPs), MTPs, soldier's manuals, operator's manuals, technical manuals (TMs), and similar publications.

# Appendix C

# Battlefield Workload Analysis

With minor adjustments in the capability block, the BWA can be used to determine MP requirements for the division. A BWA work sheet is shown in Figure C-1. A separate matrix is prepared for each EAC and corps MP brigade. Figure C-2, page C-3, is a sample of a BWA matrix for a 3-division corps MP CS company.

| Mission/Function[1] | Capability[1] | Workload[2] | Rqmt | HN[18] | US Rqmt | Risk[19] | Mission Rqmt |
|---|---|---|---|---|---|---|---|
| Divisions supported[3] | 1 | | | | | | |
| MMS | 360 km MSR and ASR | | | | | | |
| Refugees on control routes[4] | 150,000 per day | | | | | | |
| AS—corps rural terrain[5] | 2,000 sq km | | | | | | |
| AS—EAC rural terrain[5] | 3,000 sq km | | | | | | |
| AS—urban terrain[5] | 800 sq km (urban) | | | | | | |
| AS—USAF main opns base[5] | 1,000 sq km/USAF MOB | | | | | | |
| AS—convoy escort[6] | # tms per convoy | # convoys | | | | | |
| AS—ammo security[7] | 1.3 modular ammo co | | | | | | |
| | 2 modular ammo co | | | | | | |
| AS—pipeline security[8] | 400 km | | | | | | |
| AS—EAC pipeline co[8] | 3 | | | | | | |
| AS—trains[9] | 7 | | | | | | |
| AS—small critical sites[10] | 12 | | | | | | |
| AS—major HQ security[11] | 1 | | | | | | |
| AS—EAC port security[12] | 0.66 of port | | | | | | |
| AS—combat operations | When required, assets are diverted from other missions. | | | | | | |
| L&O—incident response[13] | 75,000 nondivision troops | | | | | | |
| I/R—US prisoner detention[14] | 700 | | | | | | |
| I/R—EPW escort/corps[15] | 1,900 walking | | | | | | |
| | 2,500 by vehicle | Min: 1 plt | | | | | |
| | 3,800 by train | | | | | | |
| I/R—EPW holding/corps[16] | 2,000 | Min: 1 plt | | | | | |
| Other assigned missions[17] | Mission | Force size | | | | | |
| | | Totals[20] | | | | | |
| Battalion HHD requirement = 0.199 x number of companies[21] | | | | | | | |

**Figure C-1. Sample BWA Work Sheet**

**NOTE: Workload ÷ capability = doctrinal requirement - HN offset = US requirement - acceptable risk = mission requirement**

[1]Mission/capabilities are as stated in section I of TOEs 19477L000 and 19677L000. The unit capability for each mission is what the unit can accomplish when performing no other mission (for example, if it is only doing MSRs, the unit can cover 360 kilometers in a 24-hour period). Unless otherwise specified, the capabilities are the same at corps and EAC.

[2]Workload is the amount (MSR or area) or number of supported units/facilities.

[3]Corps support to each division is covered by allocating one MP company per supported division. Any additional support requirements should be based on a detailed division PM/MP mission analysis and included under other assigned missions.

[4]Unless specifically tasked, the Army normally relies on HN support to manage refugees. Redirecting refugees from MSRs and keeping them off of MSRs is part of (as is straggler control and TCP) the mission and MSR workload. When assigned the mission, the capability is for control of refugees on established refugee control routes that are in addition to the MSRs.

[5]Compute each square kilometer mission requirement separately. Do not compute the same terrain in more than one computation (for example, the area around the Air Force's main operating base (MOB) is not included with either rural or urban square kilometers).

[6]Convoy-escort capability is not stated in the TOE and is an AS function rather than an MMS function. Determine the average number of teams per mission and the average number of simultaneous missions. Multiply the number of teams required per mission by 0.028, then multiple the result by the number of simultaneous missions to determine the total requirement. State the number of teams per convoy and the number of convoys, then convert the force size to a 3-decimal-place company equivalent. Each additional team, squad, and platoon equivalent equals 0.028, 0.083, and 0.25 respectively.

[7]Security for modular ammo companies (ASPs) supporting divisions is three platoons and is normally conducted directly behind the division (corps forward). However, METT-TC may require the ammo units to move into the DRA. In this case, corps units should be attached to the division for the duration. Security for corps storage areas (CSAs) and theater storage areas (TSAs) is one MP company per two ammo companies. Security is provided on an area-support basis in the form of screening, with limited fixed posts.

[8]The EAC petroleum terminal/pipeline operating company manages the EAC pipeline that is normally laid adjacent to MSRs. Pipeline security is inherently provided by the MP operating on the MSRs. Unless the pipeline is not near the MSR, MP support is only allocated against the EAC petroleum terminal/pipeline operating company's bulk storage facilities. Security is provided on an area-support basis in the form of screening, with limited fixed posts.

[9]Train security is normally an EAC mission. This capability equates to simultaneously providing five MP teams to each of seven trains. Actual workloads and requirements are computed in the same manner as a convoy escort.

[10]Small critical facilities are normally defined as key MSR bridges and tunnels and communication nodes that may require additional security.

[11]The major headquarters are corps, COSCOM, TSC, unified command and, if directed, the JTF. The ASCC depends on the MP to provide security for two CPs. The ASCC requirement is an additional three platoons to support a second CP.

[12]The requirement for one seaport is 1.5 companies.

[13]Teams provide incident response, crime-scene security, and traffic control. They also assist L&O detachment teams and CID investigators.

[14]For deliberate planning in support of an OPLAN, personnel to be confined will normally be held by either the unit or the EAC I/R MP battalion performing the confinement mission. This does not preclude a requirement for the echelon PM to operate a field detention facility until prisoners are released back to their units or evacuated to the EAC confinement facility.

[15]The primary mode is vehicle (back-haul doctrine). Regardless of the mode, a minimum of one platoon (0.25 requirement) per supported division is allocated for EPW escort from the division central collection point (DCCP) to the corps holding area (CHA). Because the actual number of EPWs is unknown, coupled with no time/distance study for each supported division, one platoon per division provides a minimum capability for continuous evacuation within the stated doctrinal timeline (evacuate to the CHA within 12 hours of arrival at the DCCP).

[16]At a minimum, one platoon (0.25 requirement) is allocated to operate the CHA (provides the capability to hold 500 EPWs at any given time).

[17]Add any additional continuous missions assigned by the echelon commander not covered above. Examples include VIP escort/security for other than the CG and the deputy commanding general (DCG) of major headquarters, screening exposed corps flanks, and increasing teams per mission/function which decreases doctrinal capability (for example, permanent TCPs or checkpoints). Each additional team, squad, or platoon equivalent equals 0.028, 0.083, and 0.25 respectively.

**Figure C-1. Sample BWA Work Sheet (continued)**

[18]For each mission/function, any workload either fully or partially accomplished by the HN is subtracted from the doctrinal mission requirement. Enter the MP company equivalent offset and subtract from the doctrinal requirements to determine the US requirement.

[19]For each mission/function, if the operational/MP commander will accept mission risk, subtract the MP company equivalent to determine the mission requirement.

[20]Add columns to determine the total doctrinal company requirement, HN offsets, US requirement, acceptable risk, and mission requirement to support the operation/AO. Round the requirement to the nearest whole number (0.4 or fewer rounds down, 0.5 or greater round up).

[21]Determine battalion HHD requirements by multiplying the company requirements by 0.199 and rounding.

**Figure C-1. Sample BWA Work Sheet (continued)**

| Mission/Function | Capability | Workload | Rqmt | HN | US Rqmt | Risk | Mission Rqmt |
|---|---|---|---|---|---|---|---|
| Divisions supported | 1 | 3 | 3.000 | | 3.000 | | 3.000 |
| MMS | 360 km MSR and ASR | 1,050 | 2.917 | | 2.917 | | 2.917 |
| Refugees on control routes | 150,000 per day | 25,000 | 0.167 | 0.167 | 0.000 | | 0.000 |
| AS—corps rural terrain | 2,000 sq km | 6,500 | 3.250 | | 3.250 | 0.500 | 2.750 |
| AS—EAC rural terrain | 3,000 sq km | | | | | | |
| AS—urban terrain | 800 sq km (urban) | 1,500 | 1.875 | | 1.875 | 0.500 | 1.625 |
| AS—USAF main opns base | 1,000 sq km/USAF MOB | 1,000 | 1.000 | | 1.000 | | 1.000 |
| AS—convoy escort | 5 tms per convoy | 4 | 0.560 | | 0.560 | | 0.560 |
| AS—ammo security | 1.3 modular ammo co | 3 | 2.308 | | 2.308 | | 2.308 |
| | 2 modular ammo co | 1 | 0.500 | | 0.500 | | 0.500 |
| AS—pipeline security | 400 km | | | | | | |
| AS—EAC pipeline co | 3 | | | | | | |
| AS—trains | 7 | | | | | | |
| AS—small critical sites | 12 | 9 | 0.750 | | 0.750 | 0.250 | 0.500 |
| AS—major HQ security | 1 | 2 | 2.000 | | 2.000 | | 2.000 |
| AS—EAC port security | 0.66 of port | | | | | | |
| AS—combat operations | When required, assets are diverted from other missions. | | | | | | |
| L&O—incident response | 75,000 nondivision troops | 45,000 | 0.600 | | 0.600 | 0.300 | 0.300 |
| I/R—US prisoner detention | 700 | | | | | | |
| I/R—EPW escort/corps | 1,900 walking | | | | | | |
| | 2,500 by vehicle | Min: 1 plt/div | 0.750 | | 0.750 | | 0.750 |
| | 3,800 by train | | | | | | |
| I/R—EPW holding/corps | 2,000 | Min: 1 plt/ | 0.250 | | 0.250 | | 0.250 |
| Other assigned missions | PSD for Corps CofS | 1 sqd | 0.083 | | 0.083 | | 0.083 |
| | Flank screen | 1 plt | 0.250 | | 0.250 | | 0.250 |
| | | **Totals** | 20 | 0.177 | 20 | 1.3 | 19 |
| Battalion HHD requirement = 0.199 x number of companies | | | 4 | | 4 | | 4 |



**NOTE: Workload ÷ capability = doctrinal requirement - HN offset = US requirement - acceptable risk = mission requirement**

**Figure C-2. Sample BWA Matrix**

## Appendix D

# Automated Information Systems in Support of Battle Command

As the MP Corps continues to advance into the twenty-first century, it must continue to pursue and stay in tune with the Army's initiative to acquire advanced technologies and operational concepts. This will give MP soldiers on the battlefield an information advantage that is critical for successful performance of their functions. The following information opens a window into the near future and explains some of the automation initiatives that will enhance the Army's and the MP Corps's soldiers.

## ARMY INFORMATION SYSTEMS

D-1. The Global Command and Control System (GCCS) is the midterm solution and the bridge to the concepts outlined in the concept of command, control, communications, computers, and intelligence for the warrior (C4IFTW). The GCCS's core consists of the basic functions required by the war fighter to plan, execute, and manage military operations. The Army's selected $C^2$ system is the Army Battle Command System (ABCS).

D-2. The ABCS integrates the $C^2$ systems found at each echelon—from the ground-force component commander at the theater or JTF level to the individual soldier or the weapons platform. Whether ARFOR are deployed for land combat or are conducting peace operations, providing humanitarian assistance, or giving aid to civil authorities, the ABCS supports the mission by integrating the battle-space automation systems and communications that functionally link strategic and tactical headquarters. The ABCS is interoperable with joint and multinational $C^2$ systems at upper echelons across the full range of $C^2$ functionality, and it is vertically and horizontally integrated at the tactical and operational levels. The ABCS has three major components—

- The Global Command and Control System—Army (GCCS-A).
- The Army Tactical Command and Control System (ATCCS).
- The FBCB2.

### GLOBAL COMMAND AND CONTROL SYSTEM—ARMY

D-3. The GCCS-A is the Army's component of the joint GCCS. It was built from application programs developed by the Army Worldwide Military Command and Control System [WMCCS] Information System (AWIS), namely the—

- Strategic Theater Command and Control System (STCCS).

- Theater Automated Command and Control Information Management System (TACCIMS).
- EAC portion of the Combat Service Support Control System (CSSCS) (CSSCS/EAC).

D-4. The GCCS-A is required in the MP LRP, S1, S3/NBC, and S4 sections within the MP brigade at EAC; in the Assistant Chief of Staff, G1 (Personnel) (G1), the G3, and the Assistant Chief of Staff, G4 (Logistics) (G4) sections within the MP command (I/R); and in the S1, S3, and S4 sections within the MP brigade (I/R).

## ARMY TACTICAL COMMAND AND CONTROL SYSTEM

D-5. The ATCCS embodies the commander and his staff and the doctrine, procedures, and tools used to command and control forces on the tactical battlefield. The ATCCS's functional subsystems provide tools to support the $C^2$ process across the spectrum of low- to high-intensity conflicts worldwide; for MOOTW; and for combat, CS, and CSS functional areas at corps-and-below units. The ATCCS provides a common set of $C^2$ tools and functions for commanders that extends the ATCCS to the battalion level. The component's ATCCS's are the—

- Maneuver Control System (MCS).
- All-Source Analysis System (ASAS).
- Advanced Field-Artillery Tactical Data System (AFATDS).
- Air- and Missile-Defense Planning and Control System (AMDPCS).
- CSSCS.

D-6. The FBCB2 provides situational awareness and $C^2$ to the lowest tactical echelons. It facilitates a seamless flow of battle-command information and situation-awareness information across the battle space at brigade and below on Army platforms.

## FORCE XXI BATTLE COMMAND—BRIGADE AND BELOW

D-7. The FBCB2 system, a battle-command information system, provides the technology to complete the ABCS information flow process from brigade to platform and across all platforms within the brigade TF. The FBCB2 enables commanders to remotely operate and maintain ABCS database connectivity (regardless of the command vehicle) and to digitally control and monitor their subordinate units' statuses and positions. Interoperable FBCB2 systems will enhance the battle commander's ability to have a detailed picture of the battlefield (situational awareness), to maximize force survivability, and to facilitate battlefield synchronization to reduce or destroy the enemy's capability to fight. The FBCB2 relies on integrating, processing, and moving information in a digital form. To support projections of future conflicts with an accompanying increase in the tempo of battle, the FBCB2 must be able to obtain information contained in databases from higher, lower, and adjacent elements without formally requesting that information through command channels. The war fighter requires the ability to pull data from all battlefield-system databases to keep current with the rapid changes caused by the dynamics of battle. The FBCB2 is required for divisional MP company commanders, PMs, platoon leaders, teams, and squads.

## MANEUVER CONTROL SYSTEM

D-8. The MCS is the ABCS's horizontal and vertical integrator. The MCS will enable commanders and their staffs to collect, coordinate, and act swiftly on near-real-time battlefield information. It will quickly, accurately, and automatically move tactical information on the battlefield. This allows a commander to readily access information and display current situation reports (SITREPs) that assess enemy strength and movement as well as the status of friendly forces. The MCS will then aid the battle staff in rapidly disseminating the commander's orders. It is intended to provide two major functional capabilities—maneuver functional-area control and force-level information. The MCS is the ATCCS's functional subsystem that—

- Supports the combined-arms force commander directly by providing automated support for planning, coordinating, controlling, and using maneuver functional-area assets and tasks.
- Coordinates and synchronizes supporting arms in conducting operational planning, field operations, and training.

D-9. The MCS is required for division PM operations sections, company CPs at echelons corps and below (ECB), MP battalion commanders, S3 sections at corps, MP brigade commanders, and corps MP LRP and S3 sections.

## ALL-SOURCE ANALYSIS SYSTEM—REMOTE WORK STATION

D-10. The ASAS—remote workstation (RWS) subsystem will operate within the existing intelligence and electronic warfare (IEW) battlefield operating system (BOS) and ABCS architectures to provide automation support to G2/S2 intelligence staff elements and other designated intelligence organizations from the maneuver battalion through EAC. It will operate at the security level appropriate to its echelon, mission, and functions. It assists the G2/S2 in providing IEW support to the commander and other staff elements in planning, coordinating, and conducting current and future tactical operations. It interfaces with the sensitive compartmented information (SCI) analysis and control element (ACE) subsystem to provide operational guidance and to exchange IEW information and products. It also operates as the technical control portion of the IEW $C^2$ node of the ABCS to provide current IEW and enemy-situation information to the force-level information database for access and use by ABCS users. Two RWSs are also provided to the brigade analysis control team (ACT) of the DS MI company to process data input from DS collection systems, to filter intelligence data from broadcast systems, and to pass tactically relevant enemy-situation information or situation-development data to the brigade S2's RWS. The ASAS-RWS is required in the S2 section of the CID group, the S2/S3 section of the CID battalion, the G2 section of the MP command (I/R), the S2 section of the MP battalion (I/R), and the S2 section of the MP battalion and brigade at all echelons.

## COMBAT SERVICE SUPPORT CONTROL SYSTEM

D-11. The CSSCS will function as an integral part of the ATTCS. It is designated as an information system with the capability to rapidly collect, analyze, and disseminate logistical, medical, and personnel information to accelerate and improve decision making by CSS and force commanders. It also enables the rapid exchange of $C^2$ information with other battlefield functional areas (BFAs) through automated interfaces. The system will process unclassified through secret information. The CSSCS will be employed in all divisions, corps, EAC, separate brigades, and armored cavalry regiments. The MP require CSSCS functionality in all S1 and S4 sections in battalion or higher organizations except where the GCCS-A is required as stated above.

### TRANSPORTATION COORDINATOR AUTOMATED MOVEMENT SERVER

D-12. The Transportation Coordinator Automated Movement Server (TC-AIMS II) is used by transportation agents and deploying units of each service to automate the processes of planning, organizing, coordinating, and controlling deployment and redeployment operations worldwide during peace as well as contingencies. The TC AIMS II will link all DOD-component unit movements with the installation transportation officer/traffic-management officer (ITO/TMO). This will provide the capability to plan and execute deployments rapidly to meet CINC-required arrival and departure dates. The TC AIMS II will also provide the ITO's/TMO's day-to-day traffic management functions and the theater transportation managers with the capabilities to use theater common lift assets effectively. The TC AIMS II will also allow CSS organizations and supporting commands to sustain the force and perform effective movement control, coordination, distribution, and sustainment once in theater. This is also defined as transportation movement and support of DOD personnel and cargo during all phases of military operations in all environments (including RSOI and battlefield operations). The TC AIMS II's functionality is required at all division PM operations and S3 sections within MP battalions and brigades to facilitate MSR regulation enforcement. Additionally, MP companies, battalions, brigades, and commands require the TC AIMS II to plan, organize, coordinate, and control deployments and redeployments.

### GLOBAL COMBAT SUPPORT SYSTEM—ARMY

D-13. The GCSS-A is an integrated logistical support system with accompanying hardware that will connect the service's supply, finance, transportation, maintenance, and personnel databases. Under this system, the Army will integrate 13 Legacy tactical logistics systems and will eventually interface with the service's wholesale logistics systems as well as the consolidated logistics systems of the other services and the DOD. The system is critical for modernizing automation, for reducing logistical footprints in theaters, and for expanding war fighters' capabilities. The system will also replace the Standard Army Management Information System (STAMIS). The GCSS-A is required at all MP division company CPs and MP battalion and brigade S4 sections. Additionally, it is required at all supply functions within MP detachments that support I/R missions.

## MILITARY POLICE AUTOMATED SYSTEMS

D-14. In an effort to maintain an information advantage and to acquire advance technologies and operational concepts, the MP Corps has acquired the following systems:

- The Military Police Management Information System (MPMIS) and its subsystems
- The CID's automated systems and their subsystems.

## MILITARY POLICE MANAGEMENT INFORMATION SYSTEM

D-15. The MPMIS is a DA-mandated reporting system that does not interface with the ABCS. The MPMIS provides automated tools for MP records management and reporting. It reduces the administrative burden on the PM's staff. It increases the awareness of problem areas by managing crime information. The MPMIS consists of the following elements:

- **Offense Reporting Systems Windows (ORS-2W).** The ORS-2W provides automation support to track and report complaints, criminal incidents, and traffic incidents. It incorporates the CRC, the federally mandated National Incident-Based Reporting System (NIBRS), the DOD-mandated Defense Incident-Based Reporting System (DIBRS), the military community (MILCOM), and the installation's data requirements. The ORS-2W's functionality is required at PM activities at all echelons of command within the Army.
- **Correctional Reporting System (CRS-3).** The CRS-3 provides automation support to track and report inmates and corrections staff at all Army-controlled correctional facilities.
- **Security Management System (SMS).** The SMS provides automated support to assist physical-security inspectors and force-protection officers in reporting inspections and surveys. It tracks intrusion-detection devices and MWD training and health requirements.
- **Vehicle Registration System (VRS).** The VRS provides installations with an automated means to track vehicle, weapon, and bicycle registrations.
- **Prisoner-of-War Information System (PWIS-3).** The PWIS-3 provides the automated means of tracking an EPW's status from capture through release. This system is required at all EPW management activities.

## CID AUTOMATED SYSTEMS

D-16. The Headquarters (HQ) USACIDC mandates CID automated systems. Similar to the MPMIS, it does not interface with the ABCS. The system facilitates data-collection efforts within the command. The CID's automated systems consist of the—

- **Automated Criminal Investigative Reporting System (ACIRS).** The ACIRS supports management and data-collection functions at the CRC and HQ USACIDC. It supports the following:
  - Maintaining investigative data.
  - Conducting agent investigative activities.
  - Reporting investigative data.
  - Managing the investigative processes.
  - Conducting criminal-activity analyses.

- **CRIMINTEL.** Data sources for CRIMINTEL include MP reports, CRIMINTEL reports and bulletins, reports of investigation, crime-prevention surveys, case files, collection folders, raw-data folders, significant data files, and target-analysis files. The CRIMINTEL interfaces with the ACIRS, but it does not share CRIMINTEL products.

# Appendix E

# Police Information Assessment Process

The PIAP differs somewhat from the tactical IPB. While the IPB allows the maneuver commander to see the effects that a variety of factors may have on his forces, the PIAP looks beyond this approach to determine how other relevant information may impact his forces. This information may be obtained from police or criminal actions or incidents encountered during the performance of MP functions.

## IMPLEMENTATION

E-1. The PIAP is not a substitute for the IPB. MP leaders must continue to rely on the IPB and to use its estimates as a starting point for the PIAP. However, the PIAP is a dynamic and continuous cycle that complements the IPB (Figure E-1). The following paragraphs represent the doctrinal approach to implementing the PIAP. These steps provide the basic foundation to the process. They are not exclusive; the successful development of the PIAP will depend on the MP leader's ability to apply the process to his specific environment, METT-TC, and the commander's priorities.

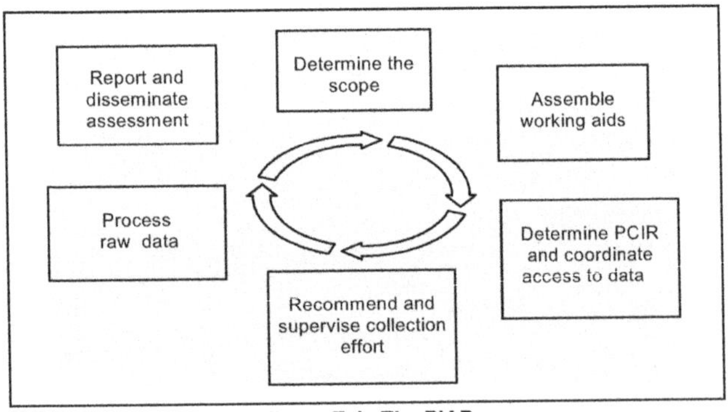


**Figure E-1. The PIAP**

### DETERMINE THE SCOPE OF THE PIAP

E-2. While determining the scope of the PIAP might sound simple, it is one of the most crucial elements of the entire process. It entails identifying what you want to be the end result of your PIAP. To help determine the scope of the PIAP, MP leaders must perform the following subtasks:

- **Conduct a mission analysis.** A detailed mission analysis will ensure that all available resources are focused toward one goal. The more specific the mission is, the more focused and productive the effort. This involves more than just looking at the mission statement, the commander's intent, and the concept of operations. It involves translating the mission and situation into easily understood, manageable goals. The following are examples of these missions:
  - Identifying criminal threats from local national to US Army logistics operations within the port of Mogadishu, Somalia.
  - Determining the impact of gang violence on US forces operating in the vicinity of Camp Bondsteel, Kosovo.
  - Determining how rioters' violence will affect force protection before deploying forces to Los Angeles, California.
  - Determining the status of public service agencies (such as the police, the fire department, and the emergency medical services [EMS]) in Homestead, Florida, after the city was struck by Hurricane Andrew.

**NOTE: As you can see from the examples, these missions are quite specific. In all likelihood, it would be necessary to establish several different PIAP missions to adequately protect Army elements operating in the MP's AOR.**

- **Prioritize missions.** MP leaders must prioritize each PIAP mission since MP resources will probably be limited. The commander's intent is the key factor for determining priorities. When the commander's most important mission has been determined, MP leaders must decide how they can contribute to the overall success of that mission.
- **Determine the AO.** Generally, the AO is a geographical area (including the airspace above) usually defined by lateral, forward, and rear boundaries assigned to a commander. MP leaders must know where major units are located and how the characteristics of the location may impact ongoing operations. For example, units in direct contact with the enemy or isolated from the local population are less likely to be the target of a PIAP mission than those units adjacent to great population centers or in rest and relaxation centers.
- **Determine key terrain.** Key terrain is any locality or area that affords a marked advantage to the retaining side.
- **Determine the AOI.** An AOI is the area of concern to the commander. This includes the area of influence and the areas adjacent thereto and extending into enemy territory and to the objectives of current or planned operations. A thorough understanding of the AOI may help predict potential threat actions. For example, how far are local agitators and ringleaders willing to travel to interfere with US operations in Pristina, Kosovo? Would they go to other US compounds?
- **Determine required information products/sources.** The desired end-state product must be decided. Should it be a report? An updated SITREP? An upgrade in force-protection measures? Or a combination of products? It should also be determined who can provide the best

information (the CID; MI; the local police chief; or joint, interagency, or multinational agencies).

## ASSEMBLE WORKING AIDS

E-3. During this step, MP leaders assemble the necessary graphic and nongraphic aids and information that will help them accomplish their mission. The CID and the MP may form an analysis team to track and analyze police information that might assist tactical operations or population-control operations. Working aids include, but are not limited to—

- **Maps.** Besides the standard military maps that depict the military AOs, the MP must attempt to procure indigenous maps of AOs. Detailed city and street maps will prove to be invaluable even if they are in a foreign language.
- **Language aids.** Phrase books and translation dictionaries cannot take the place of intensive language training, but they are better than nothing. These are essential even if the MP have assigned interpreters.
- **Open-source information.** This can cover an extensive range of invaluable information, but it should be available from a variety of sources. Examples include—
  - Demographic information on the HN population.
  - Societal information on the HN.
  - The locale and disposition of HN police forces.
  - Historical crime data of the area.
  - Environmental information (terrain and weather factors).
  - Local newspaper articles.
  - Internet sources on all of the above.

**NOTE: Sources for this type of information can come from intelligence estimates provided by the higher headquarters' S2/G2, other military units, or the HN.**

- **OPSEC information.** This information includes the disposition of US forces (to include MP, MI, CID, and other services' investigating units) deploying or being used within the AO.
- **Other working aids.** These may include items such as computers, printers, boards, and acetate that will be necessary depending on the missions the MP are undertaking.

## DETERMINE INFORMATION REQUIREMENTS AND COORDINATE ACCESS TO DATA

E-4. Once the mission analysis is completed, the next step is to determine if there are information gaps. An information gap is a missing piece of information that is critical to the analysis. MP leaders must identify and prioritize the gaps using the time, the available resources, and the commander's intent. Once an information gap is identified, it becomes PCIR. Examples of PCIR are as follows:

- Where is the nearest police station, and how did they handle a certain problem?

- What is the composition of the two gangs operating in our AO, and how can they influence US operations?
- Why is river-crossing point Bravo a danger to US crossing forces? Why are local rioters disrupting only US operations on the east bank?

E-5. The next step is to determine if PCIR can be answered with data already available. To do this, the available military and nonmilitary sources must be identified. Some sources may include—

- **USACIDC.** The USACIDC is responsible for programs such as combating terrorism, CRIMINTEL, personal-security assessment, crime analysis, and LOGSEC threat assessment. Some of the information gathered by the USACIDC (although releasable under existing controls and restraints) may help with the PIAP mission. In fact, USACIDC special agents may already be working in the AO and may be available to help the MP with the collection effort.
- **MI element.** Like the USACIDC, the MI element operating in the AO may have come across a piece of information (police or criminal) that may help accomplish the PIAP mission. The MP must make daily contact with the S2/G2 and coordinate access to invaluable information. As with the USACIDC, available intelligence information may only be releasable under strict controls and restrictions imposed by the commander.
- **Other MP units.** Coordination with an adjacent MP unit may prove to be beneficial, especially if the other MP unit has conducted a similar PIAP mission or has conducted operations in the AOI.
- **HN law-enforcement agencies.** Local law-enforcement agencies will provide the essential populace knowledge otherwise not available to US forces. Chiefs of police, sheriffs, and other key community leaders may help MP leaders fill the information need of the PIAP mission.
- **Joint, interagency, and multinational forces.** Close and effective liaison with these elements will result in quick access to information that can produce the same benefits as the agencies listed above.

**NOTE: If, after consulting with all available sources, the PCIR is not resolved, the MP leader must decide the best collection effort to solve the PCIR.**

## RECOMMEND AND SUPERVISE THE POLICE/CRIMINAL COLLECTION EFFORT

E-6. The collection effort is the means by which specific PCIR will be met. This implies developing a collection strategy, tasking specific collectors, and supervising the collection effort.

- **Develop a collection strategy.** After a through study of the availability, capability, and disposition of the potential collecting resources, MP leaders select which asset is better suitable to perform the mission. Are organic MP the best collectors for this mission? Or do we need to request support from USACIDC units? What unique or organic capabilities do each bring to the effort? Which MP function is the best? Do we conduct MMS, AS, or L&O operations? Part of the collection strategy includes coordinating with the S2/G2, the SJA, the

CID, and other agencies before launching the collection effort. This coordination will eliminate duplication of effort, interference with an ongoing effort, or stepping out of legal limits.

- **Task or request specific collectors.** The missions that will be tasked to respective collectors must be determined. Information collectors can be tasked with more than one mission at a time. However, it is imperative that their tasks be prioritized based on mission requirements and time available. The appropriate tasking or request chain must be used to request an MP team, a platoon, or USACIDC special agents.
- **Supervise the collection effort.** The collectors must be provided with reporting guidelines. How often should they report? Should they report "no information?" Specific reporting instructions should be provided, including how, when, and where they report. A collection tasking chart should also be used (Figure E-2). As more and more collectors are being used, it becomes important to track their missions, capabilities, and success. A brief note on the reporting instructions of each collector should be included.

| PIAP Missions | Unit Tasked | Echelon Employed | POC | Collection Strategy Employed | Date Collection Started | Location | Date Collection Ended |
|---|---|---|---|---|---|---|---|
| PCIR 1 | | | | | | | |
| PCIR 2 | | | | | | | |
| PCIR 3 | | | | | | | |
| PCIR 4 | | | | | | | |

**Figure E-2. Collection Tasking Chart**

## PROCESS POLICE/CRIMINAL RAW DATA

E-7. This processing step converts raw data into police intelligence through analysis. The key to processing the data is to understand it. For example, as collectors report raw data it may often appear to be meaningless information. Rarely will one collector receive all of the necessary information to answer the PCIR at one time. Instead, a variety of information must be fitted together to form a coherent, even if incomplete, picture.

E-8. Every piece of data is important. This is the cardinal rule of the entire process. Until the data is analyzed for reliability and compared with other data, all reports must be treated equally. This is important to remember, for the trend will be to use data that either fits your expectations or gives the most complete picture.

E-9. A report of "no activity" can be as important as reports with information. If the collector is capable of collecting the information and is actively trying to collect, reports of "no activity" can be critical. First, if there is no indication of a change, they can be used to reprioritize PIAP missions. Second, they can be used to retask the collectors to other areas. It is important to check the mission's original intent periodically to ensure that it is still valid and merits further efforts.

E-10. To determine the reliability of the data, the MP leader must evaluate the source. Does the information come from the police chief, local youth, or other military forces? One of the best ways to determine the reliability of the information is to have more than one collector reporting on the same PCIR. If more than one collector reports the same information, it can normally be presumed to be accurate.

E-11. Information can be analyzed in a variety of ways—chronologically, geographically, by the impact on the force, by the type of activity or the modus operandi, or by the association of participants. The MP leader must evaluate the information after it is analyzed to determine whether the information collected answers the PCIR or whether it helps accomplish the PIAP mission. If the information does not answer these questions, he must decide whether it is necessary to task other collectors, retask the same collectors, or wait until the situation develops.

### REPORT AND/OR DISSEMINATE POLICE AND CRIMINAL ASSESSMENT

E-12. Reporting and disseminating is the most important aspect of the PIAP. Collection and analysis is valueless unless the data is communicated to commanders who can best use the information.

E-13. Reporting requirements must be established based on the unit's SOP and other established processes. MP leaders must decide if the information collected must be sent using an established priority format or if it can wait to be sent using SITREPs. Does the information collected affect the force-protection measures or does it immediately impact the tactical scenario? Do we need to alert units operating in the vicinity of AA Fox to avoid sector 2? Additionally, MP leaders must ensure that any information released follows existing constraints and regulations.

E-14. It is also advantageous to provide feedback to the collectors. Collectors often receive no feedback on their performance. Informing them as to which part of their information was vital and beneficial can often increase their productivity in future operations.

## SAMPLE SCENARIO

E-15. The following example helps illustrate the implementation of all six steps of the PIAP:

E-16. While deployed in support of Operation Control Chaos, US forces assigned to TF Blue Thunder are tasked with stability and support operations in the city of Corbina. This city, although currently implementing a peace treaty, was the site of extreme civil unrest. Even though the overall situation is not as chaotic as it was before US intervention, there are some sectors that are still in turmoil. Common crimes (such as robberies, assaults, looting, and vandalism) still occur daily and there is fear that this condition may spread out of control throughout the city. The TF commander knows that in order to bring peace to the entire city, he must attempt to stabilize the situation in some of these bad sectors. He decides that the best way to deal with this situation is to assign a sector (sector 51) to an MP battalion in an attempt to control the situation. If the MP battalion succeeds, he may use the same concept throughout the city. He is aware of the five MP functions and how they can help bring this sector under some control. Under this framework, the TF commander issues the following mission: *On order, the MP battalion will conduct stability and support operations in the vicinity of sector 51, the city of Corbina, in support of TF Blue Thunder.* Upon receipt of this mission, the battalion commander gathered his staff and initiated the planning process. Before the MP staff could initiate the PIAP, they made contact with the S2/G2 and received an updated IPB estimate. Now that the IPB estimates are on-hand, the staff initiates the PIAP.

## DETERMINE THE SCOPE OF THE PIAP

E-17. The staff performs the following steps to determine the scope of the PIAP:

- **Conduct a mission analysis.** The staff looks at the given mission, the higher echelon commander's intent, and the concept of operations. After considering all of these factors, they produce a restated mission—*On order, the battalion will conduct MP operations in sector 51 to determine the type of threat and to bring this sector under control in support of TF Blue Thunder.*
- **Prioritize PIAP missions.** Since there is only one mission, this subtask is not applicable.
- **Determine the AO and the AOI.** The staff determines that the AO is the entire sector 51 and that the AOIs are the sectors adjacent to 51.
- **Determine key terrain.** The staff determines that the key terrains are the market place, the plaza, and the city park. They select these areas because most of the incidents occurred in these populated areas. Stabilization in these key areas may be advantageous to mission success.
- **Determine required information products/sources.** The end result is the stabilization of sector 51. Minimize the crime incidents and restore L&O. Information should come from the local police chief, local community leaders, civil affairs, PSYOP, and the CID.

## ASSEMBLE WORKING AIDS

E-18. The staff assembles the following working aids in support of the PIAP:

- **Maps.** The staff acquires local maps that detail the market place, the plaza, and the city park. Although the maps are in a foreign language, the staff is able to identify these three prominent features and surrounding areas.
- **Language aids.** In addition to dictionaries, the staff is able to get two translators from civil affairs.

- **Open-source information.** Open-source information includes a variety of encyclopedias and country books. Newspaper reports and historical crime data is available via the internet or through MI sources. The emphasis is in obtaining information on the history and composition of the local police force.
- **Classified intelligence.** The staff receives a briefing from the S2/G2 describing the operations of SOF units in the area.
- **Other working aids.** Computers, charts, templates, and acetate are available.

## DETERMINE INFORMATION REQUIREMENTS AND COORDINATE ACCESS TO DATA

E-19. After a careful analysis of all available information, the following information gaps were turned into PCIR:

- Why is the police force not actively engaged in L&O?
- What is the major threat to civilians in each area?
- Are local gangs involved?
- Is their local judicial system in place?
- Where are prisoners housed?

E-20. The staff consults with the USACIDC, MI, other MP units, and other sources but determines that the information provided is not clear or sufficient. Since the PCIR are not resolved, the staff decides to recommend that the commander initiate a collection effort.

## RECOMMEND AND SUPERVISE THE POLICE/CRIMINAL COLLECTION EFFORT

E-21. The staff performs the following steps to recommend and supervise the police/criminal collection effort:

- **Develop a collection strategy.** Upon coordinating with the S2/G2, the SJA, and the CID, the MP commander decides that the best collectors will come from within organic assets. He decides that he will employ the MP to go out and seek the needed information. The battalion commander will employ a company commander to interview the local police chief and an MP/interpreter will patrol around the designated AOI. The MP commander will need additional interpreters to help him with his collection effort.
- **Task or request specific collectors.** The MP staff requests an interpreter through the normal chain of command. The TF G3 assigns six interpreters for a period of 90 days.
- **Supervise the collection effort.** All collectors are instructed to report not later than (NLT) 1800 every day. The report will be in the form of an "end of day" outbrief to the S2 and the battalion commander. A collection tasking chart (Figure E-3) will also be completed.

## PROCESS POLICE/CRIMINAL RAW DATA

E-22. After receiving all of the above information, the S2 concludes that the inability of the police force to do their job has had a domino effect on all areas of peace and order. This information is confirmed through other reports from information collected by the CID (through their own efforts) and the CA. The information also reveals that the gangs have intentions of attacking US forces if they interfere with their "business." The S2 determines that although the situation may seem to be chaotic, once the locals are able to trust the police forces, everything should return to normal.

## REPORT AND/OR DISSEMINATE POLICE AND CRIMINAL ASSESSMENT

E-23. Since the collection effort identifies a potential threat to US forces, the commander decides to notify higher headquarters and increase his own force-protection measures. He requires MP elements patrolling in designated areas to be no less than a squad. If patrolling dismounted, a HMMWV or an ASV will trail the squad. The MP commander uses the information provided during the PIAP to develop his implementation plan. This plan includes joint patrols, training, and construction of a joint MP/local national police station.

| | PIAP Mission: Bring Sector 51 Under Control | | | | | | | |
|---|---|---|---|---|---|---|---|---|
| **PCIR** | **Unit Tasked** | **Echelon Employ-ed** | **POC** | **Collection Strategy Employed** | **Date Collection Started** | **Location** | **Date Collection Ended** | **PCIR Result** |
| Why is the police force not actively engaged in L&O operations? | A Co | 2 teams | CPT Smith | Direct interview | 31 Jan 00 | Sector 51, Police HQ | 2 Feb 00 | The police force does not know how to interact with the population. |
| What is the major threat to civilians in each area? | A Co | 6 teams | 1LT Thomas | Patrol and interview | 31 Jan 00 | Market place, plaza, and park | 7 Feb 00 | Muggings, robberies, and stabbings are the major threats to civilians. |
| Are local gangs involved? | A Co | 2 teams | CPT Smith | Direct interview | 31 Jan 00 | Sector 51, Police HQ | 2 Feb 00 | Two rival gangs exist. |
| Is a local judicial system in place? | A Co | 2 teams | 1LT Jones | Direct interview | 31 Jan 00 | Courthouse | 3 Feb 00 | A judicial system exists, but the judge is afraid to sentence anyone. |
| Where are the prisoners housed? | A Co | 4 teams | 1LT Lucas | AS and MMS | 31 Jan 00 | Local prison | 2 Feb 00 | The local prison is abandoned. |

SAMPLE

**Figure E-3. Sample Collection Tasking Chart**

# Glossary

**1LT** first lieutenant

**A/L** administrative and logistics

**AA** assembly area

**AATF** air-assault task force

**ABCS** Army Battle Command Systems

**ACE** analysis and control element

**ACIRS** Automated Criminal Investigative Reporting System

**ACT** analysis control team

**ADA** air-defense artillery

**ADC** area damage control

**AFATDS** Advanced Field-Artillery Tactical Data System

**ALO** authorized level of organization

**AMDPCS** Air- and Missile-Defense Planning and Control System

**AO** area of operation

**AOE** Army of Excellence

**AOI** area of interest

**AOR** area of responsibility

**APOD** aerial port of debarkation

**APOE** aerial port of embarkation

**AR** armored

**AR** Army regulation

**ARFOR** Army forces

**AS** area security

**ASAS** All-Source Analysis System

**ASCC** Army service component command

**ASG** area support group

**ASP** ammunition supply point

**ASR** alternate supply route

**ASV** armored security vehicle

**ATCCS** Army Tactical Command and Control System

**attn** attention

**AWIS** Army WMCCS Information System

**BCT** brigade combat team

**bde** brigade

**BFA** battlefield functional area

**BLD** brigade liaison detachment

**bn** battalion

**BOS** battlefield operating system

**BRT** brigade reconnaissance troop

**BWA** battlefield workload analysis

**$C^2$** command and control

**$C^3$** command, control, and communications

**C4I** command, control, communications, computers, and intelligence

**C4IFTW** command, control, communications, computers, and intelligence for the warrior

**CAS** close air support

**CASE** corps-area support element

**CCIU** computer-crime investigative unit

**cdr** commander

**CG** commanding general

**CHA** corps holding area

**CI** civilian internee

**CID** Criminal Investigation Division

**CIDR** Criminal Investigation Division Regulation

**CINC** commander in chief

**CIP** Criminal Intelligence Program

**cmd** command

**CMOC** civil-military operations center

**co** company

**COA** course of action

**CofS** Chief of Staff

**commo** communications

**COMMZ** communications zone

**conf** confinement

**CONUS** continental US

**COSCOM** corps support command

**CP** command post

**CPT** captain

**CRC** Crime Records Center

**CRIMINTEL** criminal intelligence

**CRS-3** Correctional Reporting System

**CS** combat support

**CSA** corps storage area

**CSE** corps support element

**CSG** corps support group

**CSM** command sergeant major

**CSS** combat service support

**CSSCS** Combat Service Support Control System

**cust** customs

**CZ** combat zone

**DA** Department of the Army

**DC** District of Columbia

**DCCP** division central collection point

**DCG** deputy commanding general

**DIBRS** Defense Incident-Based Reporting System

**DISCOM** division support command

**div** division

**DLEA** domestic law-enforcement agency

**DMMC** division materiel management center

**DNA** deoxyribonucleic acid

**DOD** Department of Defense

**DOJ** Department of Justice

**DOS** Department of State

**DPM** deputy provost marshal

**DRA** division rear area

**DS** direct support

**DSA** division support area

**DSB** division support battalion

**DSE** division support element

**DSS** decisive, shaping, and sustaining

**DZ** drop zone

**EAC** echelons above corps

**EAD** echelons above division

**ECB** echelons corps and below

| | |
|---|---|
| **EG** | escort guard |
| **EMS** | emergency medical services |
| **EPW** | enemy prisoner of war |
| **ERP** | engineer regulation point |
| **ES** | escort |
| **FA** | field artillery |
| **FBCB2** | Force XXI Battle Command—Brigade and Below |
| **FBI** | Federal Bureau of Investigation |
| **Feb** | February |
| **FEBA** | forward edge of the battle area |
| **FIU** | field investigative unit |
| **FM** | field manual |
| **FOB** | forward operating base |
| **FOIA** | Freedom of Information Act |
| **FORSCOM** | US Army Forces Command |
| **FSB** | forward-support battalion |
| **FSC** | forward-support company |
| **G1** | Assistant Chief of Staff, G1 (Personnel) |
| **G2** | Assistant Chief of Staff, G2 (Intelligence) |
| **G3** | Assistant Chief of Staff, G3 (Operations and Plans) |
| **G4** | Assistant Chief of Staff, G4 (Logistics) |

**G5** Assistant Chief of Staff, G5 (Civil Affairs)

**G6** Assistant Chief of Staff, G6 (Communications)

**GAC** ground-assault convoy

**GCCS** Global Command and Control System

**GCCS-A** Global Command and Control System—Army

**GD** guard

**GO** government organization

**GS** general support

**HHB** headquarters and headquarters battery

**HHC** headquarters and headquarters company

**HHD** headquarters and headquarters detachment

**HMMWV** high-mobility, multipurpose, wheeled vehicle

**HN** host nation

**HQ** headquarters

**HRP** high-risk personnel

**HVA** high-value asset

**I/R** internment and resettlement

**IBCT** initial/interim brigade combat team

**IC** information center

**IEW** intelligence and electronic warfare

**IG** inspector general

**IHO** international humanitarian organization

**IMETP** International Military Education and Training Program

**IO** international organization

**IPB** intelligence preparation of the battlefield

**IRIC** internment/resettlement information center

**ISB** intermediate staging base

**ITO** installation transportation officer

**J3** Operations Directorate

**Jan** January

**JFC** joint-force command

**JOA** joint-operations area

**JP** joint publication

**JRA** joint rear area

**JSOA** joint special operations area

**JTF** joint task force

**km** kilometer(s)

**L&O** law and order

**LID** light infantry division

**liq** liquid

**LOC** lines of communication

**LOGSEC** logistics security

**LRP** long-range plan

**LSTA** logistics-security threat assessment

**LVOSS** Light-Vehicle Obscuration Smoke System

**LZ** landing zone

**M/O** marksman/observer

**M/S** mobility/survivability

**MACOM** major Army command

**MAJ** major

**MCRP** Marine Corps reference publication

**MCS** Maneuver Control System

**MDMP** military decision-making process

**mech** mechanized

**MEDEVAC** medical evacuation

**METL** mission-essential task list

**METT-TC** mission, enemy, terrain, troops, time available, and civilian considerations

**MEVA** mission-essential or vulnerable area

**MFPU** major fraud procurement unit

**MI** military intelligence

**MILCOM** military community

**min** minimum

**MKT** mobile kitchen trailer

**MLRS** Multiple-Launch Rocket System

**MMS** maneuver and mobility support

**MOB** main operating base

**MOOTW** military operations other than war

**MP** military police

**MPBTF** military police brigade task force

**MPMIS** Military Police Management Information System

**MQS** military qualification standards

**MSR** main supply route

**MTOE** modified table of organization and equipment

**MTP** mission training plan

**MTT** mobile training team

**MWD** military working dog

**NATO** North Atlantic Treaty Organization

**naut** nautical

**NBC** nuclear, biological, and chemical

**NCO** noncommissioned officer

**NEO** noncombatant evacuation operation

**NG** national guard

**NGO** nongovernmental organization

**NIBRS** National Incident-Based Reporting System

**NLT** not later than

**NLT** nonlethal weapon

**NWP** Naval warfare publication

**OCONUS** outside the continental US

**OPCON** operational control

**OPLAN** operation plan

**opns** operations

**OPSEC** operations security

**ORS-2W** Offense Reporting Systems Windows

**PASE** port-area support element

**PCIR** police/criminal information requirements

**PDF** Panamanian Defense Force

**PEO** peace-enforcement operation

**PI** police intelligence

**PIAP** police information assessment process

**PIO** police intelligence operations

**PIR** priority intelligence requirements

**PKO** peacekeeping operation

**plt** platoon

**PM** provost marshal

**POC** point of contact

**POD** port of debarkation

**POE** port of embarkation

**POL** petroleum, oil, and lubricants

**PRC** populace and resource control

**PSD** protective security detail

**PSU** protective-service unit

**PSVA** personal-security vulnerability assessment

**PSYOP** psychological operations

**pub** publication

**PWIS-3** Prisoner-of-War Information System

**PZ** pickup zone

**R&S** reconnaissance and surveillance

**RAOC** rear-area operations center

**RC** reserve component

**ROC** rear operations center

**ROE** rules of engagement

**rqmt** requirement

**RSOI** reception, staging, onward movement, and integration

**RSTA** reconnaissance, surveillance, and target acquisition

**RWS** remote workstation

**S1** Adjutant (US Army)

**S2** Intelligence Officer (US Army)

**S3** Operations and Training Officer (US Army)

**S4** Supply Officer (US Army)

**S5** Civil-Affairs Officer (US Army)

**S6** Communications Officer (US Army)

**SALUTE** size, activity, location, unit, time, and equipment

**SAW** squad automatic weapon

**SCI** sensitive compartmented information

**sec** section

**SEP** separate

**SFC** sergeant first class

**SITREP** situation report

**SJA** staff judge advocate

**SMS** Security Management System

**SOC** security-operations cell

**SOF** special-operations forces

**SOFA** status of forces agreement

**SOI** signal operating instructions

**SOP** standing operating procedures

**SPOD** seaport of debarkation

**SPOE** seaport of embarkation

**SPOTREP** spot report

**spt** support

**sq** square

**sqd** squad

**SRT** special-reaction team

**STAMIS** Standard Army Management Information System

**stat** statute

**STCCS** Strategic Theater Command and Control System

**TA** theater Army

**TAA** tactical assembly area

**TAACOM** Theater Army Area Command

**TAC** tactical

**TACCIMS** Theater Automated Command and Control Information Management System

**TAI** traffic accident investigation

**TASE** TSC-area support element

**TC-AIMS** Transportation Coordinator Automated Movement Server

**TCF** tactical combat force

**TCP** traffic-control post

**TF** task force

**THREATCON** threat condition

**TM** technical manual

**TMO** traffic-management officer

**tms** teams

**TO** theater of operations

**TOE** tables of organization and equipment

**TRADOC** US Army Training and Doctrine Command

**TRP** traffic-regulation plan

**TSA** theater storage area

**TSC** theater support command

**TSE** tactical support element

**TTP** tactics, techniques, and procedures

**UMT** unit ministry team

**UN** United Nations

**US** United States

**USACIDC** US Army Criminal Investigation Command

**USACIL** US Army Criminal Investigations Laboratory

**USAF** US Air Force

**USAMPS** US Army Military Police School

**USC** US Code

**USCG** United States Coast Guard

**USCS** US Customs Service

**USDA** US Department of Agriculture

**VIP** very important person

**VRS** Vehicle Registration System

**WMCCS** Worldwide Military Command and Control System

# Bibliography


AR 15-6. *Procedures for Investigating Officers and Boards of Officers.* 11 May 1988.

AR 190-12. *Military Police Working Dogs.* 30 September 1993.

AR 190-45. *Law Enforcement Reporting.* 30 September 1988.

AR 190-47. *The Army Corrections System.* 15 August 1996.

AR 195-2. *Criminal Investigation Activities.* 30 October 1985.

AR 525-13. *Antiterrorism / Force Protection (AT / FP): Security of Personnel, Information, and Critical Resources.* 10 September 1998.

CIDR 195-1. *Special Agent Operations Manual.* October 1994.

DA Form 2028. *Recommended Changes to Publications and Blank Forms.* 1 February 1974.

FM 3-0. *Operations.* To be published within the next 6 months.

FM 19-1. *Military Police Support for the AirLand Battle.* 23 May 1988.

FM 19-4. *Military Police Battlefield Circulation Control, Area Security, and Enemy Prisoner of War Operations.* 7 May 1993.

FM 19-10. *The Military Police Law and Order Operations.* 30 September 1987.

FM 19-40. *Enemy Prisoners of War, Civilian Internees and Detained Persons.* 27 February 1976.

FM 34-2. *Collection Management and Synchronization Planning.* 8 March 1994.

FM 34-130. *Intelligence Preparation of the Battlefield.* 8 July 1994.

FM 41-10. *Civil Affairs Operations.* 14 February 2000.

FM 63-4. *Combat Service Support Operations—Theater Army Area Command.* 24 September 1984.

FM 71-3. *The Armored and Mechanized Infantry Brigade.* 8 January 1996.

FM 71-100. *Division Operations.* 28 August 1996.

FM 90-13-1. *Combined Arms Breaching Operations.* 28 February 1991.

FM 90-40. *Multiservice Procedures for the Tactical Employment of Nonlethal Weapons (NLW) (MCRP 3-15.8; NWP 3-07-31; USCG Pub 3-07.31).* 6 October 1998.

FM 100-1. *The Army.* 14 June 1994.

FM 100-5. *Operations.* 14 June 1993.

FM 100-8. *The Army in Multinational Operations.* 24 November 1997.

FM 100-10. *Combat Service Support.* 3 October 1995.

FM 100-15. *Corps Operations.* 29 October 1996.

FM 100-17-3. *Reception, Staging, Onward Movement, and Integration.* 17 March 1999.

FM 101-5. *Staff Organization and Operations.* 31 May 1997.

FM 101-5-1. *Operational Terms and Graphics (MCRP 5-2A).* 30 September 1997.

JP 3-10. *Joint Doctrine for Rear Area Operations.* 28 May 1996.

USC Title 18. *Crimes and Criminal Procedure.* 23 January 2000.

USC Title 32. *National Guard.* 5 January 1999.

# Index

***FM 3-19.30**

Field Manual
No. 3-19.30

Headquarters
Department of the Army
Washington, DC, 8 January 2001

# PHYSICAL SECURITY

## Contents



Distribution Restriction: Approved for public release; distribution is unlimited.

*This publication supersedes FM 19 30, 1 March 1979.

# Preface

This field manual (FM) sets forth guidance for all personnel responsible for physical security. It is the basic reference for training security personnel. It is intended to be a "one-stop" physical-security source for the Department of Defense (DOD), the Department of the Army (DA), and other proponents and agencies of physical security.

Prevention and protection are the two primary concerns of physical security. Both serve the security interests of people, equipment, and property. These interests must be supported at all staff and command levels; and this support must be unified in joint, multinational, and interagency operations.

Support to joint, multinational, and interagency operations relies on the fact that the Army will not conduct operations alone. Additionally, force-projection operations conducted by the military will involve the integration of war-fighting capabilities with stability and support operations. This manual's primary focus is the articulation of a balanced understanding of physical security for joint, multinational, and interagency operations throughout the environments of peacetime, conflict, and war (whether in the continental United States [CONUS] or outside the continental United States [OCONUS]).

Physical security must integrate the various capabilities of joint, multinational, and interagency operations in pursuit of a seamless connection between the strategic, operational, and tactical levels of war. Physical security must also address an expanded range of threats that embraces not only traditional threat components of war, but also nontraditional threats generated by guerrillas, terrorists, criminals, and natural or man-made disasters. In addition, physical security must address the concept of Homeland Defense due to the aforementioned threats.

Homeland Defense is the military's role in the United States (US) government's principal task of protecting its territory and citizens. This is accomplished by joint, interagency, and multijurisdictional organizations. Homeland Defense includes—

- Supporting domestic authorities for crisis and consequence management with regard to weapons of mass destruction (WMD).
- Protecting national-security assets (such as installations) and deploying forces and ensuring the availability, integrity, and adequacy of other critical assets.
- Deterring and defending against strategic attacks while maintaining freedom of action through antiterrorism and force-protection operations.

With this in mind, it is essential to address the five pillars of force protection—combating terrorism, physical security, personal security, law enforcement, and operations security (OPSEC). Physical security is a central component of force protection and provides an integrated venue to express support for operations. Physical security is a primary-leader task and an inherent part of all operations to protect soldiers, family members, civilians, and resources. This function directly supports the Army's universal task list.

While the effects of these changes (when viewed individually) appear revolutionary, the basic activities remain relatively unchanged, though executed under different conditions and standards. Another component that remains unchanged is our reliance upon quality soldiers and leaders well versed in physical-security fundamentals. Leaders will be challenged to ensure that they are functionally proficient; possess an understanding of physical-security operations; are educated in joint, multinational, and interagency operations; and have the ability to perform physical-security functions in support of full-dimension operations.

Appendix A contains an English-to-metric measurement conversion chart. Appendix B is a sample installation crime-prevention handbook. This handbook is designed to assist commanders in developing crime-prevention programs for their installation and units.

The proponent of this publication is HQ TRADOC. Send comments and recommendations on DA Form 2028 directly to Commandant, US Army Military Police School (USAMPS), ATTN: ATSJ-MP-TD, Directorate of Training, 401 Engineer Loop, Suite 2060, Fort Leonard Wood, Missouri 65473-8926.

Unless this publication states otherwise, masculine nouns and pronouns do not refer exclusively to men.

# Chapter 1

# Physical-Security Challenges

Physical security is defined as that part of security concerned with physical measures designed to safeguard personnel; to prevent unauthorized access to equipment, installations, material, and documents; and to safeguard against espionage, sabotage, damage, and theft. As such, all military operations face new and complex physical-security challenges across the full spectrum of operations. Challenges relative to physical security include the control of populations, information dominance, multinational and interagency connectivity, antiterrorism, and the use of physical-security assets as a versatile force multiplier.

## OVERVIEW

1-1. Reductions in manpower and funding are critical challenges to physical security. Manpower for supporting physical-security activities is reduced through deployments and cutbacks. The rapid evolution of physical-security-equipment technology also lends to physical-security challenges, which are exponentially multiplied by the introduction of the information age.

1-2. Physical-security challenges must be understood, and measures must be taken to minimize them to enhance force protection. Leaders must create order when coming upon a situation; and when they depart, some semblance of that order must remain. They must be aware of the human-dimension factors and ensure that their soldiers do not become complacent. It was human error rather than modern technology that took lives in the bombings of the African embassy. Warning was given, but not heeded. Complacency became a physical-security challenge.

## AUTOMATED INFORMATION SYSTEMS

1-3. Success on past battlefields has resulted not so much from technological advances, but from innovative ways of considering and combining available and new technologies as they apply to war fighting. Some of these technologies dealt with disseminating and processing information. For example, the telegraph, the telephone, the radio, and now the computer have redefined the fire-support paradigm.

1-4. As the armed forces move into the technological age, a greater need for physical-security measures is required. The risks associated with automated information systems (AISs) are widespread because computers are used for everything. Army Regulation (AR) 380-19 outlines the requirements that commanders and managers need for processing unclassified and classified information and for securing media, software, hardware, and different systems.

1-5. The threat to AISs and information systems security (ISS) involves deliberate, overt, and covert acts. This includes the physical threat to tangible property, such as the theft or destruction of computer hardware. Also included is the threat of electronic, electromagnetic-pulse, radio-frequency (RF), or computer-based attacks on the information or communications components that control or make up critical Army command and control ($C^2$) infrastructures. In most cases, the threat's target is the information itself rather than the system that transmits it. The threat comes from a range of sources, including the following:

- Unauthorized users (such as hackers) are the main source of today's attacks, primarily against computer-based systems. The threat they pose to AIS networks and mainframe computers is growing.
- Insiders are those individuals with legitimate access to an AIS. They pose the most difficult threat to defend against. Whether recruited or self-motivated, the AIS insider has access to systems normally protected by ISS against an attack.
- Terrorists once had to operate in the immediate vicinity of a target to gain access to or collect intelligence on that target. The proximity to the target risked exposure and detection. Today, a terrorist can accomplish most target selection, intelligence collection, and preoperational planning by gaining access through a computer network. He can increase his probability of success by using computer systems to reduce his "time on target." Terrorist access to an AIS also increases the threat of critical-data destruction or manipulation. Although his presence would be virtual, the potential for damage to Army $C^2$ systems could be equal to or greater than that achieved by physical intrusion, especially when used as a force multiplier in conjunction with a traditional terrorist attack. Therefore, while traditional preventive measures are still needed to protect unwanted access to information, the information age has added additional concerns for the commander and new opportunities for those with hostile intent.
- Non-state- and state-sponsored groups provide additional challenges. In many cases, it is difficult to confirm state sponsorship of threat activity against an AIS, no matter how apparent the affiliation might seem. Activists of all persuasions are increasingly taking advantage of information-age technology. Neither AISs nor ISS are immune from an adversary's interest in exploiting US military information systems or disrupting communication infrastructures. The availability of low-cost technology and the proliferation of an AIS increase the risk to the Army by potential adversaries.
- Foreign-intelligence services (FIS), both civil and military, are continually active and are another source of contention concerning information systems. In peacetime, they are increasingly targeted against US commercial and scientific interests, rather than military information. With little effort, this peacetime intrusiveness could easily be refocused on AISs and ISS using a wide range of information operations tactics.

- Political and religious groups are other potential adversaries to AISs and ISS. The world's political climate is diverse and complicated. It embraces traditional mainstream political values, as well as radical religious fundamentalism and political extremism. When political or religious viewpoints also incorporate anti-US sentiment, US information infrastructures (including AISs) are increasingly at risk of penetration or exploitation by these potential adversaries.

1-6. When considering an AIS, physical security is more than just safeguarding the equipment. It includes the following elements:

- Software is marked for each system and secured when not in use.
- Initial logon is password-protected (at a minimum).
- Passwords are a minimum of eight characters, using a mixture of letters and numerals.
- Access to an AIS is allowed only to authorized and cleared personnel (per AR 380-19).

1-7. Classified material is entered and transmitted only on approved devices with the following considerations:

- Approved classified devices are operated in a secured environment.
- Classified devices are secured in appropriate containers when not in use.
- Secure telephone unit–III (STU-III) keys are secured in an appropriate safe when not in use (as outlined in AR 380-19).

1-8. Additional information regarding AISs can be found in ARs 380-5 and 380-19. Required training of personnel working with an AIS is located in AR 380-19.

## OPSEC AND THE THREAT

1-9. OPSEC is a process of identifying critical information and subsequently analyzing friendly actions attendant to military operations and other activities. The threat is identified using the factors of mission, enemy, terrain, troops, time available, and civilian considerations (METT-TC). The threat defines the physical-security challenges. Implementing physical-security measures supports OPSEC. Providing soundproof rooms for conducting briefings is a simple but invaluable measure.

1-10. Another issue to consider when evaluating physical-security challenges is what actions to take in case of political implications interfering with physical-security measures. In the devastating event at Khobar Towers, a warning was given but not everyone received it. It took too long to evacuate the building after the warning was issued because a cohesive plan was not in place.

1-11. Commanders can minimize the challenges to physical security through proactive measures. They should periodically change the physical-security posture of their area of responsibility to throw off perpetrators.

## Chapter 2

# The Systems Approach

Commanders must ensure that appropriate physical-security measures are taken to minimize the loss of personnel, supplies, equipment, and material through both human and natural threats. Commanders commonly exercise those protective responsibilities through the provost marshal (PM) and/or physical-security officer and the force-protection officer. The force-protection officer must coordinate with several different agencies to complete his mission. For example, the Army's Intelligence and Counterintelligence Program (see Appendix C) provides information that will be used to complete the unit's crisis-management plan (see Appendix D).

## PROTECTIVE SYSTEMS

2-1. The approach to developing protective measures for assets should be based on a systematic process resulting in an integrated protective system. The protective system focuses on protecting specific assets against well-defined threats to acceptable levels of protection. The system is organized in-depth and contains mutually supporting elements coordinated to prevent gaps or overlaps in responsibilities and performance.

2-2. Effective protective systems integrate the following mutually supporting elements:

- Physical protective measures, including barriers, lighting, and electronic security systems (ESSs).
- Procedural security measures, including procedures in place before an incident and those employed in response to an incident. (These include procedures employed by asset owners and those applied by and governing the actions of guards.)
- Terrorism counteraction measures that protect assets against terrorist attacks.

2-3. The following determinations are made when considering system-development procedures:

- The resources available.
- The assets to be protected.
- The threat to those assets.
- The risk levels applicable to those assets.
- The applicable regulatory requirements for protecting the assets.
- The applicable level of protection for those assets against the threat.
- Additional vulnerabilities to the assets (based on the threat).

## SYSTEMS DEVELOPMENT

2-4. AR 190-51, DA Pamphlet (Pam) 190-51, and Technical Manual (TM) 5-853-1 are useful tools for developing protective systems using the systems approach. The key to applying these tools successfully is to use a team approach. A team may include physical-security, intelligence, and operations personnel; the installation engineers; and the user of the assets. It may also include representatives from the multinational, host-nation (HN), and local police as well as the regional security office from the embassy.

### ASSETS

2-5. Protective systems should always be developed for specific assets. The goal of security is to protect facilities and buildings and the assets contained inside. The risk-analysis procedure in DA Pam 190-51 is used to identify assets. This procedure is applied to all mission-essential or vulnerable areas (MEVAs) according to AR 190-13. It represents the majority of assets with which DOD is commonly concerned. These assets include—

- Aircraft and components at aviation facilities.
- Vehicle and carriage-mounted or -towed weapons systems and components at motor pools.
- Petroleum, oil, and lubricants (POL).
- Controlled medical substances and other medically sensitive items.
- Communication and electronics equipment; test, measurement, and diagnostic equipment (TMDE); night-vision devices (NVDs); and other high-value precision equipment and tool kits.
- Organizational clothing and individual equipment stored at central-issue facilities.
- Subsistence items at commissaries, commissary warehouses, and troop-issue facilities.
- Repair parts at installation-level supply activities and direct-support (DS) units with authorized stockage lists.
- Facilities-engineering supplies and construction materials.
- Audiovisual equipment, training devices, and subcaliber devices.
- Miscellaneous pilferable assets (not included above) and money.
- Mission-critical or high-risk personnel.
- General military and civilian populations.
- Industrial and utility equipment.
- Controlled cryptographic items.
- Sensitive information (included in TM 5-853-1, but not included in DA Pam 190-51).
- Arms, ammunition, and explosives (AA&E).
- Installation banks and finance offices.

### RISK LEVELS

2-6. DA Pam 190-51 provides a procedure for determining risk levels—assessing the value of the assets to their users and the likelihood of compromise. These factors are assessed by answering a series of questions leading to value and likelihood ratings.

2-7. Asset value is determined by considering the following three elements:

- The criticality of the asset for its user and the Army as a whole.
- How easily the asset can be replaced.
- Some measure of the asset's relative value.

2-8. The relative value differs for each asset. For some assets, the relative value is measured in terms of monetary cost.

2-9. The likelihood of the threat is assessed for each applicable aggressor category by considering the asset's value to the aggressor, the history of or potential for aggressors attempting to compromise the asset, and the vulnerability of the asset based on existing or planned protective measures.

## REGULATORY REQUIREMENTS

2-10. The risk level is the basis for determining the required protective measures for assets covered in AR 190-51. For each asset type, there may be physical protective measures, procedural security measures, and terrorism counteraction measures. These measures are specified by risk level. The measures identified in AR 190-51 are the minimum regulatory measures that must be applied for the identified threat level. The minimum regulatory measures for AA&E are based on the risk category established in AR 190-11.

## ANTITERRORISM/FORCE-PROTECTION CONSTRUCTION STANDARDS

2-11. In accordance with DOD Instruction 2000.16, the commanders in chief (CINCs) have developed standards for new construction and existing facilities to counter terrorism threat capabilities within the area of responsibility. These construction standards have specific requirements for such measures as standoff distance, perimeter barriers, building construction, and parking. The DOD construction standard provides for minimum standards that must be incorporated into all inhabited DOD structures regardless of the identified threat. These standards provide a degree of protection that will not preclude the direct effects of blast but will minimize collateral damage for buildings and people and will limit the progressive collapse of structures. These standards add relatively little cost, may facilitate future upgrades, and may deter acts of aggression. (All services have adopted common criteria and minimum standards to counter antiterrorism/force-protection [AT/FP] vulnerabilities and terrorism threats.) Protection to identified threat levels is described in the following paragraphs. Physical-security personnel must be familiar with the CINC and DOD AT/FP construction standards because these standards may affect elements of physical-security plans and how individual facilities are secured.

## THREAT IDENTIFICATION

2-12. The threat must be described in specific terms to help determine the assets' vulnerabilities or to establish protective measures. This description should include the tactics that aggressors will use to compromise the asset (weapons, tools, and explosives are likely to be used in an attempt). For example, the threat might be described as a moving vehicle bomb consisting of a 4,000-pound vehicle containing a 500-pound explosive. Another example would be a forced-entry threat using specific hand, power, or thermal tools. These types of threat descriptions (called the design-basis threat) can be used to design detailed protective systems to mitigate the attacks. TM 5-853-1 and DA Pam 190-51 contain procedures for establishing design-basis threat descriptions in the format described above. These procedures can be used together or separately. Threats listed in the TM will be summarized later in this chapter. When using the TM as a lone source or in conjunction with DA Pam 190-51, the following actions occur:

- When the TM process is used alone, the user goes through an identical process to that in DA Pam 190-51 up to the point where the risk level would be determined. In TM 5-853-1, the value and likelihood ratings are used differently than in DA Pam 190-51. The likelihood rating is used to determine the weapons, tools, and explosives that will be used by a particular aggressor in carrying out a specific tactic. In this procedure, higher likelihood ratings result in more severe mixes of weapons, tools, and explosives. The assumption is that the more likely the attack, the more resources the aggressor is likely to use in carrying out the attack.
- When the procedure in TM 5-853-1 is used in conjunction with the results of the DA Pam 190-51 risk analysis, the likelihood rating is taken directly from the risk analysis and applied as described above.

## LEVEL OF PROTECTION

2-13. The level of protection applies to the design of a protective system against a specified threat (for example, a bomb, breaking and entering, pilfering, and so forth). The level of protection is based on the asset's value rating from either DA Pam 190-51 or TM 5-853-1. The level increases as the asset's value rating increases. There are separate levels of protection for each tactic. TM 5-853-1 provides detailed guidance on how to achieve the levels of protection, and Chapter 3 of this manual provides a summary of the levels of protection as they apply to various tactics.

## VULNERABILITIES

2-14. Vulnerabilities are gaps in the assets' protection. They are identified by considering the tactics associated with the threat and the levels of protection that are associated with those tactics. Some vulnerabilities can be identified by considering the general design strategies for each tactic described in TM 5-853-1 and as summarized in Chapter 3 of this manual. The general design strategies identify the basic approach to protecting assets against specific tactics. For example, the general design strategy for forced entry is to provide a way to detect attempted intrusion and to provide barriers to delay the aggressors until a response force arrives. Vulnerabilities may involve inadequacies in intrusion-detection systems (IDSs) and barriers. Similarly, the general design strategy for a moving vehicle bomb is to keep the vehicle as far from the facility as possible and to harden the facility to resist the explosive at that distance. Vulnerabilities may involve limited standoff distances, inadequate barriers, and building construction that cannot resist explosive effects at the applicable standoff distance.

### PROTECTIVE MEASURES

2-15. Where vulnerabilities have been identified, protective measures must be identified to mitigate them. AR 190-13, AR 190-51, DA Pam 190-51, and TM 5-853-1 are effective tools for developing protective measures. The key to effective development of protective systems is a partnership between physical-security personnel and the installation engineers. Appendix E of this manual discusses information for office security, which should be listed in the physical-security plan (see Appendix F). Appendix G discusses personal-protection measures.

## THE INTEGRATED PROTECTIVE SYSTEM

2-16. Protective systems integrate physical protective measures and security procedures to protect assets against a design-basis threat. The characteristics of integrated systems include deterrence, detection, defense, and defeat.

### DETERRENCE

2-17. A potential aggressor who perceives a risk of being caught may be deterred from attacking an asset. The effectiveness of deterrence varies with the aggressor's sophistication, the asset's attractiveness, and the aggressor's objective. Although deterrence is not considered a direct design objective, it may be a result of the design.

### DETECTION

2-18. A detection measure senses an act of aggression, assesses the validity of the detection, and communicates the appropriate information to a response force. A detection system must provide all three of these capabilities to be effective.

2-19. Detection measures may detect an aggressor's movement via an IDS, or they may detect weapons and tools via X-ray machines or metal and explosive detectors. Detection measures may also include access-control elements that assess the validity of identification (ID) credentials. These control elements may provide a programmed response (admission or denial), or they may relay information to a response force. Guards serve as detection elements, detecting intrusions and controlling access.

2-20. Nuclear, biological, and chemical (NBC) detection systems must be used to measure and validate acts of aggression involving WMD. NBC detection systems should also be used to communicate a warning.

### DEFENSE

2-21. Defensive measures protect an asset from aggression by delaying or preventing an aggressor's movement toward the asset or by shielding the asset from weapons and explosives. Defensive measures—

- Delay aggressors from gaining access by using tools in a forced entry. These measures include barriers along with a response force.
- Prevent an aggressor's movement toward an asset. These measures provide barriers to movement and obscure lines of sight (LOSs) to assets.
- Protect the asset from the effects of tools, weapons, and explosives.

2-22. Defensive measures may be active or passive. Active defensive measures are manually or automatically activated in response to acts of aggression. Passive defensive measures do not depend on detection or a response. They include such measures as blast-resistant building components and fences. Guards may also be considered as a defensive measure.

### DEFEAT

2-23. Most protective systems depend on response personnel to defeat an aggressor. Although defeat is not a design objective, defensive and detection systems must be designed to accommodate (or at least not interfere with) response-force activities.

## SECURITY THREATS

2-24. Security threats are acts or conditions that may result in the compromise of sensitive information; loss of life; damage, loss, or destruction of property; or disruption of mission. Physical-security personnel and design teams must understand the threat to the assets they are to protect in order to develop effective security programs or design security systems. Historical patterns and trends in aggressor activity indicate general categories of aggressors and the common tactics they use against military assets. Aggressor tactics and their associated tools, weapons, and explosives are the basis for the threat to assets.

### THREAT SOURCES

2-25. There are many potential sources of threat information. Threat assessment is normally a military-intelligence (MI) responsibility. MI personnel commonly focus on such security threats as terrorists and military forces. Within the US and its territories, the Federal Bureau of Investigation (FBI) has primary responsibility for both foreign and domestic terrorists. The FBI, the US Army Criminal Investigation Command (USACIDC [CID]), and local law-enforcement agencies are good sources for physical-security personnel to obtain criminal threat information. Coordinating with these elements on a regular basis is essential to maintaining an effective security program.

### THREAT CATEGORIES

2-26. Security threats are classified as either human or natural. Human threats are carried out by a wide range of aggressors who may have one or more objectives toward assets such as equipment, personnel, and operations. Aggressors can be categorized and their objectives can be generalized as described below. (See DA Pam 190-51 and TM 5-853-1 for more information.)

## Aggressor Objectives

2-27. Four major objectives describe an aggressor's behavior. Any one of the first three objectives can be used to realize the fourth. These objectives include—

- Inflicting injury or death on people.
- Destroying or damaging facilities, property, equipment, or resources.
- Stealing equipment, materiel, or information.
- Creating adverse publicity.

## Aggressor Categories

2-28. Aggressors are grouped into five broad categories—criminals, vandals and activists, extremists, protest groups, and terrorists. Hostile acts performed by these aggressors range from crimes (such as burglary) to low-intensity conflict threats (such as unconventional warfare). Each of these categories describes predictable aggressors who pose threats to military assets and who share common objectives and tactics.

- Criminals can be characterized based on their degree of sophistication. They are classified as unsophisticated criminals, sophisticated criminals, and organized criminal groups. Their common objective is the theft of assets; however, the assets they target, the quantities they seek, their relative efficiency, and the sophistication of their actions vary significantly. Vandals and activists may also be included under this category.
- Vandals and activists are groups of protesters who are politically or issue oriented. They act out of frustration, discontent, or anger against the actions of other social or political groups. Their primary objectives commonly include destruction and publicity. Their selection of targets will vary based on the risk associated with attacking them. The degree of damage they seek to cause will vary with their sophistication.
- Extremists are radical in their political beliefs and may take extreme, violent actions to gain support for their beliefs or cause.
- Protesters are considered a threat only if they are violent. Lawful protesters have to be considered, but significant protective measures and procedures are not normally needed to control their actions. The presence of extremists or vandals/activists at a peaceful protest increases the chance of the protest becoming violent.
- Terrorists are ideologically, politically, or issue oriented. They commonly work in small, well-organized groups or cells. They are sophisticated, are skilled with tools and weapons, and possess an efficient planning capability. There are three types of terrorists—CONUS, OCONUS, and paramilitary OCONUS.
  - CONUS terrorists are typically right- or left-wing extremists operating in distinct areas of the US.
  - OCONUS terrorists generally are more organized than CONUS terrorists. They usually include ethnically or religiously oriented groups.

- Paramilitary OCONUS terrorist groups show some military capability with a broad range of military and improvised weapons. Attacks by OCONUS terrorists are typically more severe.

2-29. Natural threats are usually the consequence of natural phenomena. They are not preventable by physical-security measures, but they are likely to have significant effects on security systems and operations. They may require an increase in protective measures either to address new situations or to compensate for the loss of existing security measures. They may reduce the effectiveness of existing security measures by such occurrences as collapsed perimeter fences and barriers, inoperable protective lighting, damaged patrol vehicles, and poor visibility. Natural threats and their effects relative to security include the following:

- Floods may result in property damage, destruction of perimeter fences, and damage to IDSs. Heavy rains or snowfalls may have similar effects even if they do not result in flooding.
- Storms, tornadoes, high winds, or rain may cause nuisance alarms to activate and cause damage to IDSs. They may limit the visibility of security personnel and may affect closed-circuit television (CCTV) systems. Winds may also disrupt power or communication lines and cause safety hazards from flying debris.
- Earthquakes may cause nuisance alarms to activate or may disrupt IDSs. They may also cause broken water or gas mains, fallen electrical or communication lines, and weakened or collapsed buildings.
- Snow and ice can make travel on patrol roads difficult, may delay responses to alarms, may impede the performance of IDSs, and may freeze locks and alarm mechanisms. Heavy ice may also damage power and communication lines.
- Fires may damage or destroy perimeter barriers and buildings, possibly leaving assets susceptible to damage or theft.
- Fog can reduce the visibility of security forces, thereby requiring additional security personnel. It may also increase the response time to alarms and reduce the effectiveness of security equipment such as CCTV systems.

## Aggressor Tactics

2-30. Aggressors have historically used a wide range of offensive strategies reflecting their capabilities and objectives. These offensive strategies are categorized into 15 tactics that are specific methods of achieving aggressor goals (see TM 5-853-1). Separating these tactics into categories allows facility planners and physical-security personnel to define threats in standardized terms usable as a basis for facility and security-system design. Common aggressor tactics include—

- **Moving vehicle bomb.** An aggressor drives an explosive-laden car or truck into a facility and detonates the explosives. His goal is to damage or destroy the facility or to kill people. This is a suicide attack.
- **Stationary vehicle bomb.** An aggressor covertly parks an explosive-laden car or truck near a facility. He then detonates the explosives either by time delay or remote control. His goal in this tactic is the

same as for the moving vehicle bomb with the additional goal of destroying assets within the blast area. This is commonly not a suicide attack. It is the most frequent application of vehicle bombings.

- **Exterior attack.** An aggressor attacks a facility's exterior or an exposed asset at close range. He uses weapons such as rocks, clubs, improvised incendiary or explosive devices, and hand grenades. Weapons (such as small arms) are not included in this tactic, but are considered in subsequent tactics. His goal is to damage the facility, to injure or kill its occupants, or to damage or destroy assets.
- **Standoff weapons.** An aggressor fires military weapons or improvised versions of military weapons at a facility from a significant distance. These weapons include direct (such as antitank [AT] weapons) and indirect LOS weapons (such as mortars). His goal is to damage the facility, to injure or kill its occupants, or to damage or destroy assets.
- **Ballistics.** The aggressor fires various small arms (such as pistols, submachine guns, shotguns, and rifles) from a distance. His goal is to injure or kill facility occupants or to damage or destroy assets.
- **Forced entry.** The aggressor forcibly enters a facility using forced-entry tools (such as hand, power, and thermal tools) and explosives. He uses the tools to create a man-passable opening or to operate a device in the facility's walls, doors, roof, windows, or utility openings. He may also use small arms to overpower guards. His goal is to steal or destroy assets, compromise information, injure or kill facility occupants, or disrupt operations.
- **Covert entry.** The aggressor attempts to enter a facility or a portion of a facility by using false credentials or stealth. He may try to carry weapons or explosives into the facility. His goals include those listed for forced entry.
- **Insider compromise.** A person authorized access to a facility (an insider) attempts to compromise assets by taking advantage of that accessibility. The aggressor may also try to carry weapons or explosives into the facility in this tactic. His goals are the same as those listed for forced entry.
- **Visual surveillance.** The aggressor uses ocular and photographic devices (such as binoculars and cameras with telephoto lenses) to monitor facility or installation operations or to see assets. His goal is to compromise information. As a precursor, he uses this tactic to determine information about the asset of interest.
- **Acoustic eavesdropping.** The aggressor uses listening devices to monitor voice communications or other audibly transmitted information. His goal is to compromise information.
- **Electronic-emanations eavesdropping.** The aggressor uses electronic-emanation surveillance equipment from outside a facility or its restricted area to monitor electronic emanations from computers, communications, and related equipment. His goal is to compromise information.

- **Mail-bomb delivery.** The aggressor delivers bombs or incendiary devices to the target in letters or packages. The bomb sizes involved are relatively small. His goal is to kill or injure people.
- **Supplies-bomb delivery.** The aggressor conceals bombs in various containers and delivers them to supply- and material-handling points such as loading docks. The bomb sizes in this tactic can be significantly larger that those in mail bombs. His goal is to damage the facility, kill or injure its occupants, or damage or destroy assets. Appendix H addresses the actions to take when a bomb is suspected.
- **Airborne contamination.** An aggressor contaminates a facility's air supply by introducing chemical or biological agents into it. His goal is to kill or injure people.
- **Waterborne contamination.** An aggressor contaminates a facility's water supply by introducing chemical, biological, or radiological agents into it. These agents can be introduced into the system at any location with varying effects, depending on the quantity of water and the contaminant involved. His goal is to kill or injure people.

2-31. The aforementioned tactics are typical threats to fixed facilities for which designers and physical-security personnel can provide protective measures. However, some common terrorist acts are beyond the protection that facility designers can provide. They cannot control kidnappings, hijackings, and assassinations that take place away from facilities or during travel between facilities. Protection against these threats is provided through operational security and personal measures (see Appendices G and I), which are covered in doctrine relative to those activities and are under the general responsibility of the CID.

## TACTICAL ENVIRONMENT CONSIDERATIONS

2-32. When determining the assets and threats, the same considerations should be given to the systems approach in the tactical environment as when in the cantonment area. The same process of determining the assets, their risk level, and any regulatory guidance apply. Identifying potential threats and the level of protection required for the assets are necessary. Commanders and leaders must also identify additional vulnerabilities and other required protective measures. Commanders are not expected to have the same physical protective measures due to the impact of resources, budget, location, and situations.

2-33. Commanders must consider the various tactics used by aggressors and use their soldiers' abilities to counteract these tactics. Considerations for specific assets (such as military-working-dog [MWD] and explosive-ordnance-disposal [EOD] teams and their abilities to detect and disassemble a bomb) must be identified. Units must have the ability to improvise in a tactical environment. Their training and resourcefulness will compensate for shortcomings in the field.

2-34. The systems approach to security provides focus and integration of resources. Protective systems are mutually supporting and systematically developed to negate the threat. Commanders conduct an intelligence preparation of the battlefield (IPB) and vulnerability assessments (VAs) to determine risks. Security resources and measures are applied to mitigate risks and to deter, detect, defend, and defeat the threat.

## Chapter 3

# Design Approach

Developing protective systems to protect assets depends on an effective partnership between engineers and physical-security personnel. Physical-security personnel need to understand the basic approaches the engineers will take in laying out protective systems. Engineers must understand the issues involved with ensuring that anything they lay out is compatible with security operations and the operations of the asset users. The best way to ensure a viable design is through teamwork. This chapter provides a summary of the basic approaches to protecting assets against threats (the design strategies). Understanding these strategies is critical to being an effective team member in developing protective systems.

## DESIGN STRATEGIES

3-1. There are separate design strategies for protecting assets from each tactic described in Chapter 2. There are two types of strategies associated with each tactic—the general-design and specific-design strategies. The general-design strategy is the general approach to protecting assets against tactics. The specific-design strategy refines the general-design strategy to focus the performance of the protective system on a particular level of protection. (See TM 5-853-1 for more information.)

## PROTECTIVE MEASURES

3-2. Protective measures are developed as a result of the general- and specific-design strategies. These protective measures commonly take the form of site-work, building, detection, and procedural elements.

- Site-work elements include the area surrounding a facility or an asset. Technically, they are associated with everything beyond 5 feet from a building. They can include perimeter barriers, landforms, and standoff distances.
- Building elements are protective measures directly associated with buildings. These elements include walls, doors, windows, and roofs.
- Detection elements detect such things as intruders, weapons, or explosives. They include IDSs, CCTV systems used to assess intrusion alarms, and weapon and explosive detectors. These elements can also include the guards used to support this equipment or to perform similar functions.
- Procedural elements are the protective measures required by regulations, TMs, and standing operating procedures (SOPs). These elements provide the foundation for developing the other three elements.

## VEHICLE BOMBS

3-3. Vehicle-bomb tactics include both moving and stationary vehicle bombs. In the case of a moving vehicle bomb, the aggressor drives the vehicle into the target. This is commonly known as a suicide attack. In a stationary vehicle bomb, he parks the vehicle and detonates the bomb remotely or on a timed delay.

### GENERAL-DESIGN STRATEGY

3-4. Blast pressures near an exploding vehicle bomb are very high, but they decrease rapidly with distance from the explosion. The design strategy for these tactics is to maintain as much standoff distance as possible between the vehicle bomb and the facility and then, if necessary, to harden the facility for the resulting blast pressures. Barriers on the perimeter of the resulting standoff zone maintain the required standoff distance. The difference between moving and stationary vehicle-bomb tactics is that the aggressor using the moving vehicle bomb will attempt to crash through the vehicle barriers; the aggressor using the stationary vehicle bomb will not. Therefore, vehicle barriers for the moving vehicle bomb must be capable of stopping a moving vehicle at the perimeter of the standoff zone. For a stationary vehicle bomb, vehicle barriers must mark the perimeter of the standoff zone, but they are not required to stop the moving vehicle. They only need to make it obvious if an aggressor attempts to breach the perimeter.

### LEVELS OF PROTECTION

3-5. There are three levels of protection for vehicle bombs—low, medium, and high. The primary differences between the levels are the degree of damage allowed to the facility protecting the assets and the resulting degree of damage or injury to the assets.

- **Low.** The facility or the protected space will sustain a high degree of damage but will not collapse. It may not be economically repairable. Although collapse is prevented, injuries may occur and assets may be damaged.
- **Medium.** The facility or the protected space will sustain a significant degree of damage, but the structure will be reusable. Occupants and other assets may sustain minor injuries or damage.
- **High.** The facility or the protected space will sustain only superficial damage. Occupants and other assets will also incur only superficial injury or damage.

### SITE-WORK ELEMENTS

3-6. The two primary types of site-work elements for vehicle bombs are the standoff distance and vehicle barriers. The vehicle's speed must also be taken into consideration.

#### Standoff Distance

3-7. The standoff distance is the maintained distance between where a vehicle bomb is allowed and the target. The initial goal should be to make that distance as far from the target facility as practical. Figure 3-1 shows the distances required to limit building damage to particular levels (including the levels of protection described above) for a range of bomb weights. All bomb weights are given in terms of equivalent pounds of trinitrotoluene (TNT), which is a standard way of identifying all explosives regardless of their composition. The example in Figure 3-1 is a building of conventional construction (common, unhardened construction). Buildings built without any special construction at these standoff distances will probably withstand the explosive effects. Conventionally constructed buildings at standoff distances of less than those shown in Figure 3-1 will not adequately withstand blast effects. (Refer to TM 5-853-1 for information on hardening buildings to resist a blast.) Do not allow vehicles to park within the established standoff distances. Recognize that this restriction can result in significant operational and land-use problems.

3-8. **Exclusive Standoff Zone.** When an exclusive standoff zone is established, do not allow vehicles within the perimeter unless they have been searched or cleared for access. The zone's perimeter is established at the distance necessary to protect the facility against the highest threat explosive. All vehicles should be parked outside the exclusive standoff zone; only

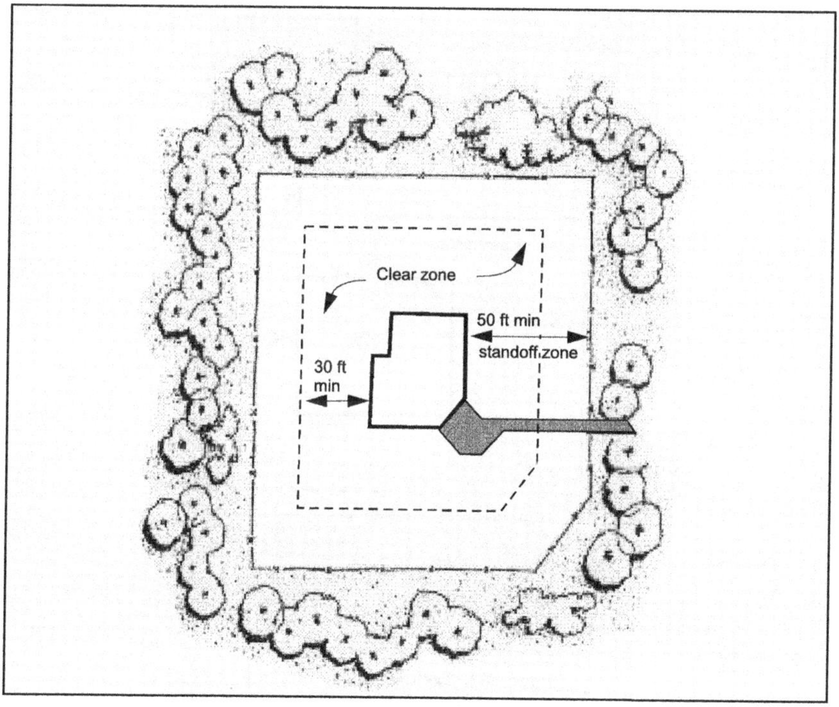


**Figure 3-1. Standoff Distance**

maintenance, emergency, and delivery vehicles should be allowed within the zone after being searched. Figure 3-2 shows an exclusive standoff zone.

3-9. **Nonexclusive Standoff Zone.** A nonexclusive standoff zone is established in a location having a mixture of cars and trucks (with relatively few trucks). A nonexclusive standoff zone takes advantage of aggressors being able to conceal a smaller quantity of explosives in a car than they can in a truck. Therefore, a nonexclusive standoff zone includes inner and outer perimeters. The inner perimeter is set at a distance corresponding to the weight of explosives that can be concealed in cars. The outer perimeter is set at a distance associated with the weight that can be placed in trucks (refer to TM 5-853-1). With these two perimeters, cars can enter the outer perimeter without being searched but they cannot enter the inner perimeter. Trucks cannot enter the outer perimeter, since it is established based on what they can carry. Figure 3-3 shows a nonexclusive standoff zone. The nonexclusive standoff zone provides the advantages of allowing better use of the parking areas and limiting the number of vehicles that need to be searched at the outer perimeter.

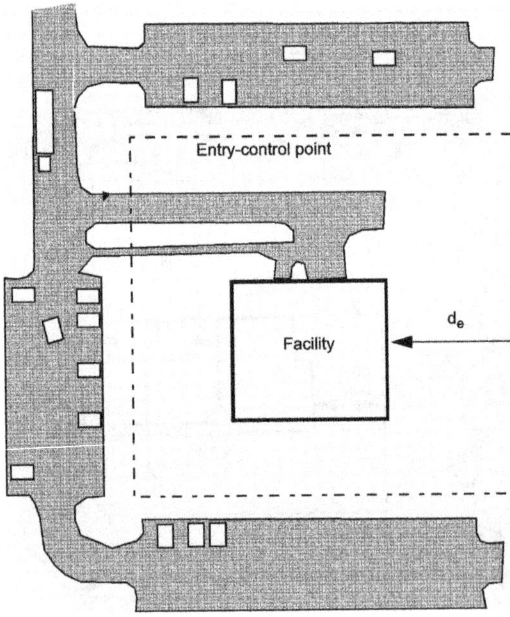


$d_e$ = exclusive standoff-zone distance

**Figure 3-2. Exclusive Standoff Zone**

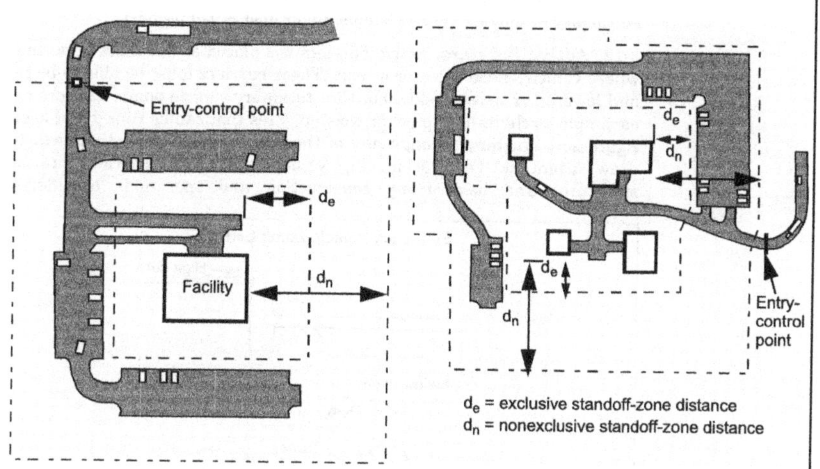


Figure 3-3. Nonexclusive Standoff Zone

## Vehicle Barriers

3-10. Two types of vehicle barriers are used for vehicle bombs—perimeter and active barriers. The type of barrier used for a moving vehicle bomb differs from the barrier used for a stationary vehicle bomb. The barrier used for a stationary vehicle bomb does not have to stop a vehicle's motion. The goal for that barrier is to make anybody driving through the barrier noticeable. The assumption is that the aggressor's goal in the stationary vehicle bomb is to park the vehicle and sneak away without being noticed. Crashing through a barrier would be noticeable. Barriers for the moving vehicle bomb need to stop the vehicle's motion; they must be much more substantial.

3-11. **Perimeter Barriers.** Perimeter barriers are fixed barriers placed around the entire perimeter of a standoff zone. Anything that presents a fixed obstacle will work for the stationary vehicle bomb. Common applications include chain-link fences, hedges made of low bushes, and high (over 8 inches) curbs. Aggressors driving through such barriers are likely to be noticed. Barriers capable of stopping moving vehicles include chain-link fences reinforced with cable, reinforced concrete "Jersey barriers", pipe bollards, planters, ditches, and berms. When barriers such as the Jersey barriers and planters are used to stop moving vehicles, they must be anchored into the ground to be effective. The cables in the reinforced fence also have to be anchored into the ground or partially buried. Spaces between barriers should be no greater than 4 feet. Figure 3-4 shows common perimeter barriers for stationary or moving vehicle bombs. Refer also to TM 5-853-1.

3-12. **Active Barriers.** Active barriers are placed at openings in perimeters where vehicles need to enter or exit. These barriers must be able to be raised and lowered or moved aside. For the stationary vehicle bomb, barriers can be as simple as chain-link, pipe, or wooden gates that can be raised and lowered. Aggressors crashing through any of these or similar obstructions will likely draw attention. For the moving vehicle bomb, the barriers are heavy structures and have many construction and operations considerations

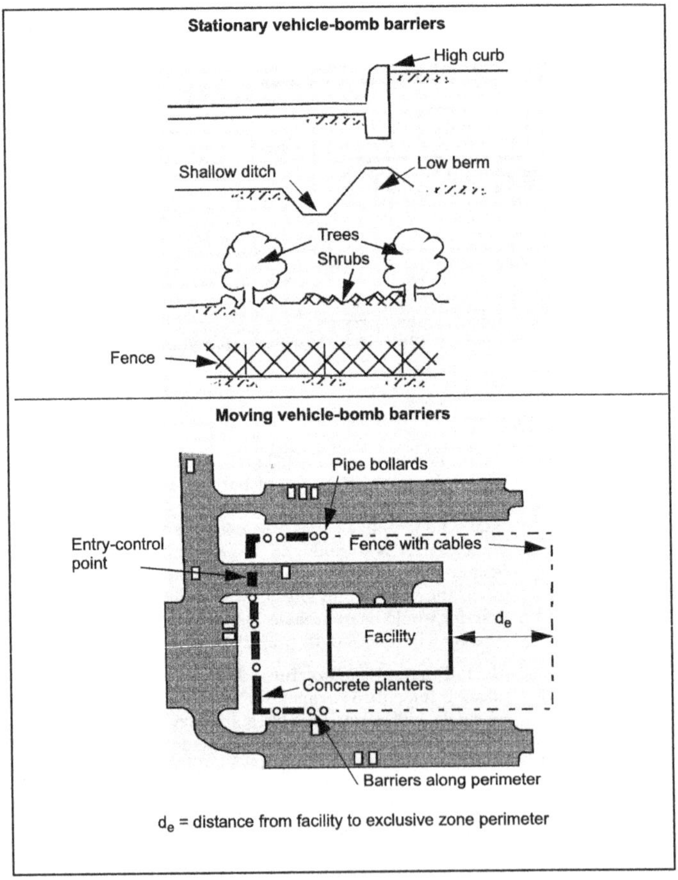


**Figure 3-4. Perimeter-Barrier Application**

associated with them. These barriers may stop vehicles weighing up to 15,000 pounds and travelling 50 miles per hour. They commonly cost tens of thousands of dollars (refer to TM 5-853-1). Some common active vehicle barriers are shown in Figure 3-5. For temporary or deployed conditions, park a vehicle across an opening and move it aside to grant access.

## Speed Control

3-13. It is important to control the speed of a vehicle approaching a barrier used for a moving vehicle bomb. The energy from a vehicle that a barrier must stop increases as its speed increases. The energy also increases with more weight, but the effect of speed is much greater. Therefore, decreasing the vehicle's speed results in smaller and less costly barriers. The best way to limit a vehicle's approach speed to perimeter barriers is to place or retain obstacles in potential approach paths. The vehicles are forced to reduce speed when going around these obstacles. The same principle applies for road approaches. Placing obstacles in a serpentine pattern on the road forces a vehicle to reduce its speed (see Figure 3-6, page 3-8). If the vehicle hits the obstacles instead of going around them, they are still slowed down. Other means to slow vehicles include forcing them to make sharp turns and installing traffic circles.

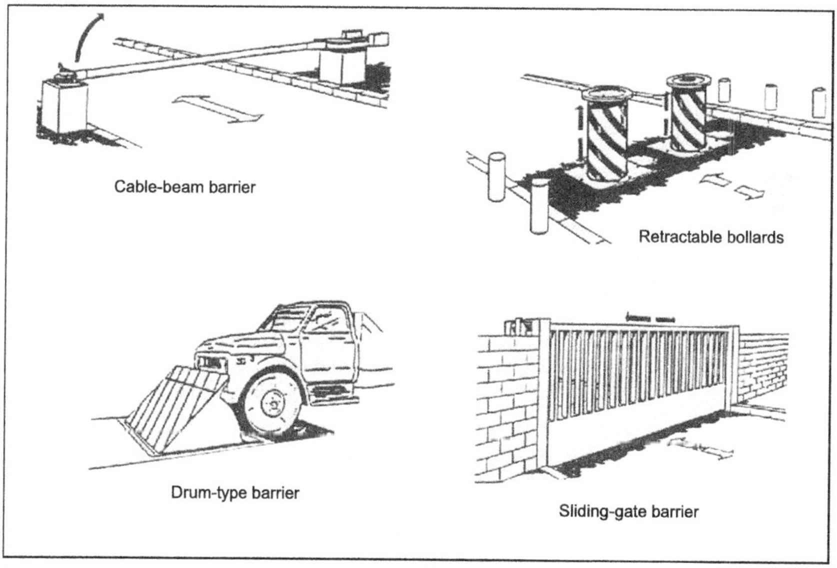


**Figure 3-5. Active Vehicle Barriers**

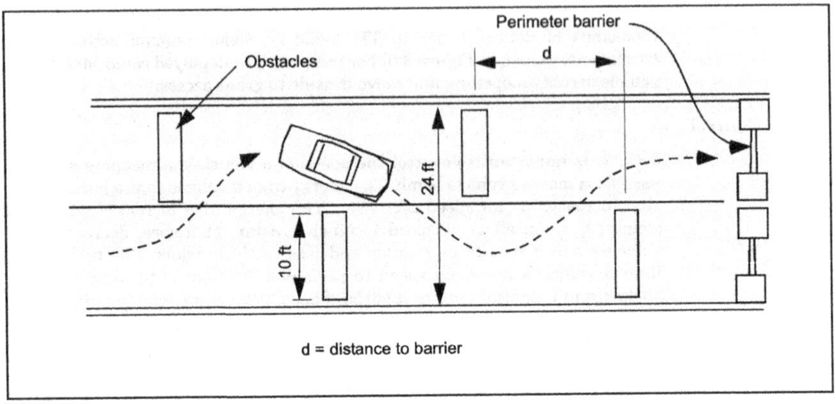


Figure 3-6. Serpentine Pattern

## BUILDING ELEMENTS

3-14. Once the standoff distance is established and the site has been laid out, the designers can select the building components necessary to protect the assets against the threat explosives at the standoff distance. The building components include the walls, roofs, doors, and windows. Detailed design issues related to these building elements are covered in TM 5-853-1.

### Walls and Roofs

3-15. If the distances shown for the desired damage levels in Figure 3-1, page 3-3, cannot be enforced, the building's walls and roofs will need to be strengthened. This can be achieved in new construction by using reinforced masonry or reinforced concrete in the walls and reinforced concrete in the roof. When the standoff distance is not available for existing construction, a more detailed analysis may be required to determine what the explosion's impact will be on the structure. When the construction is inadequate, more standoff distance should be investigated or the engineers should apply specialized techniques for retrofitting the construction to increase its strength.

### Windows

3-16. Historically, glass fragments have caused about 85 percent of injuries and deaths in bomb blasts. There are two basic approaches to mitigating the effects of bomb blasts on glass—retrofitting the windows with film or curtains and using blast-resistant glazing.

3-17. **Retrofitting Windows.** One of the most common means of decreasing the hazards from broken glass is to install fragment-retention film on the glass. The film is a plastic (polyester) sheet that adheres to the window glass with a special adhesive. The film does not strengthen the glass; but when the glass breaks, it keeps the fragments from spreading throughout the room. The glass fragments stick to the film, and the film either stays in the window frame or falls into the room in one or more large, relatively nonhazardous pieces instead of many small, lethal pieces. Another retrofit approach is to install a blast curtain or a heavy drape behind the window. The curtain or drape catches the glass fragments. The curtains are generally used with fragment-retention film. Another retrofit technique is to use fragment-retention film with a metal bar placed across the window. This "catcher bar" catches the window. The designs for this and other types of retrofit devices are complicated and require specialized engineering-analysis tools. The retrofit techniques are generally thought of as providing a lower level of protection than the glazing replacement techniques. For deployed locations, removing the windows and covering them with plywood minimizes the danger.

3-18. **Blast-Resistant Glazing.** To achieve higher levels of protection, the window glass must be replaced and the window frame should be reinforced. Because of its expense, this procedure is generally limited to new construction and major renovations. Special blast-resistant glazing and frames are available that use either tempered glass or a plastic glazing (such as polycarbonate). Another promising type of blast-resistant glazing is laminated glass, in which several layers of common glass are adhered together with a special interlayer. The resulting laminated construction is usually stronger than common glass while retaining the same thickness. The interlayer acts similarly to fragment-retention film. For deployed locations, a means of minimizing the danger of windows is to remove them and replace them with plywood.

### Doors

3-19. Doors are another building component particularly vulnerable to an explosive blast. Common metal and wood doors provide little resistance to a blast. The two ways to address the problem of doors is to install them in foyers or to replace them. Glass doors or doors containing windows should be avoided.

### Foyers

3-20. Door hazards can be reduced by installing doors in foyers during construction or by adding foyers to existing buildings. When a door is located in a foyer and the outer door fails, the outer door flies into a wall instead of the building's interior (see Figure 3-7, page 3-10). The inner door then has a greater chance of remaining intact. This option generally provides a low level of protection.

3-21. Another option is to replace the doors with specially constructed blast-resistant doors and frames. These doors are commercially available and can provide a high level of protection, but they are very expensive and heavy. The doorframe must be made of the same type of material and provide the same level of protection as the door.

## DETECTION ELEMENTS

3-22. Detection elements for vehicle bombs are limited to the use of guards to control access into standoff zones. The guards search vehicles seeking entry

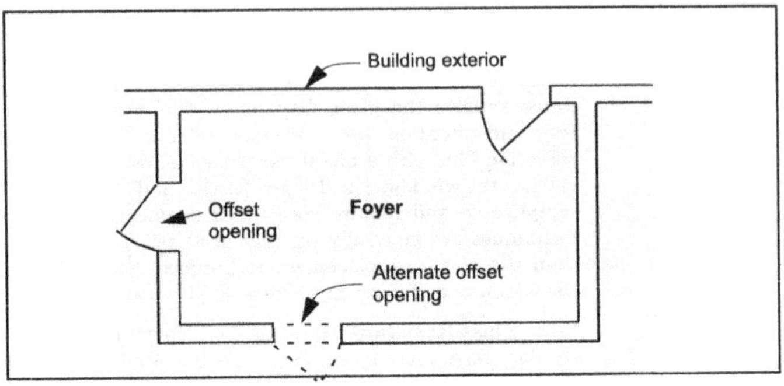


Figure 3-7. Door in Foyer

into the perimeter through an entry-control point. The recommended levels of searches depend on the required level of protection (see TM 5-853-1). Guards can be stationed at entry-control points continuously, or they can be summoned to an entry-control point when access is needed. The latter is commonly the case for the inner perimeter of exclusive standoff zones where only delivery and maintenance vehicles need access.

# EXTERIOR ATTACK

3-23. An exterior attack is a physical attack using weapons such as rocks, clubs, improvised incendiary devices (IIDs) such as Molotov cocktails, explosives such as improvised explosive devices (IEDs), and hand grenades. The explosives can be thrown at or placed near a facility's exterior. Examples of IEDs for this tactic range from pipe bombs and hand grenades to briefcase-sized explosives.

## GENERAL-DESIGN STRATEGY

3-24. Because the exterior attack is directed at a facility's exterior surfaces, the general-design strategy is to keep aggressors away from the facility (at a standoff distance) and, if necessary, to harden the facility's exterior components to resist the effects of weapons and explosives. A standoff distance from the facility reduces the degree of hardening required to resist weapons effects. When briefcase-sized bombs are a threat, an obstacle-free zone should be established around the facility and the explosives placed within should be detected and disarmed.

## LEVELS OF PROTECTION

3-25. The levels of protection for exterior attacks are similar to those for vehicle bombs. Levels of protection vary based on the level of building damage and asset injury or damage allowed. However, due to the limited sizes of explosives involved in this tactic, the damage to the building will be much more localized and injuries or damage to assets will be confined to smaller areas.

## SITE-WORK ELEMENTS

3-26. Site-work elements for exterior attacks are relatively limited because the explosive weights are more limited. Large standoff distances are not a consideration. The common approach to site-work elements is to lay out a standoff zone of about 50 feet and to provide a fence or perimeter barrier about 7 feet high. The purpose of the standoff is to make it harder for aggressors to throw pipe bombs and hand grenades at targets inside the perimeter. Trees can be left around the perimeter to make it harder for aggressors to throw explosives over the fence. The remaining component of site-work elements is a clear zone around the facility. A clear zone is applied so that anything placed in that area can be detected visually. This limits the aggressor's ability to place explosives near the target facility.

## BUILDING ELEMENTS

3-27. Building elements for exterior attacks are similar to those for vehicle bombs. For small IEDs and IIDs, the building-element requirements do not increase the cost of the building significantly. For larger, briefcase-sized bombs, the measures are more significant than for incendiary devices but less than for vehicle bombs.

### Walls and Roofs

3-28. Walls and roofs are not a problem with small explosives. Conventional construction normally provides adequate protection. Walls with 6-inch reinforced concrete or 8-inch, grout-filled, reinforced masonry will withstand the effects of typical pipe bombs or hand grenades. The corresponding roof construction is 6-inch reinforced concrete. In the case of briefcase-sized bombs, considerations similar to those discussed for vehicle bombs need to be employed.

### Windows

3-29. A significant goal when constructing windows is to make them difficult to throw an explosive or incendiary device through, especially when considering smaller explosives. This is accomplished by constructing smaller windows or making narrow windows (see Figure 3-8, page 3-12). For existing windows, parts of the windows can be covered to achieve a narrow effect. These windows still may be susceptible to breakage due to explosive effects, even from the smaller explosives. This problem is solved by installing 3/4-inch-thick plastic (polycarbonate) glazing or by raising the windows over 6 feet high to develop a small standoff distance (as shown in Figure 3-9, page 3-12). A 3/4-inch glazing will also stop grenade fragments. Fragment-retention film, a blast curtain, or a heavy drape as described in vehicle-bomb tactics are also good applications for small bombs.

### Conventionally Constructed Doors

3-30. Doors are not a significant problem with small bombs and incendiary devices. Generally, metal doors are adequate for incendiary devices, and doors placed in foyers (as shown in Figure 3-7) are adequate for pipe bombs and hand grenades. A similar application for briefcase-sized bombs would provide

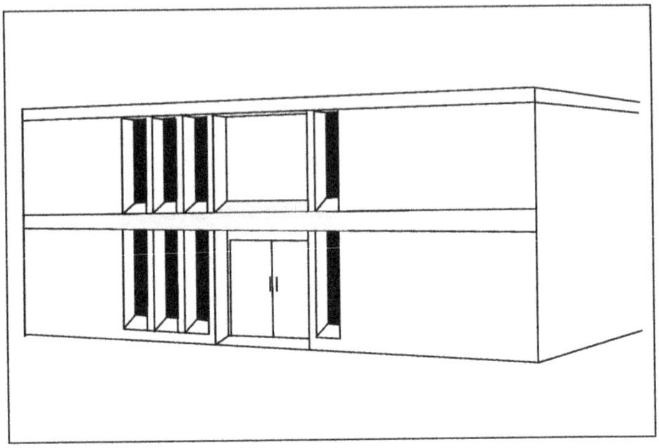

Figure 3-8. Narrow Recessed Windows

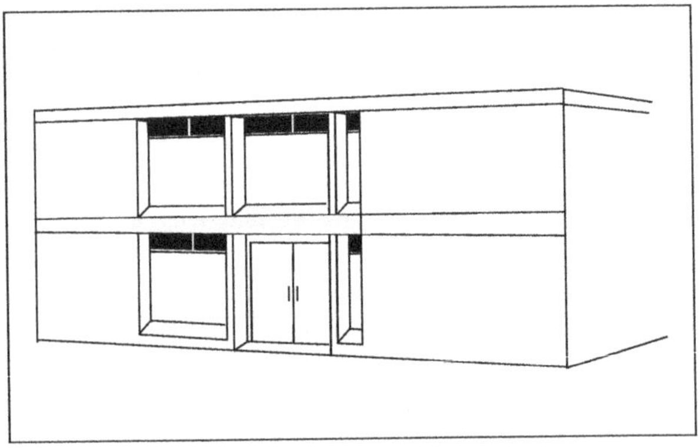

Figure 3-9. Narrow Raised Windows

only a low level of protection. To achieve higher levels of protection for briefcase-sized bombs, blast-resistant doors must be installed.

3-31. The requirements to meet the levels of protection for larger explosives are similar to those described for vehicle bombs, but they will not stop grenade fragments. Fragment-retention film and drapes or curtains can provide a low level of protection, but blast-resistant glazing is required to achieve a higher level of protection.

### DETECTION ELEMENTS

3-32. Other than awareness of aggressor activity on or outside the site, detection is only a specific design goal where briefcase-sized bombs are anticipated. When that is the case, the clear zone around the building must be visually monitored so that any objects placed in it are detected. At higher levels of protection, visual surveillance is augmented by IDSs.

## STANDOFF WEAPONS

3-33. The standoff-weapons tactic includes the use of AT weapons and mortars. In both of these tactics, the aggressor fires weapons at assets located in the protected facility from a distance. An AT-weapon attack requires a clear LOS to the target, while mortars can fire over obstacles and only need a clear line of flight.

### GENERAL-DESIGN STRATEGY

3-34. Standoff-weapons attacks cannot be detected reliably before they occur. Protective design to resist these tactics relies on blocking LOSs to protected areas of a facility or hardening the facility to resist the particular weapon's effects. The approaches to protection against mortars and AT weapons differ from each other and will be discussed separately. Detection measures are not applicable for these tactics.

### LEVELS OF PROTECTION

3-35. There are two levels of protection against both mortars and AT weapons. For AT weapons, the low level of protection depends on detonating the AT round before it hits the target facility. The high level of protection avoids the risk associated with that and hardens the building to resist the direct impact of the AT round.

3-36. For mortars, the low level of protection involves allowing some areas of the facility to be sacrificed. Those spaces provide a buffer to the assets to be protected. The assets within the sacrificial areas and the areas themselves may be destroyed. At the high level of protection, the building's exterior fully resists the mortar rounds and there are no sacrificial areas.

### SITE-WORK ELEMENTS

3-37. The primary site-work element for standoff weapons is to obstruct LOSs from vantage points outside of the site. With AT weapons, the aggressor cannot hit what he cannot see. This is not true with mortars, but blocking LOSs from mortar firing points helps to make targeting more difficult. The LOSs are blocked by using trees, other buildings, vehicle parking areas, or fences. Another site-work element, a predetonation screen, applies only to an AT weapon. When using a predetonation screen, the AT round is detonated on the screen and its effects are dissipated in the distance between the screen and the target (see Figure 3-10, page 3-14). Any screen material (such as a

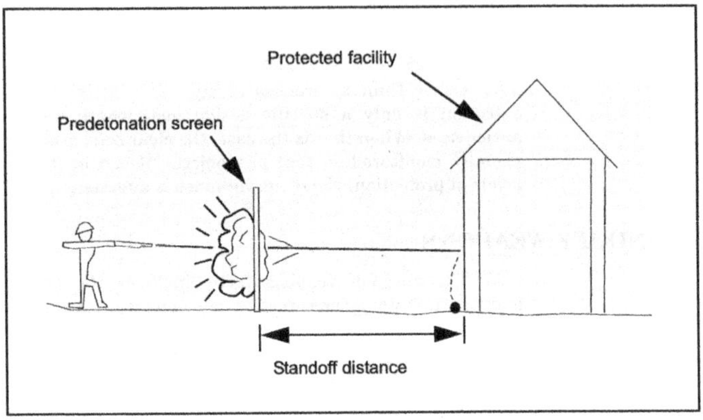


Figure 3-10. Predetonation Screen

wooden fence) will detonate the round unless it has spaces in it. The screen distances vary from less than 10 feet to almost 40 feet, depending on the building construction (see TM 5-853-1). This measure only applies to the low level of protection.

## BUILDING ELEMENTS

3-38. Building elements for AT weapons and mortars involve the building's layout. This includes the materials used in the construction.

### Layout

3-39. A building's interior layout is only an issue for the low level of protection against a mortar round. The layout issue involves designating sacrificial areas in which unimportant assets are located. The assets to be protected are located in a hardened interior layer. Figure 3-11 includes a plan view (from above). The sacrificial area has to be both around and above the protected area in case a mortar round comes from above. If such a layout is not feasible, other options include going to a higher level of protection and either hardening the entire building or building the facility underground (which are both very expensive).

### Walls and Roofs

3-40. Walls and roofs must offer protection against both AT weapons and mortar rounds. The design of walls that protect against AT weapons varies with the level of protection. For the low level of protection where the round is predetonated, the walls can be of conventional construction, varying with the standoff distance from the predetonation screen to the wall. For higher levels of protection, the walls must resist the full effect of the round, requiring the walls to be 24-inch-thick reinforced concrete. Roofs are not an issue in

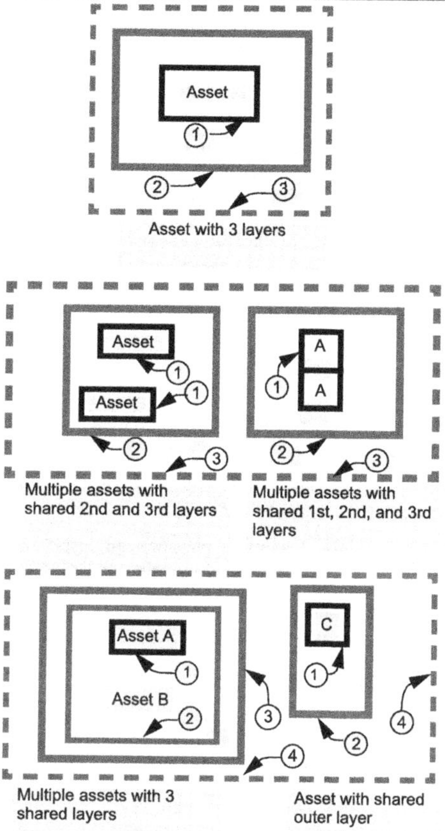


**Figure 3-11. Assets Protected by Hardened Interior Layer**

protecting against AT weapons because it is difficult to get direct LOSs to roofs. If such LOSs are possible, the roof should be designed like the walls.

3-41. To provide protection against mortar rounds, walls and roofs should be designed to resist the explosive effects in the rounds at the standoff distance that the sacrificial space provides. In the case of sacrificial areas, the walls can be of common construction. The interior protected-area walls are then designed of reinforced concrete or reinforced masonry for the standoff distance those sacrificial walls provide. When the walls must resist the full effect of the rounds (as in the higher level of protection), they are likely to be very thick (up to 30 inches of reinforced concrete for some improvised mortars). Similar considerations should be made for roofs. Roofs are designed to take the direct effects of the round or to take the round at the standoff distance provided by the sacrificial area.

### Doors and Windows

3-42. It is impractical to provide doors and windows that are resistant to mortar rounds and AT weapons. Windows should only be used in sacrificial areas where there is a mortar threat. When there is an AT weapon threat, windows can only be used where the round is predetonated. The windows should be narrowed or raised to present a smaller target (see Figures 3-8 and 3-9, page 3-12). Doors should be placed in foyers (see Figure 3-7, page 3-10) for protection against AT rounds and to achieve a low level of protection against mortars. Blast-resistant doors are necessary to achieve a high level of protection against mortar rounds.

# BALLISTICS

3-43. In a ballistics tactic, aggressors fire small arms at assets from vantage points outside of the target facility's control. Ballistic attacks cannot be detected reliably before they occur.

## GENERAL-DESIGN STRATEGY

3-44. Protective measures to resist these tactics rely on blocking LOSs to protected areas of a facility or by hardening the facility to resist the ballistic effects. This strategy focuses on assets within buildings. Protecting people or property in the open is difficult and can only be addressed through operational measures. Detection measures are not applicable for this tactic.

## LEVELS OF PROTECTION

3-45. There are only two levels of protection for this tactic. The low level of protection depends on blocking LOSs to assets. This strategy assumes that the aggressor cannot hit what he cannot see. The risk of an aggressor firing into a building randomly and hitting something is what makes this the low level of protection. The high level of protection involves hardening building components to resist the ballistic effects. These strategies can be thought of as either hardening or hiding.

## SITE-WORK ELEMENTS

3-46. Site-work elements are of limited use for the ballistics tactic. When they are applied, they are used to obstruct LOSs from vantage points outside of the site, which is consistent with the low level of protection. The LOSs can be blocked using trees, other buildings, motor pools, or fences.

## BUILDING ELEMENTS

3-47. Building elements are the principal means of protecting assets against a ballistics attack. They can be applied to achieve either the low or high level of protection.

### Walls and Roofs

3-48. Walls and roofs are inherently opaque, so it is easy to achieve the low level of protection (hiding) with them. Achieving the high level of protection (hardening) for walls and roofs can be done within conventional construction using reinforced concrete, concrete-masonry units (CMUs), or clay brick. The material's required thickness is shown in Table 3-1. The thicknesses of CMUs and clay brick are nominal, meaning they do not represent the actual thickness of the material; they represent the thicknesses at which those materials are commercially available. Steel plates (mild steel and armor steel) and bullet-resistant fiberglass can be used to retrofit existing building components that would not provide the needed bullet resistance.

**Table 3-1. Required Thicknesses, in Inches**

| Ballistics Type | Reinforced Concrete | Grouted CMU* | Clay Brick* | Steel Plate | | Bullet-Resistant Fiberglass |
|---|---|---|---|---|---|---|
| | | | | Mild | Armor | |
| .38 special | 2 | 4 | 4 | 1/4 | 3/16 | 5/16 |
| 9 mm | 2 1/2 | 4 | 4 | 5/16 | 1/4 | 7/16 |
| 7.62 and 5.56 mm | 4 | 6 | 6 | 9/16 | 7/16 | 1 1/8 |
| 7.62-mm AP | 6 1/2 | 8 | 8 | 13/16 | 11/16 | N/A |
| *Nominal thicknesses | | | | | | |

### Windows

3-49. Windows can include openings in walls and skylights, although skylights are only an issue where there are LOSs to them. When skylights require protection, treat them like windows. Achieving the low level of protection (hiding) for windows requires making it difficult to see through them, such as installing reflective film on the glass. An aggressor cannot see through the windows during daylight while it is lighter outside than inside, but he may see through them at night when the opposite might be true. Drapes or blinds that can be closed at night address that vulnerability. To achieve the high level of protection requires bullet-resistant window assemblies. These are commercially available for a wide range of ballistics types. They are purchased as manufactured-and-tested assemblies (including glazing and frames, both of which are equally bullet-resistant). The glazing materials and thicknesses and the framing details are proprietary to their manufacturers. The manufacturers make them according to industry test standards to ensure an effective product.

### Doors

3-50. Doors without glass easily meet the requirements for the low level of protection. Meeting the high level of protection requires the installation of bullet-resistant door assemblies. Doors can be installed in foyers so that there is no direct LOS into assets within the building (see Figure 3-7, page 3-10).

## FORCED ENTRY

3-51. In the forced-entry tactic, an aggressor tries to forcibly gain access to assets. He may use tools or explosives to breach building components or other barriers.

## GENERAL-DESIGN STRATEGY

3-52. The general-design strategy for forced entry is to detect the aggressor early in the forced-entry attempt and delay him long enough for a response force to intercept him. The combination of detection and defensive measures must provide sufficient time for a response force to intercept the aggressor before he reaches the asset or before he escapes with it, depending on the protective goals for the asset. The first goal would apply where the asset is likely to be destroyed or where access to it is not acceptable. The second goal would be applied when the idea is to prevent it from being stolen.

## LEVELS OF PROTECTION

3-53. Several levels of protection apply to forced entry. These levels vary in terms of system design, delay time, and response-force arrival time.

## SITE-WORK ELEMENTS

3-54. Site-work elements do not normally play a major role in protecting against a forced entry. However, the site should be laid out and maintained so that an aggressor does not have a hiding place nearby that will conceal his attempts to break into the building. Another site-work element is the application of perimeter barriers, most commonly fences. Fences are effective at delineating a boundary and at keeping honest people honest, but they are ineffective for preventing a forced entry. The design strategy for forced entry is based on delaying the aggressor, and any serious aggressor could climb a fence in less than 4 seconds or can cut through a fence in less than 10 seconds. Therefore, fences are not used as delay elements, but they are used to establish boundaries and as platforms on which to hang sensors. The final site-work consideration is securing utility-access ports such as manholes. If there are utility tunnels through which aggressors can enter a building, those accesses should be locked using padlocks or locking bolts.

## BUILDING ELEMENTS

3-55. Building elements are the principal construction elements of a system for protecting against a forced entry. The building elements are used to provide delay. The process for designing to resist forced entry involves laying out concentric "rings" of delay (called defensive layers). These defensive layers can include the facility's exterior, interior rooms within that layer, and containers within the interior rooms. The individual building components for each of the layers (walls, doors, windows, floors, ceilings, and roofs) provide the delay time (see TM 5-853-1).

## DETECTION ELEMENTS

3-56. For a protective system to be effective against a forced entry, the aggressors must be detected at a point of adequate delay. Detection at that point can be achieved by using an IDS. Once a sensor detects an aggressor, the alarm annunciator communicates that event to security personnel, who then dispatch a response force. The alarm can be assessed through a guard response or via CCTV. Chapter 6 and TM 5-853-4 provide detailed discussion of IDSs, CCTV systems, and other elements of ESSs.

## COVERT ENTRY AND INSIDER COMPROMISE

3-57. In the covert-entry tactic, an aggressor who is not authorized to be in the facility attempts to enter using false credentials. In the insider-compromise tactic, personnel with legitimate access to a facility try to compromise an asset. The insider may or may not have legitimate access to the asset itself. The purpose of the entry in either case can be to steal or otherwise compromise the asset or to destroy it. In the latter case, the aggressor may bring IEDs or IIDs.

### GENERAL-DESIGN STRATEGY

3-58. The general-design strategy for both the insider-compromise and covert-entry tactics is to keep people from entering areas they are not authorized to enter. For covert entry, aggressors are denied access to controlled areas. For insider compromise, aggressors are denied access to assets within controlled areas based on their need to have access to them. The general-design strategy also includes detecting aggressors removing assets from protected areas and detecting aggressors carrying tools, weapons, and explosives into protected areas.

### LEVELS OF PROTECTION

3-59. The levels of protection for these tactics address different issues, depending on whether the aggressor's goal is to steal or otherwise compromise an asset or to destroy it. When the goal is to steal or compromise an asset, the levels of protection vary with the number and sophistication of the access controls required to verify personnel access into a controlled area. When the goal is to destroy the assets, the levels of protection vary with the amount of damage the building (and the assets inside) are allowed to sustain and the sophistication of detecting weapons or explosives at entry points.

### BUILDING ELEMENTS

3-60. Building elements vary with an aggressor's goal. To protect against theft or compromise of assets, building elements are used to establish and maintain controlled areas into which only authorized personnel can enter. For insider compromise, there may be an additional requirement that access be further limited among personnel otherwise authorized access to the controlled area. That access is based on the need to have access to a specific asset. The result is that the controlled area may be compartmentalized, and each compartmentalized area may have separate access requirements. There are no special construction requirements for these tactics if the goal is theft or compromise. The only requirement is that the building elements of controlled areas should provide enough resistance to require aggressors to force their way through them to gain entry and to provide evidence of the forced entry if it is attempted. Forcing entry would be contrary to the aggressor's assumed goal to be covert. In addition, a common design goal would be to limit the number of entrances into controlled areas because there will need to be access control at each entry.

3-61. To protect against the destruction of assets, building elements are used to shield assets from the effects of explosives going off at access-control points.

The basic approach is to lay out areas at access points in which guards can search for carried-in weapons, explosives, or incendiary devices. The construction of that area is designed to limit damage to the rest of the building if an explosive is detonated in that area. Those levels of damage are similar to those discussed in relation to vehicle bombs. The walls and doors between the access point and the protected area will be hardened, and the walls and doors to the outside will be of lightweight construction so that they may fail and vent the blast pressure away from the building. At the higher level of protection, the access-control area is located in a separate facility and the target building is hardened to resist an explosion in that separate facility.

### DETECTION ELEMENTS

3-62. Detection elements for these tactics also vary based on the aggressor's goal. For theft, the detection elements are mainly related to access control. For destruction, the detection elements are used to detect weapons, explosives, or incendiary devices.

3-63. The main detection elements for theft or compromise are access-control devices. These can include procedural systems (such as guards checking ID), mechanical systems (such as keyed or combination locks), or electronic entry-control elements (such as electronic card readers, keypads, and biometric devices). Chapter 6 provides detailed discussion of electronic devices. The sophistication of these elements and the number used varies with the level of protection. For example, achieving the higher levels of protection requires the application of multiple forms of access-control elements such as a card reader and an electronic keypad for electronic-entry control or a badge check and badge exchange for a procedural system.

3-64. When destruction of the assets is the goal, detection is oriented toward detecting weapons, explosives, or incendiary devices. At the lower levels of protection, it is sufficient for guards to search for carried-in items. Achieving higher levels of protection requires the application of such equipment as metal detectors, X-ray machines, and explosive detectors.

## SURVEILLANCE AND EAVESDROPPING

3-65. Surveillance and eavesdropping tactics include visual surveillance, acoustic eavesdropping, and electronic-emanations eavesdropping. In these tactics, aggressors remain outside of controlled areas and try to gather information from within those areas. The tools used for these tactics include ocular devices for the visual-surveillance tactic and listening devices and electronic-emanations-eavesdropping equipment for the eavesdropping tactic.

### GENERAL-DESIGN STRATEGY

3-66. The general-design strategy for these tactics is to deny aggressors access to information assets. The kind of information (objects, operations, or files; secure conversations; or electronically processed data) and how it can be compromised differs for each tactic as do the specific protective strategies. Therefore, each tactic is addressed separately.

## LEVELS OF PROTECTION

3-67. Each of these tactics has only one level of protection. Either one protects or fails to protect against these tactics.

## SITE-WORK ELEMENTS

3-68. Site-work elements play a minor role in protecting assets from all surveillance or eavesdropping tactics. The main issue is to eliminate or control vantage points from which aggressors can surveil or eavesdrop on assets or operations. In addition, for the visual-surveillance tactic, a design goal can be to block LOSs from vantage points. Items used to block LOSs include trees, bushes, fences, and other buildings (see Figure 3-12).

## BUILDING ELEMENTS

3-69. Building elements are the principal components of the protective strategies for surveillance and eavesdropping tactics. For visual surveillance,

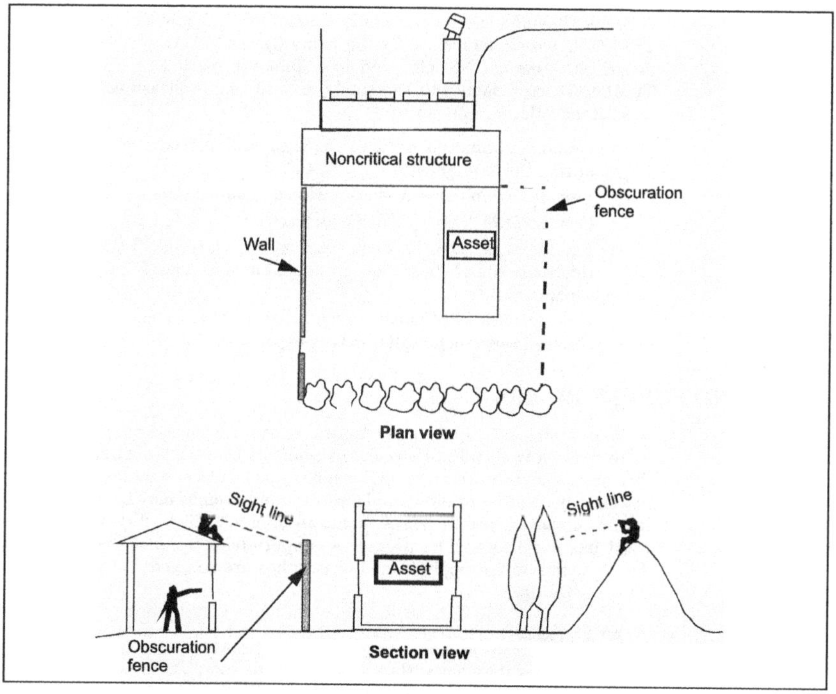


**Figure 3-12. LOS Blocked From Potential Vantage Points**

the building elements must block LOSs from outside the building. Walls and roofs perform this function effectively. Doors are only a problem when they have windows in them or are made of transparent materials. When this is the case, they can be treated like windows or they can be placed in foyers so that there are no direct LOSs through them. Windows can be treated with reflective film and drapes or blinds as described in the ballistics tactics. When there are LOSs through skylights, they should be treated like windows.

3-70. Building elements for acoustic eavesdropping relate to the construction of areas (preferably separated from the building exterior) that minimize the sound that can be transmitted through them. This requires specialized construction that has a sound-transmission-coefficient (STC) rating. Walls, floors, and ceilings can be constructed to achieve specific STC ratings using conventional construction materials as described in TM 5-853-1. Doors and windows that are STC rated are commonly manufactured and tested as assemblies. This type of design and construction can be expensive.

3-71. Protection against electronic-emanations eavesdropping involves the application of Terminal Electromagnetic-Pulse Emanation Standard (TEMPEST) guidance, most of which is classified. The protection is based on a TEMPEST assessment done for the Army by the US Army Intelligence and Security Command (INSCOM) and on guidance in AR 380-19. The results of a TEMPEST assessment will commonly lead to countermeasures from one or more of the following categories:

- Follow information security policies and procedures recommended during the assessments.
- Provide controlled space both inside and outside the facility.
- Provide TEMPEST-shielded equipment.
- Provide separation between electronic circuits that handle classified information and those that do not. This is commonly called red/black separation.
- Provide TEMPEST-shielded enclosures. This is specialized, metal-shielded construction that is very expensive.

## MAIL AND SUPPLY BOMBS

3-72. In mail- and supply-bomb tactics, aggressors place bombs in materials delivered to a facility. Explosives used in supply bombs are significantly larger (briefcase size) than those in mail bombs (pipe bombs or smaller). Mail bombs are usually directed at individuals, while supply bombs may be used to target larger numbers of people. These tactics assume that the facility containing the asset has a mail-handling area or a supplies-handling and -receiving area. These tactics do not apply if mail or supplies are handled and screened in a different facility.

### GENERAL-DESIGN STRATEGY

3-73. A bomb exploding within a building has more severe effects than the same size bomb exploding outside of the facility because the blast pressures cannot dissipate inside. Also, there is no standoff distance between the explosive and the facility to mitigate blast effects. The general-design strategy for mail and supply bombs is to detect delivered bombs before they explode and to harden the area where the explosion takes place. This minimizes the damage to the remainder of the facility. Occupants and contents within the mail room or supplies-handling area are likely to be killed or destroyed if an undetected bomb explodes.

## LEVELS OF PROTECTION

3-74. The levels of protection for mail and supply bombs are based on the amount of damage allowed to the building and, therefore, the occupants of the building. They also vary based on the sophistication of the detection measures used.

## BUILDING ELEMENTS

3-75. The purpose of building elements in relation to these bomb tactics is to shield assets from the effects of explosives going off at supply areas, receiving points, or mail rooms. The basic approach is to lay out either a mail room or a supplies-receiving area in which people can search suspicious packages for explosives or incendiary devices. Constructing this type of area will limit the damage to the rest of the building if an explosive is detonated there. Those levels of damage are similar to those discussed in relation to vehicle bombs.

### Mail Rooms

3-76. Mail rooms should be located on the facility's exterior, away from any critical assets. The walls and ceiling between the mail room and the remainder of the building are hardened to keep the blast effects out of the facility. The exterior walls and doors should be of lightweight construction so that they may fail and vent the blast pressure away from the building. There may be an explosives container in the mail room where suspicious packages can be placed. If the package explodes, the container will keep its effects from causing damage or injury. The hardened construction will protect assets outside of the mail room if the explosion occurs outside of the container. Check with EOD personnel to determine the local policy for using explosive containers. At higher levels of protection, the mail room is constructed to completely contain the effects of an explosion either through hardened construction or by using a specialized construction called vented suppressive shielding. Mail rooms should not have windows into protected areas. Doors between the mail room and the rest of the building should be avoided, placed in foyers, or replaced with blast-resistant doors, depending on the desired level of protection.

### Supplies-Handling Areas

3-77. Supplies-handling areas should also be on the building's exterior, away from critical areas of the facility. Walls and doors between the handling area and the protected area should be hardened, and the exterior walls and doors should be of lightweight construction so that they may fail and vent the blast pressure away from the building. There should be no windows between the handling area and the protected area. At the higher level of protection, the handling area is located in a separate facility and the target building is hardened to resist an explosion in that separate facility.

### DETECTION ELEMENTS

3-78. Detection for these assets varies with the level of protection. At the lower levels of protection, bombs are detected by inspection. As the level of protection goes up, the sophistication of the detection increases. At the higher levels of protection, equipment such as X-ray examining devices, metal detectors, and explosives detectors can be used. Explosive-detection dogs are an alternative to explosive detectors.

## CHEMICAL AND BIOLOGICAL CONTAMINATION

3-79. When using chemical- and biological-contamination tactics, aggressors introduce contaminants into the air or water supply to a facility or a group of facilities. Both airborne and waterborne contaminants include chemical, biological, and radiological agents. Aggressors may also forcibly enter a facility to contaminate water or air using the forced-entry tactic.

### GENERAL-DESIGN STRATEGY

3-80. Both chemical and biological agents are difficult to detect in water and air supplies. Radiological agents are relatively easy to detect in water, but they are not commonly included in water-quality examinations. It is unlikely that all agents will be detected, so the general-design strategy for these tactics is to filter out suspected airborne contaminants or to shut off suspected waterborne contaminants. Also, because contaminants can easily be entered into the environment from inside a facility, the strategy includes limiting access to the facility (especially mechanical rooms, water intakes, and so forth).

### LEVELS OF PROTECTION

3-81. The levels of protection for each of these tactics differ only in the frequency with which some protective measures are exercised. For the low level of protection, they are exercised only in response to a known threat. In the high level of protection, they are exercised continuously.

### SITE-WORK ELEMENTS

3-82. Site-work elements are only significant for waterborne contamination. They include protecting water-treatment plants and water-storage structures. This protection may include constructing perimeter barriers (such as chain-link fences) and controlling access to the plant site. These measures are used because most contaminants require quantities on the order of truckloads to contaminate a water supply, so the focus of security is to keep such large vehicles under control. The perimeter barriers do not need to stop the vehicles because the assumption is that the aggressor wants to be covert. An overt act would alert people to avoid the water supply.

### BUILDING ELEMENTS

3-83. Building elements for both tactics include controlling access so that aggressors cannot sneak in and plant devices in the building. Protection against airborne contamination at a facility involves making elements of the air-handling system (including air intakes) inaccessible and laying out toxin-free areas for people to be protected. A toxin-free area is an area in which the internal air pressure is higher than the external air pressure. Therefore, if a chemical, biological, or radiological device is set off outside, its contaminant will not be able to penetrate the protected area. Achieving that "net positive pressure" requires a significant air-handling system with air filters to filter contaminants out of the air. It also requires an air-lock entrance into the area so contaminants cannot enter through the door. At the low level of protection, the filters and the air-handling system are only used in response to a credible threat. At the high level of protection, that risk is not acceptable and the filters are run continuously.

3-84. The building-element issues for waterborne contamination are limited to providing protection against forced and covert entries into water-treatment plants and water-storage areas. These methods have been previously described. The only additional issue is the provision for alternative water sources. If it is suspected or detected that the water is contaminated, a backup water source should be in place (such as bottled water). For the high level of protection, bottled water should always be used for drinking.

# Chapter 4

# Protective Barriers

Protective barriers are used to define the physical limits of an installation, activity, or area. Barriers restrict, channel, or impede access and are fully integrated to form a continuous obstacle around the installation. They are designed to deter the worst-case threat. The barriers should be focused on providing assets with an acceptable level of protection against a threat.

## OVERVIEW

4-1. Protective barriers form the perimeter of controlled, limited, and exclusion areas. Utility areas (such as water sources, transformer banks, commercial power and fuel connections, heating and power plants, or air-conditioning units) may require these barriers for safety standards. Protective barriers consist of two major categories—natural and structural.

- Natural protective barriers are mountains and deserts, cliffs and ditches, water obstacles, or other terrain features that are difficult to traverse.
- Structural protective barriers are man-made devices (such as fences, walls, floors, roofs, grills, bars, roadblocks, signs, or other construction) used to restrict, channel, or impede access.

4-2. Barriers offer important benefits to a physical-security posture. They create a psychological deterrent for anyone thinking of unauthorized entry. They may delay or even prevent passage through them. This is especially true of barriers against forced entry and vehicles. Barriers have a direct impact on the number of security posts needed and on the frequency of use for each post.

4-3. Barriers cannot be designed for all situations. Considerations for protective structural barriers include the following:

- Weighing the cost of completely enclosing large tracts of land with significant structural barriers against the threat and the cost of alternate security precautions (such as patrols, MWD teams, ground sensors, electronic surveillance, and airborne sensors).
- Sizing a restricted area based on the degree of compartmentalization required and the area's complexity. As a rule, size should be kept to a minimum consistent with operational efficiency. A restricted area's size may be driven by the likelihood of an aggressor's use of certain tactics. For example, protecting assets from a vehicle bomb often calls for a substantial explosives standoff distance. In these cases, mitigating the vehicle bomb would often be more important than minimizing the restricted area to the extent necessary for operational efficiency. Protective barriers should be established for—
  - Controlling vehicular and pedestrian traffic flow.

- Providing entry-control points where ID can be checked.
- Defining a buffer zone for more highly classified areas.
- Precluding visual compromise by unauthorized individuals.
- Delaying forced entry.
- Protecting individual assets.

4-4. If a secured area requires a limited or exclusion area on a temporary or infrequent basis, it may not be possible to use physical structural barriers. A temporary limited or exclusion area may be established where the lack of proper physical barriers is compensated for by additional security posts, patrols, and other security measures during the period of restriction. Temporary barriers (including temporary fences, coiled concertina wire, and vehicles) may be used. Barriers are not the only restrictive element, and they may not always be necessary. They may not be ideal when working with limited or exclusion areas or when integrated with other controls.

4-5. Because barriers can be compromised through breaching (cutting a hole through a fence) or by nature (berms eroded by the wind and rain), they should be inspected and maintained at least weekly. Guard-force personnel should look for deliberate breaches, holes in and under barriers, sand dunes building up against barriers, and the proper functioning of locks.

# FENCING

4-6. Three types of fencing are authorized for use in protecting restricted areas—chain link, barbed wire, and barbed tape or concertina. The type used for construction depends primarily on the threat and the degree of permanence. It may also depend on the availability of materials and the time available for construction. Fencing may be erected for other uses besides impeding personnel access. It can impede observation, can serve as a means to defeat standoff-weapon systems (such as rocket-propelled grenades [RPGs]), and can serve as a barrier to hand-thrown weapons (such as grenades and firebombs).

4-7. Generally, chain-link fencing will be used for protecting permanent limited and exclusion areas. All three types of fencing may be used to augment or increase the security of existing fences that protect restricted areas. Examples would be to create an additional barrier line, to increase existing fence height, or to provide other methods that effectively add to physical security. It is important to recognize that fencing provides very little delay when it comes to motivated aggressors, but it can act as a psychological deterrent.

## CHAIN LINK

4-8. Chain-link fence (including gates) must be constructed of 6-foot material, excluding the top guard. Fence heights for conventional arms and ammunition security must be 6 feet for standard chain-link, wire-mesh fencing. Chain-link fences must be constructed with 9-gauge or heavier wire. They must be galvanized with mesh openings not larger than 2 inches per side and have twisted and barbed selvages at the top and the bottom. The wire must be taut and securely fastened to rigid metal or reinforced-concrete posts set in concrete. It must reach within 2 inches of hard ground or pavement. On soft ground, it must reach below the surface deep enough to compensate for shifting soil or sand. Materials and construction must meet with the US Army Corps of Engineers (USACE) guide specifications shown in the USACE Standard (STD) 872-90 series. Weaknesses in the chain-link fence occur as a result of weather (rusting) or failure to keep it fastened to the post that affects the desired tightness. Damage to the fence and fence fabric may be the result of allowing vegetation and trees to grow on or near the fence. The interaction between the fence and the overgrowth often leads to fence damage and reduces the integrity and continuity of the fence as a perimeter boundary and barrier. The perimeter fence is the most obvious protective measure. A well-maintained fence indicates that the asset owner is dedicated to physical security.

## BARBED WIRE

4-9. Standard barbed wire is twisted, double-strand, 13.5-gauge wire, with four-point barbs spaced an equal distance apart. Barbed-wire fencing (including gates) intended to prevent human trespassing should not be less than 6 feet high and must be affixed firmly to posts not more than 6 feet apart. The distance between strands should not exceed 6 inches, and at least one wire should be interlaced vertically and midway between posts. The ends must be staggered or fastened together, and the base wire must be picketed to the ground.

## BARBED TAPE OR CONCERTINA

4-10. A barbed-taped obstacle (BTO) is fabricated from 0.025-inch stainless steel and is available in 24-, 30-, 40-, and 60-inch-diameter coils. The barbs shall have a minimum length of 1.2 inches, and the barb cluster's width shall be 1.21 inches. A BTO deploys tangle-free for fast installation. It may be recovered and used again. Fifty feet (plus or minus 2 inches) can be covered by 101 coil loops. Handling barbed tape requires the use of heavy barbed-tape gauntlets instead of standard barbed-wire gauntlets.

### Barbed-Tape Concertina

4-11. Barbed-tape concertina (standard concertina barbed tape) is a commercially manufactured wire coil of high-strength-steel barbed wire that is clipped together at intervals to form a cylinder. When opened, it is 50 feet long and 3 feet in diameter. When used as the perimeter barrier for a restricted area, the concertina must be laid between poles with one roll on top of another or in a pyramid arrangement (with a minimum of three rolls).

4-12. Reinforced barbed-tape concertina consists of a single strand of spring-steel wire and a single strand of barbed tape. The sections between barbs of the barbed tape are securely clinched around the wire. Each coil is about 37 1/2 inches in diameter and consists of 55 spiral turns connected by steel clips to form a cylindrical diamond pattern when extended to a coil length of 50 feet. One end turn is fitted with four bundling wires for securing the coil when closed and each end turn is fitted with two steel carrying loops. The concertina extends to 50 feet without permanent distortion. When released, it can be retracted into a closed coil.

4-13. When possible, a top guard should be constructed on all perimeter fences and may be added on interior enclosures for additional protection. A top guard is an overhang of barbed wire or tape along the top of a fence, facing outward and upward at about a 45-degree angle. Placing barbed wire or tape above it can further enhance the top guard. Top-guard supporting arms will be permanently affixed to the top of fence posts to increase the overall height of the fence by at least 1 foot. (Due to liability issues in some locations, the top guards will not be allowed to face outward where the fence is adjacent to public areas.) Three strands of barbed wire spaced 6 inches apart must be installed on the supporting arms. The number of strands of wire or tape may be increased when required. The top guard of fencing adjoining gates may range from a vertical height of 18 inches to the normal 45-degree outward protection but only for sufficient distance along the fence to open the gates adequately. Bottom and top tension wires should be used in lieu of fence rails. A concrete sill may be cast at the bottom of the fence to protect against soil erosion. A bottom rail is used on high-security fences to prevent intruders from lifting the fence.

### Gates and Entrances

4-14. The number of gates and perimeter entrances must be the minimum required for safe and efficient operation of the facility. Active perimeter entrances must be designed so that the guard force maintains full control. Semiactive entrances, such as infrequently used vehicular gates, must be locked on the inside when not in use. When closed, gates and entrances must provide a barrier structurally comparable to their associated barriers. Care must be afforded against the ability to crawl under gates. Top guards, which may be vertical, are required for all gates.

### Triple-Standard Concertina (TSC) Wire

4-15. This type of fence uses three rolls of stacked concertina. One roll will be stacked on top of two rolls that run parallel to each other while resting on the ground, forming a pyramid. In many situations, this fence has been used effectively in place of a chain-link fence. (If perimeter fencing consists of TSC, a top guard is not feasible.)

### Tangle-Foot Wire

4-16. Barbed wire or tape may be used in appropriate situations to construct a tangle-foot obstruction either outside a single perimeter fence or in the area between double fences to provide an additional deterrent to intruders. The wire or tape should be supported on short metal or wooden pickets spaced at irregular intervals of 3 to 10 feet and at heights between 6 and 12 inches. The wire or tape should be crisscrossed to provide a more effective obstacle. The space and materials available govern the depth of the field.

## AIRCRAFT CABLE

4-17. Although not used very often, aircraft cable can be used as a temporary barrier. Refer to FM 5-34 for information required for determining the barrier's strength. The barrier is created using wire rope. Clips are spaced six times the diameter of the wire rope. Aircraft cable (deployed as described above or attached to a chain-link fence) can also be made to act as a barrier to moving vehicles. To do so, the cable must be anchored into the ground at both ends at about 200-foot intervals (see TM 5-853-1).

## UTILITY OPENINGS

4-18. Sewers, air and water intakes and exhausts, and other utility openings of 10 inches or more in diameter that pass through perimeter barriers must have security measures equivalent to that of the perimeter (see TM 5-820-4). Specific requirements of various openings are discussed below:

- Manhole covers 10 inches or more in diameter must be secured to prevent unauthorized opening. They may be secured with locks and hasps, by welding them shut, or by bolting them to their frame. Ensure that hasps, locks, and bolts are made of materials that resist corrosion. Keyed bolts (which make removal by unauthorized personnel more difficult) are also available.
- Drainage ditches, culverts, vents, ducts, and other openings that pass through a perimeter and that have a cross-sectional area greater than 96 square inches and whose smallest dimension is greater than 6 inches will be protected by securely fastened welded bar grilles (refer to TM 5-853-3, Figure 8-1). As an alternative, drainage structures may be constructed of multiple pipes, with each pipe having a diameter of 10 inches or less. Multiple pipes of this diameter may also be placed and secured in the inflow end of a drainage culvert to prevent intrusion into the area. Ensure that any addition of grilles or pipes to culverts or other drainage structures is coordinated with the engineers so that they can compensate for the diminished flow capacity and additional maintenance that will result from the installation.

## OTHER PERIMETER BARRIERS

4-19. Buildings less than two stories high that form part of a perimeter must have a top guard along the outside edge to deny access to the roof. When using masonry walls as part of a perimeter barrier, they must be at least 7 feet high and have a barbed-wire top guard. The top guard should be sloped outward at a 45-degree angle and carry at least three strands of barbed wire. This will increase the vertical height of the barrier by at least 1 foot.

4-20. Protect windows, active doors, and other designated openings by securely fastening bars, grilles, or chain-link screens. Fasten window barriers from the inside. If hinged, the hinges and locks must be on the inside. Building elements that provide delay against forced entry have stringent requirements. These elements should be designed according to TM 5-853-1.

## SECURITY TOWERS

4-21. It is not acceptable to observe a perimeter from towers only. However, all towers should be located to provide maximum observation and should be constructed for protection from small-arms fire.

4-22. Mobile towers are useful in some temporary situations such as a large, open storage area where receiving and storing activities take place. All facilities using towers must have a support force available for emergencies. Tower personnel should be rotated at frequent intervals.

4-23. The height of a tower increases the range of observation during daylight hours and at night with artificial illumination. However, during inclement weather and during a blackout, towers lose this advantage and must be supplemented by on-ground observation.

4-24. The following considerations should be made when planning for the use of towers:

- Hardening the tower against small-arms effects by using sandbags, salvaged armor, or commercially fabricated bullet-resistant construction. This may require strengthening the tower supports, which should be performed only under the supervision of an engineer. The level of protection required must equate to the threat level identified during the IPB or the military decision-making process (MDMP). The best approach is to design for the worst identified threat rather than to try and modify the tower at a later date on short notice.
- Installing communications and alarm systems, both audible and visual (primary and alternate).
- Using appropriate surveillance, target-acquisition, and night-observation (STANO) equipment with the tower and perimeter barriers being surveilled. Infrared (IR) items may be especially valuable. Considerations for the selection and use of STANO equipment must be made while evaluating the effects of perimeter protective lighting.
- Providing security lighting for route protection to the tower. Security lighting also allows for support of the guard force entering or exiting the perimeter.
- Ensuring that the tower's height is determined according to the area of observation.
- Ensuring that towers have overlapping, mutually supporting fields of observation and fire.
- Providing towers with a backup fortified defensive fighting position, as appropriate.

## INSTALLATION ENTRANCES

4-25. The number of installation or activity gates and perimeter entrances in active use should be limited to the minimum number required for safe and efficient operations. When necessary, install vehicle barriers in front of vehicle gates. Security lighting should be considered at entry points (see Chapter 5). Refer to TM 5-853-1 for the application and selection of these barriers.

4-26. Plans to use guards for controlling entry to an installation or activity must be predetermined based on the threat conditions (THREATCON). The construction of the guard post must be included in the security plan.

## PERIMETER ENTRANCES

4-27. Active perimeter entrances should be designated so that security forces maintain full control without an unnecessary delay in traffic. This is accomplished by having sufficient entrances to accommodate the peak flow of pedestrian and vehicular traffic and having adequate lighting for rapid and efficient inspection. When gates are not operational during nonduty hours, they should be securely locked, illuminated during hours of darkness, and inspected periodically by a roving patrol. Additionally, warning signs should be used to warn drivers when gates are closed. Doors and windows on buildings that form a part of the perimeter should be locked, lighted, and inspected.

## ENTRY-CONTROL STATIONS

4-28. Entry-control stations should be provided at main perimeter entrances where security personnel are present. Considerations for construction and use should be based on the information outlined in USACE STD 872-50-01.

4-29. Entry-control stations should be located as close as practical to the perimeter entrance to permit personnel inside the station to maintain constant surveillance over the entrance and its approaches. Additional considerations at entry-control stations include—

- Establishing a holding area for unauthorized vehicles or those to be inspected further. A turnaround area should be provided to keep from impeding other traffic.
- Establishing control measures such as displaying a decal on the window or having a specially marked vehicle.

4-30. Entry-control stations that are manned 24 hours each day should have interior and exterior lighting, interior heating (where appropriate), and a sufficient glassed area to afford adequate observation for personnel inside. Where appropriate, entry-control stations should be designed for optimum personnel ID and movement control. Each station should also include a telephone, a radio, and badge racks (if required).

4-31. Signs should be erected to assist in controlling authorized entry, to deter unauthorized entry, and to preclude accidental entry. Signs should be plainly displayed and be legible from any approach to the perimeter from a reasonable distance. The size and coloring of a sign, its letters, and the interval of posting must be appropriate to each situation.

4-32. Entry-control stations should be hardened against attacks according to the type of threat. The methods of hardening may include—

- Reinforced concrete or masonry.
- Steel plating.
- Bullet-resistant glass.
- Sandbags, two layers in depth.
- Commercially fabricated, bullet-resistant building components or assemblies.

## WARNING SIGNS

4-33. A significant amount of warning signs should be erected to ensure that possible intruders are aware of entry into restricted areas. Warning signs augment control signs. They warn intruders that the area is restricted and that trespassing may result in the use of deadly force.

4-34. Warning signs should be installed along the limited area's physical barriers and at each entry point where they can be seen readily and understood by anyone approaching the perimeter. In areas where English is one of two or more languages commonly spoken, warning signs must contain the local language in addition to English. The wording on the signs will denote warning of a restricted area. The signs should be posted at intervals of no more than 100 feet. They must not be mounted on fences equipped with intrusion-detection equipment. Additionally, the warning signs prescribed in AR 190-13 should be posted at all entrances to limited, controlled, and exclusion areas. See Chapter 7 for more details.

## OTHER SIGNS

4-35. Signs setting forth the conditions of entry to an installation or area should be plainly posted at all principal entrances. The signs should be legible under normal conditions at a distance not less than 50 feet from the point of entry. Such signs should inform the entrant of the provisions (search of the person, the vehicle, packages, and so forth) or prohibitions (such as against cameras, matches, and lighters and entry for reasons other than official business) that may be prescribed by the installation commander.

4-36. Signs or notices legibly setting forth the designation of restricted areas and provisions of entry should be plainly posted at all entrances and at other points along the perimeter line as necessary. The wording of these signs or notices is prescribed in AR 190-13.

## INSTALLATION PERIMETER ROADS AND CLEAR ZONES

4-37. When the perimeter barrier encloses a large area, an interior all-weather perimeter road should be provided for security-patrol vehicles. Clear zones should be maintained on both sides of the perimeter barrier to provide an unobstructed view of the barrier and the ground adjacent to it. Roads within the clear zone should be as close to the perimeter barrier as possible without interfering with it. The roads should be constructed to allow effective road barriers to deter motor movement of unauthorized personnel during mobilization periods.

4-38. Clear zones should be kept clear of weeds, rubbish, or other material capable of offering concealment or assistance to an intruder attempting to breach the barrier. A clear zone of 20 feet or more should exist between the perimeter barrier and exterior structures, parking areas, and natural or man-made features. When possible, a clear zone of 50 feet or more should exist between the perimeter barrier and structures within the protected area, except when a building's wall constitutes part of the perimeter barrier. Ammunition supply points (ASPs) will have clear zones 12 feet outside of the ASP and 30 feet inside, and the vegetation will not exceed 8 inches (4 inches for high-threat and highly controlled areas). Refer to AR 190-11 and DOD 0-2000.12-H, Appendix EE, for further information.

4-39. When it is impossible to have adequate clear zones because of property lines or natural or man-made features, it may be necessary to increase the height of the perimeter barrier, increase security-patrol coverage, add more security lighting, or install an intrusion-detection device along that portion of the perimeter.

4-40. When considering the construction of a new site or perimeter, ensure that the plans include a fence located well inside the property line, thus permitting control of enough space outside the fence to maintain at least a minimal clear zone. The following considerations apply:

- On a large installation (such as a proving ground), it is unreasonable to construct an expensive perimeter fence and keep it under constant observation. Such an installation is usually established in a sparsely inhabited area. Its comparative isolation and the depth of the installation give reasonable perimeter protection. Under these circumstances, it is usually sufficient to post warning signs or notices, reduce access roads to a minimum, and periodically patrol the area between the outer perimeter and the conventionally protected vital area of the installation.
- An alternative to erecting new or replacing old chain-link fence involving an entire installation perimeter is to relocate or isolate the sensitive area or item by—
  - Relocating the item within a safe perimeter.
  - Consolidating the item with other items.
  - Erecting a chain-link fence (regulations permitting) around individual assets rather than the installation's perimeter.

## ARMS-FACILITY STRUCTURAL STANDARDS

4-41. It is next to impossible to build a protective barrier that cannot be penetrated by a human or heavy armor. Therefore, as opposed to protecting a facility using only one barrier, enhance security by using a combination of barriers to increase delay. Multiple barriers also cause aggressors to expend more energy trying to breach all of the barriers. They also provide the appearance of additional security and may further deter some aggressors.

4-42. The interest of security must be kept in mind when constructing walls, ceilings, floors, and roofs. Facilities that house arms and ammunition are constructed as security barriers in the interest of deterring and delaying penetration. Construction guidelines for arms facilities are outlined in AR 190-11. AR 190-11 requires coordination with the engineer office, the safety office, the provost marshal office (PMO), or the security-force office when definitive drawings and specifications for new construction or upgrades or modifications of AA&E storage structures are proposed. This coordinated effort ensures that safety and physical-security requirements are met. AR 190-11 also addresses waivers and exceptions for AA&E storage structures, as well as the requirements for a tactical (training or operational) or shipboard environment. Waivers and exceptions are not discussed in this manual. The following guidelines are provided for securing AA&E in tactical and shipboard environments:

- The criteria and standards for protecting AA&E will be developed by the major Army command (MACOM) according to AR 190-11.
- The deploying commander will establish and enforce procedures for securing deployed AA&E based on the assessment of the threat, the objectives, the location, and the duration of the deployment.
- The AA&E in the tactical environment will be secured at all times.
- The AA&E will be under continuous positive control.
- Persons charged with the custody of AA&E will have the capability to sound the alarm if a forceful theft is attempted.
- A response force will be available to protect the AA&E.
- A system of supervisory checks will be established to ensure that all personnel comply with security measures. Supervisory checks of the AA&E holding area will be made to ensure that the AA&E being guarded have not been tampered with.
- All officers, noncommissioned officers (NCOs), or civilian equivalents will closely monitor the control of ammunition and explosives during field training or range firing.
- Selection of personnel to perform guard duties at AA&E holding areas will be closely monitored by commanders to ensure that only responsible individuals are assigned duties.

# Chapter 5

# Physical-Security Lighting

Security lighting allows security personnel to maintain visual-assessment capability during darkness. When security-lighting provisions are impractical, additional security posts, patrols, MWD patrols, NVDs, or other security means are necessary.

## OVERVIEW

5-1. Security lighting should not be used as a psychological deterrent only. It should also be used along perimeter fences when the situation dictates that the fence be under continuous or periodic observation.

5-2. Lighting is relatively inexpensive to maintain and, when properly used, may reduce the need for security forces. It may also enhance personal protection for forces by reducing the advantages of concealment and surprise for a determined intruder.

5-3. Security lighting is desirable for those sensitive areas or structures within the perimeter that are under observation. Such areas or structures include pier and dock areas, vital buildings, storage areas, motor pools, and vulnerable control points in communication and power- and water-distribution systems. In interior areas where night operations are conducted, adequate lighting facilitates the detection of unauthorized persons approaching or attempting malicious acts within the area. Security lighting has considerable value as a deterrent to thieves and vandals and may make the job of the saboteur more difficult. It is an essential element of an integrated physical-security program.

5-4. A secure auxiliary power source and power-distribution system for the facility should be installed to provide redundancy to critical security lighting and other security equipment. During deployed operations, primary power may not exist or may be subject to constraints or interruptions due to poor infrastructure or hostile activity. Auxiliary power sources must be available for critical electrical loads and must be secured against direct and indirect fires as well as sabotage. If automatic-transfer switches are not installed, security procedures must designate the responsibility for the manual start of the source.

## COMMANDER'S RESPONSIBILITY

5-5. Commanders determine perimeter lighting needs based on the threat, site conditions along the perimeter, surveillance capabilities, and available guard forces. Commanders ensure that security lighting is designed and used to discourage unauthorized entry and to facilitate the detection of intruders approaching or attempting to gain entry into protected areas.

## PLANNING CONSIDERATIONS

5-6. Security lighting usually requires less intensity than working lights, except for ID and inspection at entry-control points. Each area of a facility presents its own unique set of considerations based on physical layout, terrain, atmospheric and climatic conditions, and security requirements. Information is available from the manufacturers of lighting equipment and from the installation's director of public works, who will assist in designing a lighting system. This information includes—

- Descriptions, characteristics, and specifications of various lighting fixtures, arc, and gaseous-discharge lamps.
- Lighting patterns of various fixtures.
- Typical layouts showing the most efficient height and spacing of equipment.
- Minimum levels of illumination and lighting uniformity required for various applications.

5-7. In planning a security-lighting system, the physical-security manager considers the—

- Cost of replacing lamps and cleaning fixtures, as well as the cost of providing the required equipment (such as ladders and mechanical buckets) to perform this maintenance.
- Provision of manual-override capability during a blackout, including photoelectric controls. These controls may be desirable in a peacetime situation but undesirable when a blackout is a possibility.
- Effects of local weather conditions on lighting systems.
- Fluctuating or erratic voltages in the primary power source.
- Grounding requirements.
- Provisions for rapid lamp replacement.
- Use of lighting to support a CCTV system.
- Limited and exclusion areas. Specific lighting requirements are referenced in AR 190-59 and TM 5-853-2. TM 5-853-4 provides guidance for facility applications that include CCTV cameras.
  - Lighting in these areas must be under the control of the guard force.
  - For critical areas (such as weapons storage areas), instantaneous lighting with a backup source is required. Any period without lighting in a critical area is unacceptable. Therefore, these areas generally have a requirement for backup power (such as diesel-engine generators, uninterrupted power supplies, and batteries) in case of power loss.
  - Security-lighting systems are operated continuously during hours of darkness.
  - Protective lights should be used so that the failure of one or more lights will not affect the operation of the remaining lights.
- Lighting requirements for adjoining properties and activities.
- Restrike time (the time required before the light will function properly after a brief power interruption).

- Color accuracy.
- Other facilities requiring lighting, such as parking areas.

## PRINCIPLES OF SECURITY LIGHTING

5-8. Security lighting enables guard-force personnel to observe activities around or inside an installation while minimizing their presence. An adequate level of illumination for all approaches to an installation will not discourage unauthorized entry; however, adequate lighting improves the ability of security personnel to assess visually and intervene on attempts at unauthorized entry. Lighting is used with other security measures (such as fixed security posts or patrols, fences, and ESSs) and should never be used alone. Other principles of security lighting include the following:

- Optimum security lighting is achieved by adequate, even light on bordering areas; glaring lights in the eyes of an intruder; and little light on security-patrol routes. In addition to seeing long distances, security forces must be able to see low contrasts (such as indistinct outlines of silhouettes) and must be able to detect an intruder who may be exposed to view for only a few seconds. Higher levels of illumination improve these abilities.
- High brightness contrast between an intruder and the background should be the first consideration when planning for security lighting. With predominantly dark, dirty surfaces or camouflage-type painted surfaces, more light is needed to produce the same brightness around installations and buildings than when clean concrete, light brick, and grass predominate. When the same amount of light falls on an object and its background, the observer must depend on contrasts in the amount of light reflected. His ability to distinguish poor contrasts is significantly improved by increasing the illumination level.
- The observer primarily sees an outline or a silhouette when the intruder is darker than his background. Using light finishes on the lower parts of buildings and structures may expose an intruder who depends on dark clothing and darkened face and hands. Stripes on walls have also been used effectively, as they provide recognizable breaks in outlines or silhouettes. Providing broad-lighted areas around and within the installation against which intruders can be seen can also create good observation conditions.

5-9. To be effective, two basic systems or a combination of both may be used to provide practical and effective security lighting. The first method is to light the boundaries and approaches; the second is to light the area and structures within the property's general boundaries. Protective lighting should—

- Discourage or deter attempts at entry by intruders. Proper illumination may lead a potential intruder to believe detection is inevitable.
- Make detection likely if entry is attempted.
- Prevent glare that may temporarily blind the guards.

## TYPES OF LIGHTING

5-10. The type of lighting system used depends on the installation's overall security requirements. Four types of lighting units are used for security-lighting systems—continuous, standby, movable (portable), and emergency.

5-11. Continuous lighting is the most common security-lighting system. It consists of a series of fixed lights arranged to flood a given area continuously during darkness with overlapping cones of light. Two primary methods of using continuous lighting are glare projection and controlled lighting.

- The glare security-lighting method is used when the glare of lights directed across the surrounding territory will not be annoying nor interfere with adjacent operations. It is a strong deterrent to a potential intruder because it makes it difficult to see inside of the area. Guards are protected by being kept in comparative darkness and being able to observe intruders at a considerable distance beyond the perimeter.
- Controlled lighting is best when it limits the width of the lighted strip outside the perimeter, such as along highways. In controlled lighting, the width of the lighted strip is controlled and adjusted to fit the particular need. This method of lighting may illuminate or silhouette security personnel.

5-12. Standby lighting has a layout similar to continuous lighting. However, the luminaries are not continuously lit but are either automatically or manually turned on when suspicious activity is detected or suspected by the security force or alarm systems.

5-13. Movable lighting consists of manually operated, movable searchlights that may be lit during hours of darkness or only as needed. The system normally is used to supplement continuous or standby lighting.

5-14. Emergency lighting is a system of lighting that may duplicate any or all of the above systems. Its use is limited to times of power failure or other emergencies that render the normal system inoperative. It depends on an alternative power source such as installed or portable generators or batteries.

### FENCED PERIMETERS

5-15. Fenced perimeters require the lighting specifications indicated in TM 5-853-2. Specific lighting requirements are based on whether the perimeter is isolated, semi-isolated, or nonisolated.

- Isolated fenced perimeters are fence lines around areas where the fence is 100 feet or more from buildings or operating areas. The approach area is clear of obstruction for 100 or more feet outside of the fence. Other personnel do not use the area. Use glare projection for these perimeters and keep patrol routes unlit.
- Semi-isolated fenced perimeters are fence lines where approach areas are clear of obstruction for 60 to 100 feet outside of the fence. The general public or installation personnel seldom have reason to be in the area. Use controlled lighting for these perimeters and keep patrol routes in relative darkness.

- Nonisolated fenced perimeters are fence lines immediately adjacent to operating areas. These areas may be within an installation or public thoroughfares. Outsiders or installation personnel may move about freely in this approach area. The width of the lighted strip depends on the clear zones inside and outside the fence. Use controlled lighting for these perimeters. It may not be practical to keep the patrol area dark.

### ENTRANCES

5-16. Entrances for pedestrians will have two or more lighting units providing adequate illumination for recognition of persons and examination of credentials. Vehicle entrances will have two lighting units located to facilitate the complete inspection of passenger cars, trucks, and freight cars as well as their contents and passengers. Semiactive and inactive entrances will have the same degree of continuous lighting as the remainder of the perimeter, with standby lighting to be used when the entrance becomes active. Gatehouses at entrances should have a low level of interior illumination, enabling guards to see approaching pedestrians and vehicles.

### OTHER

5-17. Areas and structures within the installation's property line consist of yards; storage spaces; large, open working areas; piers; docks; and other sensitive areas and structures.

- Open yards (unoccupied land only) and outdoor storage spaces (material storage areas, railroad sidings, motor pools, and parking areas) should be illuminated. An open yard adjacent to a perimeter (between guards and fences) will be illuminated according to the perimeter's illumination requirements. Where lighting is necessary in other open yards, illumination will not be less than 0.2 foot-candle at any point.
- Lighting units are placed in outdoor storage spaces to provide an adequate distribution of light in aisles, passageways, and recesses to eliminate shadowed areas where unauthorized persons may hide.
- Illuminating both water approaches and the pier area safeguards piers and docks located on an installation. Decks on open piers will be illuminated to at least 1 foot-candle and the water approaches (extending to a distance of 100 feet from the pier) to at least 0.5 foot-candle. The area beneath the pier floor will be lit with small wattage floodlights arranged on the piling. Movable lighting is recommended as a part of the protective lighting system for piers and docks. The lighting must not in any way violate marine rules and regulations (it must not be glaring to pilots). Consult the US Coast Guard (USCG) for approval of protective lighting adjacent to navigable waters.

## WIRING SYSTEMS

5-18. The wiring circuit should be arranged so that failure of any one lamp will not leave a large portion of the perimeter line or a major segment of a critical or vulnerable position in darkness. Feeder lines will be placed underground (or sufficiently inside the perimeter in the case of overhead wiring) to minimize the possibility of sabotage or vandalism from outside the perimeter. Another advantage to underground wiring is reduced effects from adverse weather conditions.

# MAINTENANCE

5-19. Periodic inspections will be made of all electrical circuits to replace or repair worn parts, tighten connections, and check insulation. Keep fixtures clean and properly aimed.

## POWER SOURCES

5-20. Primary and alternate power sources must be identified. The following is a partial list of considerations:

- The primary source is usually a local public utility.
- An alternate source (standby batteries or diesel-fuel-driven generators may be used) is provided where required and should—
  - Start automatically upon failure of primary power.
  - Be adequate to power the entire lighting system.
  - Be equipped with adequate fuel storage and supply.
  - Be tested under load to ensure efficiency and effectiveness.
  - Be located within a controlled area for additional security.

## CCTV-CAMERA LIGHTING REQUIREMENTS

5-21. TM 5-853-4 provides a detailed discussion of CCTV-camera lighting requirements and guidelines for minimum lighting levels and lighting uniformity. The following considerations apply when lighting systems are intended to support CCTV assessment or surveillance:

- The camera's field of view.
- Lighting intensity levels.
- Maximum light-to-dark ratio.
- Scene reflectance.
- Daylight-to-darkness transitions.
- Camera mounting systems relative to lighting.
- The camera's spectral response.
- The cold-start time.
- The restrike time.

## Chapter 6

# Electronic Security Systems

An overall site-security system is comprised of three major subelements—detection, delay, and response. The detection subelement includes intrusion detection, assessment, and entry control. An ESS is an integrated system that encompasses interior and exterior sensors; CCTV systems for assessing alarm conditions; electronic entry-control systems (EECSs); data-transmission media (DTM); and alarm reporting systems for monitoring, controlling, and displaying various alarm and system information. Interior and exterior sensors and their associated communication and display subsystems are collectively called IDSs.

## OVERVIEW

6-1. Many Army and DOD regulations specify protective measures, policies, and operations related to security. Although the regulations specify minimum requirements, it is possible that more stringent requirements will be necessary at specific sites. A designer will use a previously performed site survey to determine which regulations apply and to determine whether circumstances require more stringent measures. Refer to TM 5-853-4 for additional detailed information.

6-2. AR 190-13 requires the use of a standardized ESS, if practical and available. The receiving element must determine whether a standardized system can meet the requirements and whether it is available. After coordinating with the product manager for physical-security equipment to verify that a standardized system is available, the associated MACOM can issue approval to procure a commercial system in lieu of a standardized system.

### USE OF ESS

6-3. An ESS is used to provide early warning of an intruder. This system consists of hardware and software elements operated by trained security personnel.

6-4. A system is configured to provide one or more layers of detection around an asset. Each layer is made up of a series of contiguous detection zones designed to isolate the asset and to control the entry and exit of authorized personnel and materials.

### GENERAL ESS DESCRIPTION

6-5. An ESS consists of sensors interfaced with electronic entry-control devices, CCTV, alarm reporting displays (both visual and audible), and security lighting. The situation is assessed by sending guards to the alarm point or by using CCTV. Alarm reporting devices and video monitors are located in the security center. The asset's importance will determine whether multiple or redundant security centers are required and, ultimately, the required sophistication of all elements in the ESS. Digital and analog data are transmitted from local (field) interior and exterior locations to the security center for processing. Reliability and accuracy are important functional requirements of the data-transmission system.

### ESS IMPLEMENTATION PROCESS

6-6. The ESS implementation process is shown in Figure 6-1. Implementing an ESS is based on general requirements tailored to a site-specific mission and physical profile. The process begins with a site survey that includes a top-down view of basic needs and classic configurations that are tailored to such site-specific characteristics as terrain, site geography, climatic conditions, the type of asset, and priorities. This data is used to determine the hardware and software requirements, taking into account the additional capacity that should be factored into the design system for future expansion. Once the requirements for an ESS have been identified, the user must determine whether an existing standardized system is suitable for the application. (AR 190-13 outlines the process for gaining approval to use nonstandard equipment.) The user must also secure funding for the equipment (refer to Appendix J). Depending on the current funding regulations, operation-and-maintenance, procurement, or other funds may be required. For example, operations and procurement, Army (OPA) funds may be required for IDS devices; and operations and maintenance, Army (OMA) funds may be required for installation items. A contract is normally awarded to procure and install the equipment. The procurement or installation must be overseen. This may be accomplished by reviewing submittals, inspecting the contractor's work, or responding to the contractor's requests for information. Once the equipment is installed, the acceptance-testing activities must be witnessed and verified. Site conditions during acceptance testing affect the demonstrated detection capability of an exterior IDS. As feasible, acceptance testing should be designed to determine a sensor system's probability of detection (PD) under a range of conditions. For some types of sensor systems, this may be as straightforward as conducting both daytime and nighttime trials to experience differences in temperature and solar heating. After the ESS has been accepted, it must be operated and maintained throughout the remainder of its life cycle. Planning for manpower to operate the system and forecasting the funding and personnel to properly maintain the system is critical for success.

## ESS DESIGN CONSIDERATIONS

6-7. A facility may require interior and exterior ESS elements, depending on the level of protection required. The applicable regulations, threat, and design criteria will define the ESS's general requirements. For an existing ESS, hardware and software may need to be supplemented, upgraded, or completely replaced. A site layout (in which all assets are identified and located) is required. It is a useful design tool for such tasks as configuring the DTM.

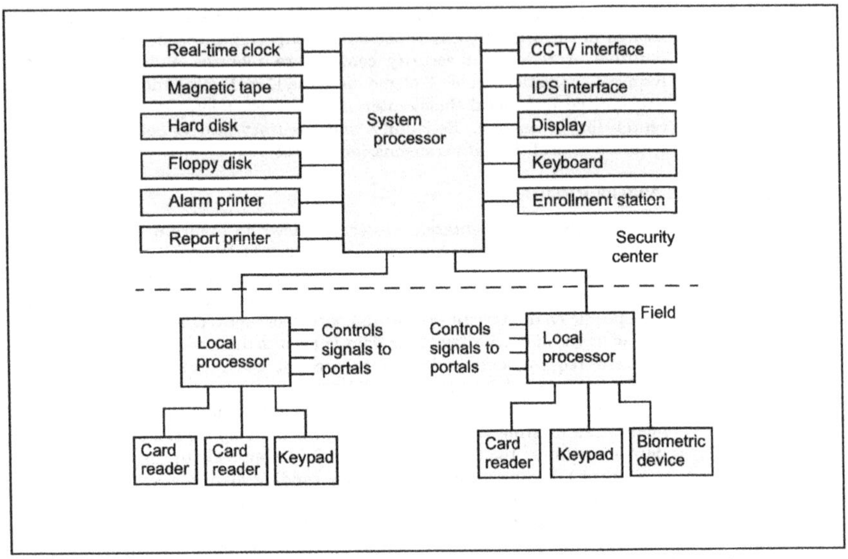


**Figure 6-1. Entry-Control System Configured With Distributed Control**

6-8. The exterior and interior IDSs should be configured as layers of unbroken rings concentrically surrounding the asset. These rings should correspond to defensive layers that constitute the delay system. The first detection layer is located at the outermost defensive layer necessary to provide the required delay. Detection layers can be on a defensive layer, in the area between two defensive layers, or on the asset itself, depending on the delay required. For example, if a wall of an interior room provides sufficient delay for effective response to aggression, detection layers could be between the facility exterior and interior-room wall or on the interior-room wall. These would detect the intruder before penetration of the interior wall is possible.

## RESPONSE AND DELAY

6-9. When dealing with an ESS, the response time is defined as the time it takes the security force to arrive at the scene after an initial alarm is received at the security center. The total delay time is defined as the sum of all of the barriers' delay times, the time required to cross the areas between barriers after an intrusion alarm has been reported, and the time required to accomplish the mission and leave the protected area.

6-10. An ESS's basic function is to notify security personnel that an intruder is attempting to penetrate, or has penetrated, a protected area in sufficient time to allow the response force to intercept and apprehend him. To accomplish this, there must be sufficient physical delay between the point where the intruder is first detected and his objective. This provides delay time equal to or greater than the response time (refer to TM 5-853-1).

6-11. When dealing with interior sensors, boundary sensors that detect penetration (such as structural-vibration sensors or passive ultrasonic sensors) provide the earliest warning of an attempted intrusion. This alarm is usually generated before the barrier is penetrated. This gives the security force advance notification of an attempted penetration, thus allowing the barrier's delay time to be counted as part of the total delay time. Door-position sensors and glass-breakage sensors do not generate an alarm until the barrier has been breached; therefore, the delay time provided by the barrier cannot be counted as part of the total delay time.

6-12. Volumetric motion sensors do not generate an alarm until the intruder is already inside the area covered by the sensors. Therefore, if these sensors are to be used to provide additional response time, additional barriers must be placed between the volumetric motion sensors and the protected asset. Point sensors, such as capacitance sensors and pressure mats, provide warning of attempted penetration only if they detect the intruder before access is gained to the protected area.

## BASIC GUIDANCE

6-13. An IDS is deployed in and around barriers (as detailed in TM 5-853-1). Voice communication links (radio, intercom, and telephone) with the response force are located in the security center. Security personnel will man the center and will alert and dispatch response forces in case of an alarm.

6-14. The barrier should always be deployed behind the IDS to ensure that integrity is maintained against intruders. An intruder will then activate the alarm sensor before penetrating or bypassing the barriers, thus providing delay for alarm assessment and response. The delay time is the determining factor in whether an assessment is conducted by dispatching a guard or by observing the CCTV. Normally, an intruder can climb a fence before a guard can be dispatched; therefore, a CCTV is usually required with an exterior IDS. Barriers can be located ahead of an alarm sensor as a boundary demarcation and can serve to keep people and animals from causing nuisance alarms by inadvertently straying into a controlled area. These barriers provide no additional response time because the barrier could be breached before the IDS sensors could be activated.

6-15. Data for monitoring and controlling an ESS are gathered and processed in the security center where the operator interacts with information from the ESS components located at remote facilities. The ESS's alarm-annunciation computer and its DTM line-termination equipment should be located in a controlled area and provided with tamper protection. Supervisory personnel should permit changes to software only, and these changes should be documented. If redundant DTM links connect the central computer to the local processor, diverse paths should be used to route these links.

6-16. The preferred medium for transmitting data in an ESS is a dedicated fiber-optics system. It provides for communications not susceptible to voltage transients, lightning, electromagnetic interference, and noise. Additionally, the fiber optics will provide a measure of communication-line security and wide bandwidth for video signals and increased data-transmission rates.

## ESS EFFECTIVENESS

6-17. An ESS has a degree-of-protection effectiveness that is based on its probability of detecting intruders attempting to go over, under, around, or through the physical-security system. The intruder may use forced-entry, covert-entry, or insider-compromise tactics. A well-designed system will minimize the possibility of a successful penetration through covert entry or insider compromise. Interior and exterior alarm sensors have a PD based on the capability to detect an intruder passing through a sensing field. An intruder disturbs the steady-state quiescent condition of a sensor for a finite period. Sensors are designed to detect a person of minimum stature moving within a specific range of speeds and distances from the sensor, and any target outside of those parameters will probably not be detected. The PD for a specific sensor is usually specified at 0.9 or greater, but the designer must be aware that the PD is based on certain constraints and environmental conditions.

6-18. Manufacturer specifications usually do not discuss environmental or nuisance alarms that can be caused by climatic conditions (such as wind or rain) or by the intrusion of animals (including birds). The alarm annunciation is valid because the sensor's thresholds have been exceeded; however, the alarm does not represent a valid penetration attempt. If the assessment system is slow, the operator may not be able to determine the cause of the alarm and must, therefore, treat an environmental or nuisance alarm as real.

6-19. Another type of false alarm is caused by electronic-circuit tolerances being exceeded, resulting in the sensor's actuation. False alarms may also result from improper installation of the sensor or from effects of other equipment in the immediate area.

6-20. After an alarm is sensed and information is displayed in the security center, the console operator must determine the cause of the alarm (intrusion, nuisance, environmental, or false). Timely assessment is required when determining its cause. For example, if an intruder scales a fence in 10 seconds and runs 20 feet per second, the intruder will have overcome the barrier and be 2,200 feet from the point of penetration in 2 minutes. To conduct an accurate assessment of the alarm after 2 minutes, guards will have to search an area of about 200 acres. A fixed-television camera properly located and integrated with the alarm processor can assess the situation while the intruder is still in the controlled area.

6-21. For a CCTV camera to be effective, the area it views must be adequately lighted. To correlate the alarms and cameras in a large system (more than 10 cameras) in a timely manner, a computer-based processing system must be used to select and display alarms and camera scenes for the operator. A complex ESS has the following basic components:

- Intrusion-detection sensors.
- Electronic entry-control devices.
- CCTV.

- Alarm-annunciation system.
- DTM.

6-22. The intrusion-detection sensors are normally deployed in a series of concentric layers. The overall PD improves with each added layer of sensors. The layers (interior and exterior) should be functionally uniform; however, their overall effectiveness and cost are different. The exterior zones significantly differ from the interior zones due to the following considerations:

- The consistency of the PD.
- The PD.
- The cost per detection zone.
- The number of zones.
- The overall sensor coverage.

6-23. Exterior IDSs usually have PDs equal to those of interior IDSs. However, exterior sensors are more likely to experience weather-related situations that cause the system's PD to vary. Sensor phenomenology (passive infrared [PIR], microwave radar, and so forth) determines which environmental factors may alter the system's PD. The frequency of occurrence, severity, and duration of a weather event jointly determine whether it represents security vulnerability with the IDS in use. Typically, sophisticated intruders will attempt their penetration and challenge an ESS under conditions most favorable to themselves. Inclement weather (fog, snow, and rain) affects the usefulness of CCTVs and security lighting such that the capability for remote assessment of alarm events may be lost. Exterior IDSs are not necessarily less likely to detect a penetration attempt during fog, rain and snow; the effect of such site conditions on the IDS depends on sensor phenomenology. For example, fence motion caused by rain impact may drive the response of a fence-mounted sensor closer to satisfying the system's alarm criteria, with the result that the margin of disturbance available to the intruder is less. Also, certain buried sensors are more likely to detect an intruder when the ground is wet because of rain or melting snow. Since interior sensor systems are less influenced by environmental conditions, their PD is typically more consistent than that of some types of exterior sensor systems. Other considerations in comparing an interior and exterior ESS are the cost, the number and size of detection zones required, and the detection height.

- Because of environmental conditions, the exterior electronics must be designed and packaged for extremes of temperature, moisture, and wind. The result is that exterior electronic packages are more costly than equivalent packages for interior applications.
- State-of-the-art exterior sensors do not detect penetration attempts above the height of a fence (typically 8 feet). Fence-mounted sensors are usually limited to this height because the fence fabric or poles are used to support the sensor. For aboveground sensors in the controlled area between the fences, the sensor's mounting brackets and posts limit the detection height. In some applications of field sensors (especially buried sensors), the detection height is no more than 3 feet. For a facility, interior sensors can be deployed on walls, floors, or ceilings, thus permitting complete protection of the asset.

6-24. An interior ESS may be far less costly than that of a comparable exterior ESS. This comparison indicates to the designer the value of selecting and deploying a well-planned, well-designed, layered system. The basic rule in overall design of an ESS is to design from the inside out; that is, layered from the asset to the site boundary.

## INTERIOR ESS CONSIDERATIONS

6-25. An interior ESS is typically deployed within a boundary in the immediate vicinity of the asset being protected. If the interior ESS operates in a controlled environment, its PD will be independent of any weather-induced variation in exterior conditions. Also, the physical-security system's effectiveness is enhanced by the interior barriers (walls, ceiling, and floor) that inherently impose a longer delay than exterior barriers (fences and gates).

6-26. Functionally, an interior asset should be viewed as being contained within a cube with sensors protecting all six faces. Interior sensors can be deployed at the cube's perimeter, in its interior space, or in the space immediately outside of the cube.

6-27. If an increased level of protection is dictated by the threat, and if the building is large enough, multiple layers of interior sensors may be deployed for a given asset. A multilayered interior IDS will improve the overall PD. Tamper protection and access-/secure-mode capabilities must be considered when planning and laying out interior sensors.

### TAMPER PROTECTION

6-28. To minimize the possibility of someone tampering with circuitry and associated wiring, all sensor-related enclosures must be equipped with tamper switches. These switches must be positioned so that an alarm is generated before the cover has been moved enough to permit access to the circuitry of adjustment controls. In addition, several types of sensors should be equipped with tamper switches to protect against being repositioned or removed. Security screens containing grid-wire sensors and vibration sensors that can be easily removed from a wall are examples of sensors that require tamper switches.

### ACCESS/SECURE MODE

6-29. During regular working hours, many of the interior sensors must be deactivated by placing the area in the access mode. For example, door-position sensors and volumetric sensors in occupied areas must be deactivated to prevent multiple nuisance alarms caused by the normal movement of people. This can be done locally or remotely. With local control, a switch is used to bypass or shunt alarm contacts when the sensor is placed in the access mode. When done remotely, the security-center operator usually enters a command that causes the processor software to ignore incoming alarms from those sensors placed in the access mode. However, when a sensor is placed in the access mode, its tamper-protection circuitry must remain in the activated or secure mode. During nonworking hours when the facility is unoccupied, all sensors must be placed in the secure mode. Certain devices (such as duress-alarm switches, tamper switches, grid-wire sensors covering vent openings, and glass-breakage sensors) should never be placed in the access mode. The designer must ensure that selected sensors can be placed in an access mode (if required) and that certain types of sensors (such as duress and tamper switches) are configured so that they cannot be put in the access mode under any condition.

# EXTERIOR ESS CONSIDERATIONS

6-30. An exterior ESS is typically deployed at a site's boundary or some other significant boundary such as the demarcation fence for a group of bunkers. An exterior ESS has the advantage that it remains in the secure mode at all times.

6-31. The ideal configuration for an exterior ESS is a rectangle or a polygon, with all sides being straight. The ESS is located in and around barriers that typically include a dual fence. The outside fence is used for demarcation, and the interior fence is used to aid in detection and provide some delay. If dual fences are not used, the sensors should be deployed on the fence or inside it.

## DESIGN GUIDELINES

6-32. The general-design criteria of a perimeter IDS involves primarily the selection and layout of exterior sensors that are compatible with the physical and operational characteristics of a specific site. Important factors to consider during the selection process include physical and environmental conditions at the site, the sensor's performance, and the overall cost of the system. Refer to TMs 5-853-1 and 5-853-2 for additional guidance on the requirements for and placement of exterior sensor systems. Since exterior barriers provide very little delay, exterior sensor systems generally do not provide a significant increase in the available response time.

### Physical and Environmental Considerations

6-33. Physical and environmental considerations are often the determining factors for selecting exterior sensors. The site's characteristics can significantly affect a sensor's operational performance, both in terms of PD and the susceptibility to nuisance alarms. Exterior sensor systems should be selected on the basis of the frequency and duration of weather-related periods of poor detection capability. An exterior IDS may have an unacceptably low PD during a particular weather event or site condition, yet otherwise be superior to other IDSs in terms of good detection capability and a low nuisance-alarm rate. It may be appropriate to select that IDS in spite of its known vulnerability, precisely because the circumstances of its vulnerability are known and precautionary measures can be taken at those times. The overall performance of that IDS, together with its cost, may justify its selection.

6-34. Weather and climatic conditions at a specific site can significantly influence sensor selection. For example, IR detectors are not very effective in heavy rain, fog, dust, or snow. Deep snow can affect detection patterns and performance of both IR and microwave sensors. High winds can cause numerous false alarms in fence-mounted sensors. Electrical storms can cause alarms in many types of sensors and may also damage the equipment.

6-35. Vegetation can be a significant cause of nuisance alarms. Tall grass or weeds can disturb the energy pattern of microwave and both thermal IR and near-IR beam-break sensors. Vegetation growing near electric-field sensors and capacitance sensors can cause nuisance alarms. Large weeds or bushes rubbing against a fence can produce nuisance alarms from fence-mounted sensors. Large trees and bushes moving within the field of view of video motion sensors can cause nuisance or environmental alarms. A clear area must be established for exterior sensors. This area must be void of vegetation or contain vegetation of carefully controlled growth.

6-36. Topographic features are extremely important. Ideally, perimeter terrain should be flat, although gently sloping terrain is acceptable. Irregular terrain with steep slopes may preclude the use of LOS sensors and make CCTV assessment difficult. Gullies and ditches crossing the perimeter represent a vulnerability to LOS sensors and may be a source of false alarms (from flowing water) for buried line sensors. Large culverts can provide an intruder with an entry or exit route across the perimeter without causing an alarm. Likewise, overhead power and communication lines may permit an intruder to bridge the perimeter without causing an alarm.

6-37. Large animals (such as cows, horses, and deer) can cause nuisance alarms in both aboveground and buried sensors. Sensors sensitive enough to detect a crawling or rolling intruder are susceptible to nuisance alarms from small animals such as rabbits, squirrels, cats, and dogs. To minimize the interference from animals, a dual chain-link-fence configuration may be established around the site perimeter with the sensors installed between the fences.

## Sensor Performance

6-38. Exterior sensors must have a high PD for all types of intrusion and have a low unwanted-alarm rate for all expected environmental and site conditions. Unfortunately, no single exterior sensor that is presently available meets both these criteria. All are limited in their detection capability, and all have high susceptibility to nuisance and environmental conditions. Table 6-1, page 6-10, provides estimates of PDs for various types of intrusions. Table 6-2, page 6-10, lists the relative susceptibility of various types of sensors to nuisance and environmental alarms.

## Economic Considerations

6-39. Exterior sensor costs are usually given in cost per linear foot per detection zone (typically 300 feet). These costs include both equipment and installation. Fence-mounted sensors (such as strain-sensitive cable, electromechanical, and mechanical) are generally less costly than stand-alone and buried line sensors. Installation costs can vary significantly, depending on the type of sensor. Table 6-3, page 6-11, provides a comparison of relative costs for procuring and installing various types of exterior sensor systems. It should be remembered that the sensor system's cost is only a portion of the total cost for employing a perimeter IDS. Additional costs include fencing, site preparation, CCTV assessment, and perimeter lighting.

### Table 6-1. Estimate of PD by Exterior Sensors

| Type of Sensor | Intruder Technique | | | | | | | | | | | |
|---|---|---|---|---|---|---|---|---|---|---|---|---|
| | Slow walk | Walking | Running | Crawling | Rolling | Jumping | Tunnelling | Trenching | Bridging | Cutting | Climbing | Lifting |
| Fence mounted | N/A | N/A | N/A | N/A | N/A | VH | VL | L | VL | M/H | H | M/H |
| Taut wire | N/A | N/A | N/A | N/A | N/A | VH | VL | VL | VL | H | H | H |
| Electric field | VH | VH | VH | H | VH | VH | VL | L | L | N/A | N/A | N/A |
| Capacitance | VH | VH | VH | H | H | VH | VL | L | L | N/A | N/A | N/A |
| Ported cable | H | VH | VH | VH | VH | H | M | VH | L | N/A | N/A | N/A |
| Seismic | H | VH | H | M | M | M | L | M | L | N/A | N/A | N/A |
| Seismic/magnetic | H | VH | H | M | M | M | L | M | L | N/A | N/A | N/A |
| Microwave | H | VH | H | M/H | M/H | M/H | VL | L/M | L | N/A | N/A | N/A |
| IR | VH | VH | VH | M/H | M/H | H | VL | L | VL | N/A | N/A | N/A |
| Video motion | H | VH | VH | H | H | H | VL | L/M | M | N/A | N/A | N/A |

**VL = very low, L = low, M = medium, H = high, VH = very high, N/A = not applicable**

### Table 6-2. Relative Susceptibility of Exterior Sensors to False Alarms

| Type of Sensor | Intruder Technique | | | | | | | | | | | |
|---|---|---|---|---|---|---|---|---|---|---|---|---|
| | Wind | Rain | Standing water/runoff | Snow | Fog | Small animals | Large animals | Small birds | Large birds | Lightning | Overhead power lines | Buried power lines |
| Fence mounted | H | M | L | L | VL | L | M | L | L | L | VL | VL |
| Taut wire | VL | VL | VL | VL | VL | VL | L | VL | VL | VL | VL | VL |
| Electric field | M | L/H | VL | M | VL | M | VH | L | M | M | L | VL |
| Capacitance | M | M | VL | M | VL | M | VH | L | M | M | L | VL |
| Ported cable | VL | M | H | L | VL | VL | M | VL | VL | M | VL | L |
| Seismic | M | L | L | L | VL | L | VH | VL | VL | L | L | M |
| Seismic/magnetic | M | L | L | L | VL | L | VH | VL | VL | H | M | H |
| Microwave | L | L | M/H | L/M | L | M/H | VH | VL | M | L/M | L | VL |
| IR | L | L | L | M | M | M | VH | L | M | L | VL | VL |
| Video motion | M | L | L | L | M/H | L | VH | VL | M | L | L | VL |

**VL = very low, L = low, M = medium, H = high, VH = very high**

Table 6-3. Exterior IDS Sensor Cost Comparison

| Type of Sensor | Equipment | Installation | Maintenance |
|---|---|---|---|
| Fence mounted | L | L | L |
| Taut wire | H | H | M |
| Electric field | H | M | M |
| Capacitance | M | L | M |
| Ported cable | H | M | M |
| Seismic | M | M | L |
| Seismic/magnetic | H | M | L |
| Microwave | M | M | L |
| IR | M | L | M |
| Video motion | M | L | M |
| **L = low, M = medium, H = high** | | | |

## PERIMETER LAYOUT AND ZONING

6-40. A protected area's perimeter is usually defined by an enclosing wall or fence or a natural barrier such as water. For exterior sensors to be effective, the perimeter around which they are to be deployed must be precisely defined. In most applications, a dual chain-link-fence configuration will be established around the perimeter. Typically, fences should be between 30 and 50 feet apart; as the distance increases, it is harder for an intruder to bridge the fences. If fence separation is less than 30 feet, some microwave and ported-coax sensors cannot be used. The area between fences (called the controlled area or isolation zone) may need to be cleared of vegetation and graded, depending on the type of sensor used. Proper drainage is required to preclude standing water and to prevent the formation of gullies caused by running water after a heavy rain or melting snow. Cleared areas are required inside and outside of the controlled area. These areas enhance routine observation, as well as sensor-alarm assessment, and minimize the protective cover available to a would-be intruder.

6-41. After the perimeter has been defined, the next step is to divide it into specific detection zones. The length of each detection zone is determined by evaluating the contour, the existing terrain, and the operational activities along the perimeter. Detection zones should be long and straight to minimize the number of sensors or cameras necessary and to aid guard assessment if cameras are not used. It may be more economical to straighten an existing fence line than to create numerous detection zones in accommodating a crooked fence line. If the perimeter is hilly and LOS sensors or CCTV assessment are used, the length of individual detection zones will be commensurate with sensor limitations. Entry points for personnel and vehicles must be configured as independent zones. This enables deactivation of the sensors in these zones; that is, placing them in the access mode during customary working hours (assuming the entry points are manned) without having to deactivate adjacent areas.

6-42. The specific length of individual zones can vary around the perimeter. Although specific manufacturers may advertise maximum zone lengths exceeding 1,000 feet, it is not practical to exceed a zone length of 300 feet. If the zone is longer, it will be difficult for an operator using CCTV assessment or for the response force to identify the location of an intrusion or the cause of a false alarm.

6-43. When establishing zones using multiple sensors, the designer should establish coincident zones where the length and location of each individual sensor will be identical for all sensors within a given zone. If an alarm occurs in a specific zone, the operator can readily determine its approximate location by referring to a map of the perimeter. This also minimizes the number of CCTV cameras required for assessment and simplifies the interface between the alarm-annunciation system and the CCTV switching system.

## ESS ALARM-ANNUNCIATION SYSTEM

6-44. Status information from the various intrusion-detection sensors and entry-control terminal devices must be collected from the field and transmitted to the alarm-annunciation system in the security center, where it is processed, annunciated, and acted on by security personnel. The alarm-annunciation system may also interface with a CCTV system. There are typically two types of alarm-annunciation configurations available. The simplest configuration, which is suitable for small installations, is the point-to-point configuration. With this configuration, a separate transmission line is routed from the protected area to the security center (see Figure 6-2). The Joint-Service Interior Intrusion-Detection System (J-SIIDS) is typical of this type of configuration but will not be further discussed in this manual. The second, and more popular type, is a digital multiplexed configuration that allows multiple protected areas to communicate with the security center over a common data line. A block diagram of a typical multiplexed alarm-annunciation system is shown in Figure 6-3.

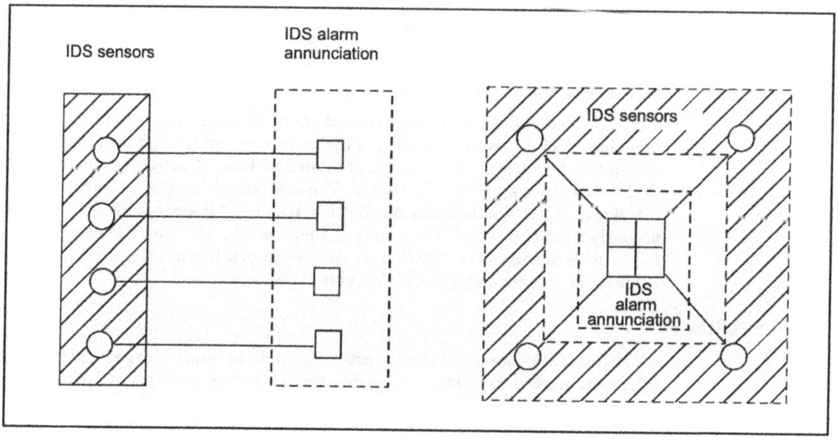


**Figure 6-2. Typical Point-to-Point IDS**

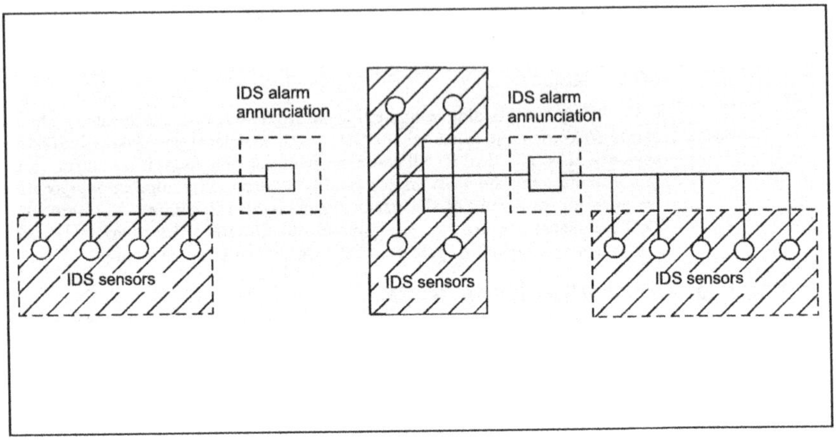


Figure 6-3. Typical Multiplexed IDS

## ALARM-ANNUNCIATION CONFIGURATION

6-45. A block diagram of a typical alarm-annunciation system is shown in Figure 6-4, page 6-14. As shown in the figure, the central computer is the hub of the information flow. The central computer receives and displays alarm and device status information and sends operator-control commands to the ESS's local processors. It also interfaces with the CCTV system. For larger facilities, the management of the DTM communications tasks may be delegated to a separate communication processor so that the central computer can turn its full attention to interpreting the incoming information and updating the control and display devices located at the security console (display, logging, control, and storage devices).

6-46. The central computer may consist of one or more digital computers. The real-time clock is usually integral to the central computer and provides a time stamp for alarms and other events. It allows for time synchronization with the CCTV and other systems, if included. The console operator must be able to set the clock, which should include a battery backup. All system events must be properly time-correlated. For example, there will be an exact time correlation for an ESS alarm event reported on the alarm printer and the corresponding video scene recorded by the CCTV's video processor.

## DATA STORAGE

6-47. Computer-based systems are required to store large amounts of information such as system software, application programs, data structures, and system events (alarm transactions and status changes). Therefore, a large amount of nonvolatile memory is required. The semiconductor memory provided with a central computer is designed for rapid storage and retrieval

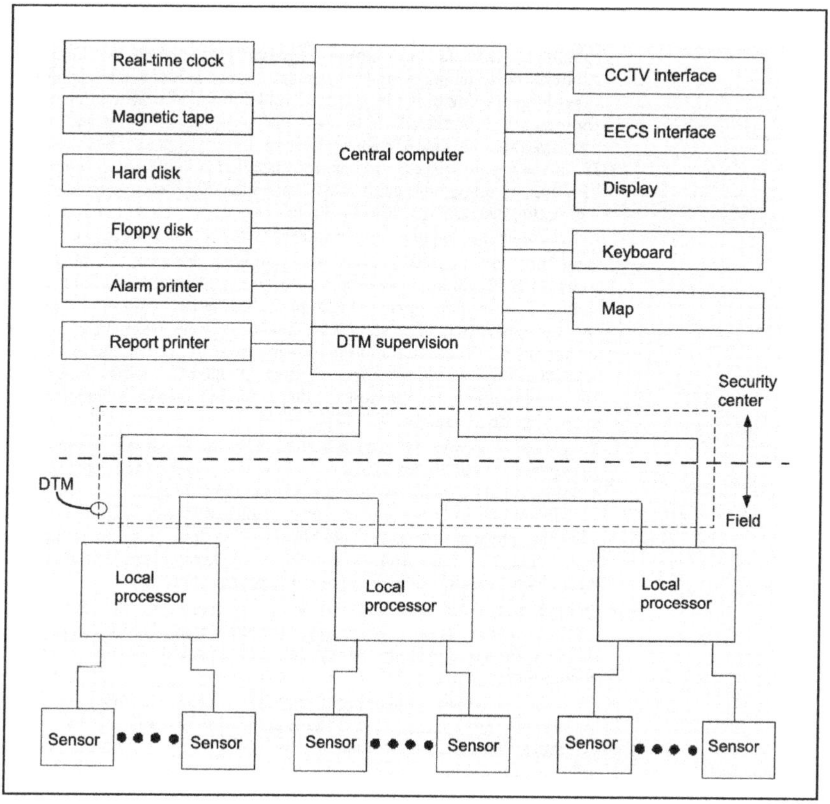


**Figure 6-4. Typical IDS Alarm-Annunciation System**

and possesses extremely fast access times. The most commonly used media for archival storage are magnetic tape; compact-disk, read-only memory (CD-ROM); and magnetic disk. These media are capable of economically storing large amounts of data.

## OPERATOR INTERFACE

6-48. The operator interacts with the alarm-annunciation system through devices that can be seen, heard, or touched and manipulated. Visual displays and printers can be used to inform the operator of an alarm or the equipment's status. Audible devices are used to alert an operator to an alarm or the equipment's failure. Devices such as push buttons and keyboards permit an operator to acknowledge and reset alarms, as well as change operational parameters.

- **Visual displays.** The type of display used to inform the operator visually of the ESS's status is determined primarily by the system's complexity. Status information is usually displayed on monitors. Alphanumeric displays and map displays are seldom used. Monitors provide great flexibility in the type and format of alarm information that may be displayed. Both text and graphic information can be displayed in a variety of colors. Multiple alarms may also be displayed. If alarms are prioritized, higher-priority alarms may be highlighted by blinking, by using bold print or reverse video, or by changing colors. To assist the operator in determining the correct response, alarm-specific instructions may be displayed adjacent to the alarm information.
- **Audible alarm devices.** In conjunction with the visual display of an alarm, the alarm-annunciation system must also generate an audible alarm. The audible alarm may be produced by the ringing of a bell or by the generation of a steady or pulsating tone from an electronic device. In any case, the audible alarm serves to attract the operator's attention to the visual-alarm display. A silence switch is usually provided to allow the operator to silence the bell or tone before actually resetting the alarm.
- **Logging devices.** All alarm-system activity (such as a change of access/secure status, an alarm event, an entry-control transaction, or a trouble event) should be logged and recorded. Logged information is important not only for security personnel investigating an event, but also for maintenance personnel checking equipment performance for such causes as false and nuisance alarms. Most alarm-annunciation systems are equipped with logging and alarm printers.
- **Alarm printers.** Alarm printers are typically of the high-speed, continuous-feed variety. The printer provides a hard-copy record of all alarm events and system activity, as well as limited backup in case the visual display fails.
- **Report printers.** Most ESSs include a separate printer (report printer) for generating reports using information stored by the central computer. This printer will usually be typical of those found in modern office environments.
- **Operator control.** A means is required to transmit information from the operator to the system. The type of controls provided usually depends on the type of display provided. The following are consistent with the controls:
  - Keypads consist of a numeric display system that will generally be provided with a 12-digit keypad and several function keys such as access, secure, acknowledge, and reset. The keypad enables an operator to key in numeric requests for the status of specific zones.
  - Monitor-based systems are usually provided with a typewriter-type keyboard that enables an operator to enter more information using a combination of alphanumeric characters and function keys.
  - An ESS may be equipped with enhancement hardware/devices to help the operator enter information or execute commands quickly. A mouse or a trackball are typical examples.

## FIELD-DATA COLLECTION

6-49. Sensor and terminal device data must be transmitted to the central alarm monitor located in the security center using a selected DTM. The following are DTM methods that may be used:

### Local Processors

6-50. Multiplexing techniques can be used to minimize the number of data links needed to communicate field-device status to the security center. This is done through devices called local processors. The following is descriptive of a local processor's capabilities:

- A local processor may have very few device inputs, or it may have many (depending on the manufacturer). Rather than having a fixed number of inputs, many local processors are expandable. For example, a basic local processor may be provided with eight device inputs with additional blocks of eight inputs available by using plug-in modules.
- The local processor must provide line supervision for all communication links to sensors, terminal devices, and so forth. Usually, direct-current (DC) line supervision is supplied as the standard with more secure techniques available as options. The data communication links between the local processor and the central alarm monitor must also be supervised.
- Local processors can also provide output signals that can be used for such functions as activating sensor remote test features, light control, or portal control or activating a deterrent (such as a loud horn).
- The local processor contains a microprocessor, solid-state memory, and appropriate software. It has the capability to perform a number of functions locally (such as access-/secure-mode selection, alarm reset, card or keypad electronic-entry control, portal control, and device testing). If the communication link to the security center is temporarily lost, local processors can continue to operate in a stand-alone mode, storing data for transmission after the link is restored.
- The number of local processors required for a specific site depends on the number of protected areas and their proximity to each other and the number of sensors within a protected area. For example, a small building may require one local processor, whereas a large building may require one or more for each floor. An exterior IDS perimeter with two or three different sensors may require one local processor for every two perimeter zones. All local processors may be linked to the central computer using one common DTM link, or the DTM may consist of several links. The designer should note that the temporary loss of a DTM link would render all local processors on that link inactive for the duration of the loss.

### Central Computer and Local-Processor Data Exchange

6-51. When the ESS is powered up or reset at the security center, the central computer will download all necessary operational information over the DTM to all local processors. After the download is complete, the central computer will automatically begin polling the local processors for ESS device status. In addition to alarm status, tamper indications, and local-processor status, the DTM may be required to convey security-center console-operator commands to field devices. Examples include security-area access-/secure-mode changes and initiation of the intrusion-sensor self test.

### CCTV Interface

6-52. If a CCTV assessment system is deployed with the ESS, an interface between the two is required. This interface allows CCTV system alarms (such as loss of video) to be displayed by the ESS's alarm-annunciation system. The interface also provides IDS alarm signals to the CCTV's video switcher so that the correct CCTV camera will be displayed on the CCTV monitors to allow real-time alarm assessment and video recording as required.

## ESS SOFTWARE

6-53. The software provided with computer-based ESS alarm-annunciation systems consists of three types—a standard operating system (such as the Microsoft®-disk operating system [MS-DOS]); vendor-developed application programs; and user-filled, site-specific data structures.

- **System software.** The designer will ensure that system software provided by the vendor conforms to accepted industry standards so that standard, follow-on maintenance and service contracts can be negotiated to maintain the central computer system.
- **Application software.** The vendor-developed application programs are typically proprietary and include ESS monitoring, display, and entry-control capabilities.
- **User-filled data structures.** These data structures are used to populate the site-specific database. Specific electronic address information, personnel access schedules, and normal duty hours are typically included in the site-specific database. The information may include preferred route descriptions for the response force, the phone number of the person responsible for the alarmed area, and any hazardous material that may be located in the alarmed area.

6-54. ESS software functions typically include the following:

- **Alarm monitoring and logging.** The software should provide for monitoring all sensors, local processors, and data communication links and notifying the operator of an alarm condition. All alarm messages should be printed on the alarm printer, archived, and displayed at the console. As a minimum, printed alarm data should include the date and time (to the nearest second) of the alarm and the location and type of alarm.
- **Alarm display.** The software should be structured to permit several alarms to be annunciated simultaneously. A buffer or alarm queue should be available to store additional alarms until they are annunciated and, subsequently, acted upon and reset by the console operator.
- **Alarm priority.** A minimum of five alarm-priority levels should be available. Higher-priority alarms should always be displayed before

lower-priority alarms. This feature permits an operator to respond quickly to the more important alarms before those of lesser importance. For example, the priority of alarm devices may be as follows:

- Duress.
- Intrusion detection.
- Electronic-entry control.
- Tamper.
- CCTV alarms and equipment-malfunction alarms.

- **Reports.** The application software should provide for generating, displaying, printing, and storing reports.

## PASSWORDS

6-55. Software security will be provided by limiting access to personnel with authorized passwords assigned by a system manager. A minimum of three password levels shall be provided. Additional security can be provided by programmed restrictions that limit the keyboard actions of logged-in passwords to the user ranks of system managers, supervisors, and console operators, as appropriate.

## OPERATOR INTERFACE

6-56. The software should enable an operator with the proper password to enter commands and to obtain displays of system information. As a minimum, an operator should be able to perform the following functions through the keyboard or the keypad:

- Log on by password to activate the keyboard.
- Log off to deactivate the keyboard.
- Request display of all keyboard commands that are authorized for the logged-in password.
- Request display of detailed instructions for any authorized keyboard command.
- Acknowledge and clear alarm messages.
- Display the current status of any device in the system.
- Command a status change for any controlled device in the system.
- Command a mode change for any access/secure device in the system.
- Command printouts of alarm summaries, status summaries, or system activity on a designated printer.
- Add or delete ESS devices or modify parameters associated with a device.

# INTERIOR INTRUSION-DETECTION SENSORS

6-57. Interior intrusion-detection sensors are devices used to detect unauthorized entry into specific areas or volumetric spaces within a building. These sensors are usually not designed to be weatherproof or rugged enough to survive an outdoor environment. Therefore, this type of sensor should not be used outdoors unless described by the manufacturer as suitable for outdoor use.

6-58. Interior intrusion-detection sensors generally perform one of three detection functions—detection of an intruder penetrating the boundary of a protected area, detection of intruder motion within a protected area, and detection of an intruder touching or lifting an asset within a protected area. Therefore, interior sensors are commonly classified as boundary-penetration sensors, volumetric motion sensors, and point sensors. Although duress switches are not intrusion-detection sensors, they are included in this discussion because they are usually wired to the same equipment that monitors the interior intrusion-detection sensors.

## BOUNDARY-PENETRATION SENSORS

6-59. Boundary-penetration sensors are designed to detect penetration or attempted penetration through perimeter barriers. These barriers include walls, ceilings, duct openings, doors, and windows.

### Structural-Vibration Sensors

6-60. Structural-vibration sensors detect low-frequency energy generated in an attempted penetration of a physical barrier (such as a wall or a ceiling) by hammering, drilling, cutting, detonating explosives, or employing other forcible methods of entry. A piezoelectric transducer senses mechanical energy and converts it into electrical signals proportional in magnitude to the vibrations. To reduce false alarms from single accidental impacts on the barrier, most vibration sensors use a signal processor that has an adjustable pulse-counting accumulator in conjunction with a manual sensitivity adjustment. The count circuit can be set to count a specific number of pulses of specific magnitude within a predefined time interval before an alarm is generated. However, the circuitry is usually designed to respond immediately to large pulses, such as those caused by an explosion. The sensitivity adjustment is used to compensate for the type of barrier and the distance between transducers. Typically, several transducers can be connected together and monitored by one signal processor. Figure 6-5, page 6-20, shows an example of wall-mounted, structural-vibration sensors.

### Glass-Breakage Sensors

6-61. Glass-breakage sensors detect the breaking of glass. The noise from breaking glass consists of frequencies in both the audible and ultrasonic range. Glass-breakage sensors use microphone transducers to detect the glass breakage. The sensors are designed to respond to specific frequencies only, thus minimizing such false alarms as may be caused by banging on the glass.

### Passive Ultrasonic Sensors

6-62. Passive ultrasonic sensors detect acoustical energy in the ultrasonic frequency range, typically between 20 and 30 kilohertz (kHz). They are used to detect an attempted penetration through rigid barriers (such as metal or masonry walls, ceilings, and floors). They also detect penetration through windows and vents covered by metal grilles, shutters, or bars if these openings are properly sealed against outside sounds.

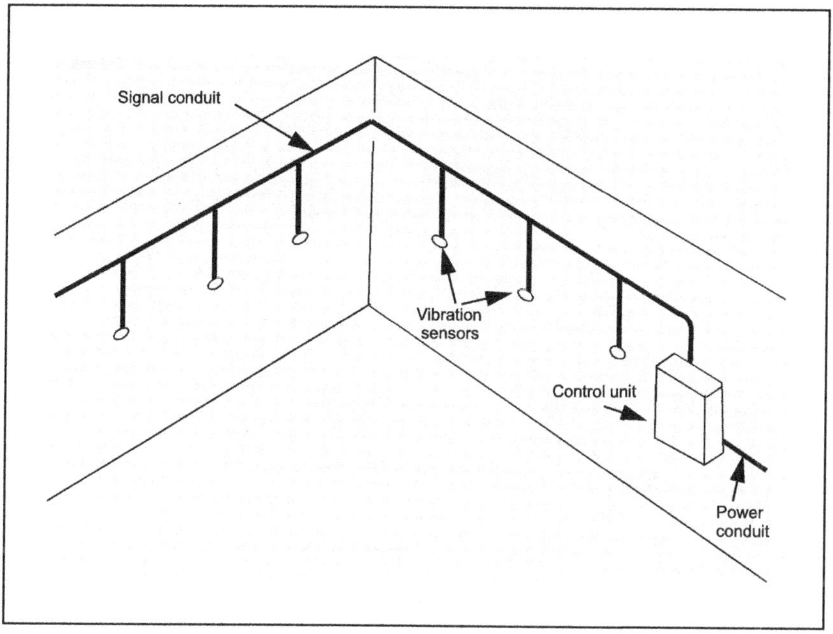


Figure 6-5. Wall-Mounted, Structural-Vibration Sensors

6-63. **Detection Transducer.** The detection transducer is a piezoelectric crystal that produces electrical signals proportional to the magnitude of the vibrations. A single transducer provides coverage of an area about 15 by 20 feet in a room with an 8- to 12-foot ceiling. A typical detection pattern is shown in Figure 6-6. Ten or more transducers can be connected to a signal processor. As with vibration sensors, the signal processor for a passive ultrasonic sensor has manual sensitivity adjustment and an adjustable pulse-counting accumulator.

6-64. **Sensors.** Passive ultrasonic sensors detect ultrasonic energy that results from the breaking of glass, the snipping of bolt cutters on metal barriers, the hissing of an acetylene torch, and the shattering of brittle materials (such as concrete or cinderblock). However, the sensors will not reliably detect drilling through most material nor attacks against soft material such as wallboard. Their effective detection range depends largely on the barrier material, the method of attempted penetration, and the sensitivity adjustment of the sensor. Examples of maximum detection distances for a typical sensor for different types of attempted penetration are shown in Table 6-4.

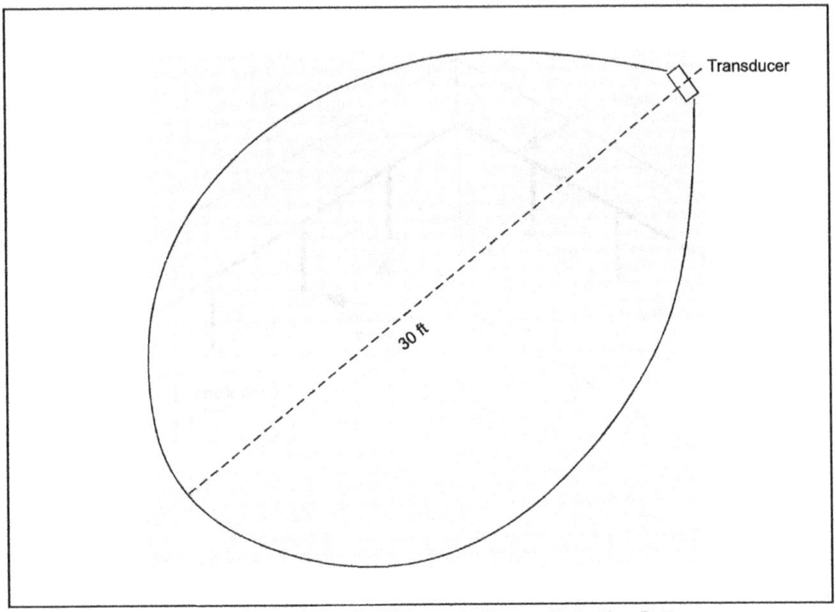


Figure 6-6. Typical Passive-Ultrasonic-Sensor Detection Pattern

Table 6-4. Detection Range for Passive Ultrasonic Sensors

| Penetration | Distance (in Feet) |
|---|---|
| Cut 1/4-inch-thick expanded metal with bolt cutters | 55 |
| Cut 5/8-inch reinforcing bar with bolt cutters | 45 |
| Use acetylene cutting torch | 39 |
| Cut wood with circular saw | 30 |
| Cut 5/8-inch reinforcing bar with hacksaw | 19 |
| Drill through brick | 15 |
| Drill through 1/8-inch steel plate | 6 |
| Cut 1/8-inch steel plate with hacksaw | 4 |
| Drill through cinderblock | 3 |

6-65. **Balanced Magnetic Switches.** Balanced magnetic switches (BMSs) are typically used to detect the opening of a door. These sensors can also be used on windows, hatches, gates, or other structural devices that can be opened to gain entry. When using a BMS, mount the switch mechanism on the door frame and the actuating magnet on the door. Typically, the BMS has a three-position reed switch and an additional magnet (called the bias magnet) located adjacent to the switch. When the door is closed, the reed switch is held in the balanced or center position by interacting magnetic fields. If the door is opened or an external magnet is brought near the sensor in an attempt to defeat it, the switch becomes unbalanced and generates an alarm. A BMS must be mounted so that the magnet receives maximum movement when the door or window is opened. Figure 6-7 shows several configurations for mounting BMSs.

### Grid-Wire Sensors

6-66. The grid-wire sensor consists of a continuous electrical wire arranged in a grid pattern. The wire maintains an electrical current. An alarm is generated when the wire is broken. The sensor detects forced entry through walls, floors, ceilings, doors, windows, and other barriers. An enamel-coated number 24 or 26 American wire gauge (AWG) solid-copper wire typically forms the grid. The grid's maximum size is determined by the spacing between the wires, the wire's resistance, and the electrical characteristics of the source providing the current. The grid wire can be installed directly on the barrier, in a grille or screen that is mounted on the barrier, or over an opening that requires protection. The wire can be stapled directly to barriers made of wood or wallboard. Wood panels should be installed over the grid to protect it from day-to-day abuse and to conceal it. When used on cinder, concrete, and

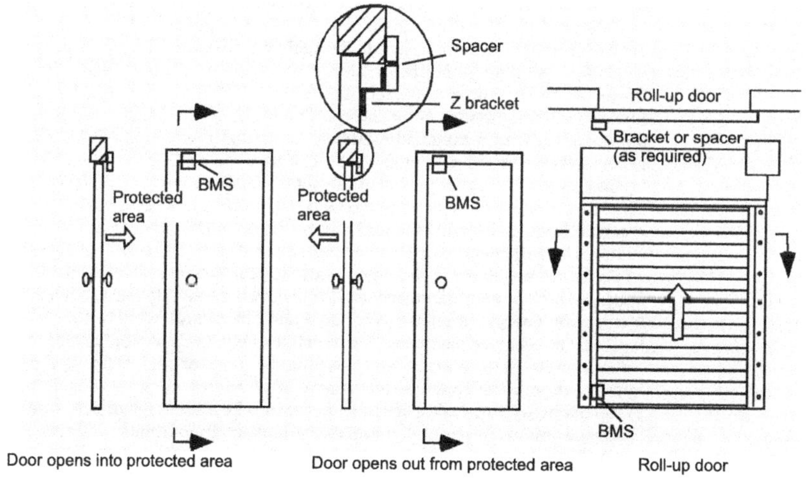


**Figure 6-7. BMS Mounting Configurations**

masonry surfaces, these surfaces must first be covered with plywood or other material to which the wire can be stapled. An alternative method is to staple the wire grid to the back side of a panel and install the panel over the surface.

## VOLUMETRIC MOTION SENSORS

6-67. Volumetric motion sensors are designed to detect intruder motion within the interior of a protected volume. Volumetric sensors may be active or passive. Active sensors (such as microwave) fill the volume to be protected with an energy pattern and recognize a disturbance in the pattern when anything moves within the detection zone. Whereas active sensors generate their own energy pattern to detect an intruder, passive sensors (such as IR) detect energy generated by an intruder. Some sensors, known as dual-technology sensors, use a combination of two different technologies, usually one active and one passive, within the same unit. If CCTV assessment or surveillance cameras are installed, video motion sensors can be used to detect intruder movement within the area. Since ultrasonic motion sensors are seldom used, they will not be discussed here.

### Microwave Motion Sensors

6-68. With microwave motion sensors, high-frequency electromagnetic energy is used to detect an intruder's motion within the protected area. Interior or sophisticated microwave motion sensors are normally used.

6-69. **Interior Microwave Motion Sensors.** Interior microwave motion sensors are typically monostatic; the transmitter and the receiver are housed in the same enclosure (transceiver). They may each be provided with a separate antenna or they may share a common antenna. The high-frequency signals produced by the transmitter are usually generated by a solid-state device, such as a gallium arsenide field-effect transistor. The power generated is usually less than 10 milliwatts, but it is sufficient to transmit the signal for distances up to about 100 feet. The shape of the transmitted beam is a function of the antenna configuration. The range of the transmitted beam can be controlled with a range adjustment. A variety of detection patterns can be generated (see Figure 6-8, page 6-24). The frequency of the transmitted signal is compared with the frequency of the signal reflected back from objects in the protected area. If there is no movement within the area, the transmitted and received frequencies will be equal and no alarm will be generated. Movement in the area will generate a Doppler frequency shift in the reflected signal and will produce an alarm if the signal satisfies the sensor's alarm criteria. The Doppler shift for a human intruder is typically between 20 and 120 hertz (Hz). Microwave energy can pass through glass doors and windows as well as lightweight walls or partitions constructed of plywood, plastic, or fiberboard. As a result, false alarms are possible because of the reflection of the microwave signals from the movement of people or vehicles outside of the protected area. The designer can sometimes take advantage of this when the protected area is large and contains a number of partitions, but this is not normally done.

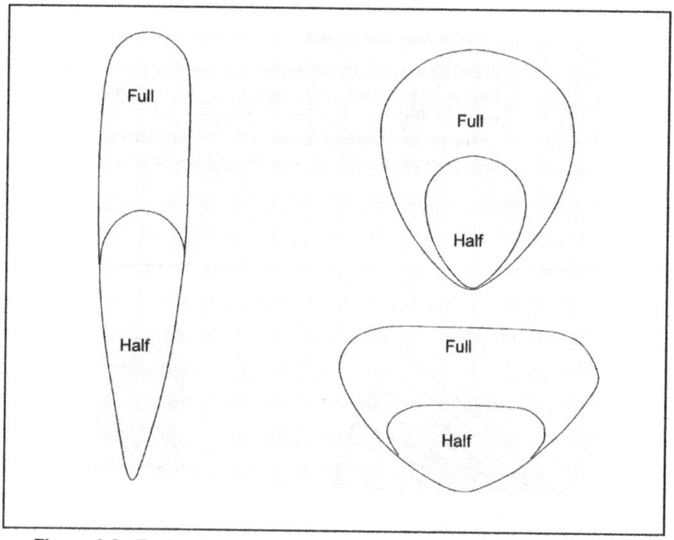


**Figure 6-8. Typical Detection Patterns for Microwave Motion Sensors**

6-70. **Sophisticated Microwave Motion Sensors.** Sophisticated microwave motion sensors may be equipped with electronic range gating. This feature allows the sensor to ignore the signals reflected beyond the settable detection range. Range gating may be used to effectively minimize unwanted alarms from activity outside the protected area.

## PIR Motion Sensors

6-71. PIR motion sensors detect a change in the thermal energy pattern caused by a moving intruder and initiate an alarm when the change in energy satisfies the detector's alarm criteria. These sensors are passive devices because they do not transmit energy; they monitor the energy radiated by the surrounding environment.

6-72. All objects with temperatures above absolute zero radiate thermal energy. The wavelengths of the IR energy spectrum lie between 1 and 1,000 microns. Because the human body radiates thermal energy of between 7 and 14 microns, PIR motion sensors are typically designed to operate in the far IR wavelength range of 4 to 20 microns.

6-73. The IR energy must be focused onto a sensing element, somewhat as a camera lens focuses light onto a film. Two techniques are commonly used. One technique uses reflective focusing; parabolic mirrors focus the energy. The other uses an optical lens. Of the various types of optical lenses, Fresnel lenses are preferred because they can achieve short focal lengths with minimal thickness. Because IR energy is severely attenuated by glass, lenses are usually made of plastic.

6-74. The sensor's detection pattern is determined by the arrangement of lenses or reflectors. The pattern is not continuous but consists of a number of rays or fingers, one for each mirror or lens segment. Numerous detection patterns are available, several of which are shown in Figure 6-9. The PIR is not provided with a range adjustment, but the range can be adjusted

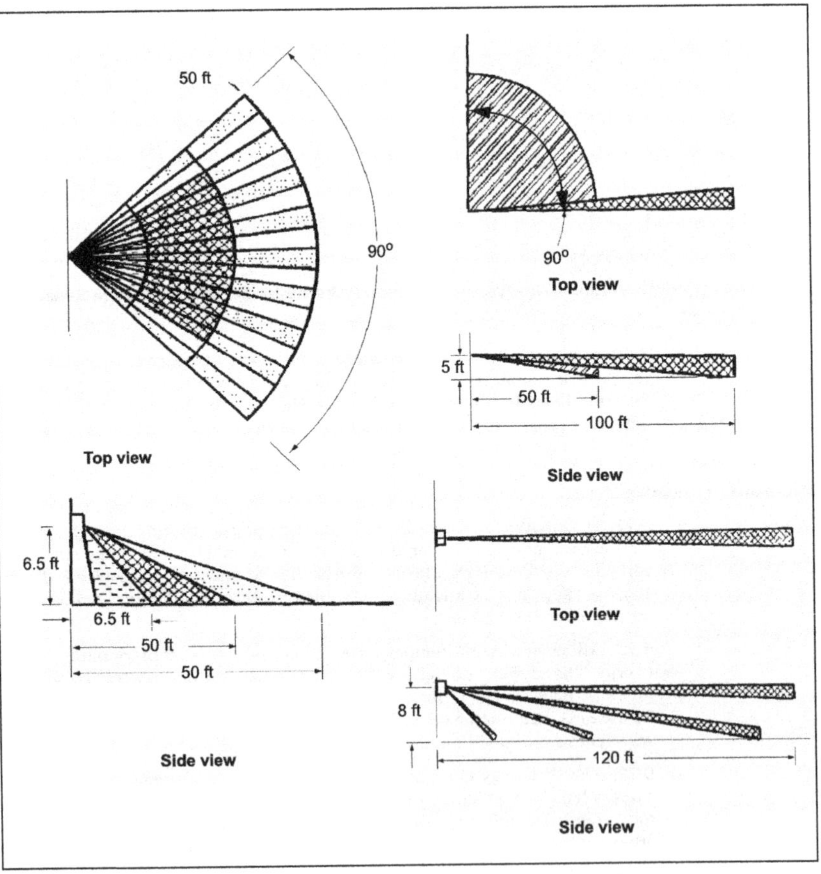


Figure 6-9. Typical Detection Patterns for a PIR Motion Detector

somewhat by manipulating the sensor's position; therefore, careful selection of the appropriate detection pattern is critical to proper sensor performance.

6-75. Most manufacturers use a pyroelectric material as the thermal sensing element. This material produces a change in electric charge when exposed to changes in temperature. To minimize false alarms caused by changes in ambient temperature, most manufacturers use a dual-element sensor. The sensing element is split into halves, one that produces a positive voltage pulse and the other a negative pulse when a change in temperature changes. An intruder entering one of the detection fingers produces an imbalance between the two halves, resulting in an alarm condition. Quadelement sensors that combine and compare two dual-element sensors are also in use. Pulse-count activation, a technique in which a predefined number of pulses within a specific interval of time must be produced before an alarm is generated, is also used.

### Dual-Technology Sensors

6-76. To minimize the generation of alarms caused by sources other than intruders, dual-technology sensors combine two different technologies in one unit. Ideally, this is achieved by combining two sensors that individually have a high PD and do not respond to common sources of false alarms. Available dual-technology sensors combine an active ultrasonic or microwave sensor with a PIR sensor. The alarms from each sensor are logically combined in an "and" configuration; that is, nearly simultaneous alarms from both active and passive sensors are needed to produce a valid alarm. Although combined technology sensors have a lower false-alarm rate than individual sensors, the PD is also reduced. For example, if each individual sensor has a PD of 0.95, the PD of the combined sensors is the product of individual probabilities (0.9). Also, ultrasonic and microwave motion sensors have the highest probability of detecting movement directly toward or away from the sensor, whereas PIR motion sensors have the highest probability of detecting movement across the detection pattern. Therefore, the PD of sensors combined in a single unit is less than that obtainable if the individual sensors are mounted perpendicular to each other with overlapping detection patterns. Because of the lower false-alarm rate, the reduced PD can be somewhat compensated for by increasing the sensitivity or detection criteria of each individual sensor.

### Video Motion Sensors

6-77. A video motion sensor generates an alarm when an intruder enters a selected portion of a CCTV camera's field of view. The sensor processes and compares successive images between the images against predefined alarm criteria. There are two categories of video motion detectors—analog and digital. Analog detectors generate an alarm in response to changes in a picture's contrast. Digital devices convert selected portions of the analog video signal into digital data that are compared with data converted previously; if differences exceed preset limits, an alarm is generated. The signal processor usually provides an adjustable window that can be positioned anywhere on the video image. Available adjustments permit changing horizontal and vertical window size, window position, and window sensitivity. More sophisticated units provide several adjustable windows that can be individually sized and positioned. Multiple windows permit concentrating on several specific areas of an image while ignoring others. For example, in a scene containing six doorways leading into a long hallway, the sensor can be set to monitor only two critical doorways.

## POINT SENSORS

6-78. Point sensors are used to protect specific objects within a facility. These sensors (sometimes referred to as proximity sensors) detect an intruder coming in close proximity to, touching, or lifting an object. Several different types are available, including capacitance sensors, pressure mats, and pressure switches. Other types of sensors can also be used for object protection.

### Capacitance Sensors

6-79. Capacitance sensors detect an intruder approaching or touching a metal object by sensing a change in capacitance between the object and the ground. A capacitor consists of two metallic plates separated by a dielectric medium. A change in the dielectric medium or electrical charge results in a change in capacitance. In practice, the metal object to be protected forms one plate of the capacitor and the ground plane surrounding the object forms the second plate. The sensor processor measures the capacitance between the metal object and the ground plane. An approaching intruder alters the dielectric value, thus changing the capacitance. If the net capacitance change satisfies the alarm criteria, an alarm is generated.

6-80. The maximum capacitance that can be monitored by this type of sensor is usually between 10,000 and 50,000 picofarads. The minimum detectable change in capacitance can be as low as 20 picofarads. The signal processor usually has a sensitivity adjustment that can be set to detect an approaching intruder several feet away or to require that the intruder touch the object before an alarm is generated.

6-81. Because air forms most of the dielectric of the capacitor, changes in relative humidity will affect the sensor's sensitivity. An increase in humidity causes the conductivity of the air to increase, lowering the capacitance. Moving a metal object (such as a file cabinet) closer to or away from the protected object can also affect the sensitivity of a capacitance sensor. Figure 6-10, page 6-28, illustrates a typical application using a capacitance sensor.

### Pressure Mats

6-82. Pressure mats generate an alarm when pressure is applied to any part of the mat's surface, as when someone steps on the mat. One type of construction uses two layers of copper screening separated by soft-sponge rubber insulation with large holes in it. Another type uses parallel strips of ribbon switches made from two strips of metal separated by an insulating material and spaced several inches apart. When enough pressure is applied to the mat, either the screening or the metal strips make contact, generating an alarm. Pressure mats can be used to detect an intruder approaching a protected object, or they can be placed by doors or windows to detect entry. Because pressure mats are easy to bridge, they should be well concealed, such as placing them under a carpet.

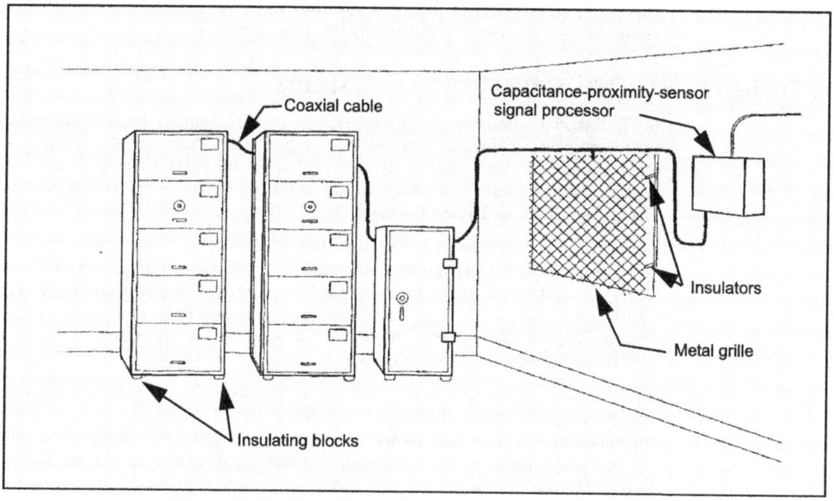


**Figure 6-10. Capacitance-Proximity-Sensor Application**

### Pressure Switches

6-83. Mechanically activated contact switches or single ribbon switches can be used as pressure switches. Objects that require protection can be placed on top of the switch. When the object is moved, the switch actuates and generates an alarm. In this usage, the switch must be well concealed. The interface between the switch and the protected object should be designed so that an adversary cannot slide a thin piece of material under the object to override the switch while the object is removed.

## DURESS-ALARM DEVICES

6-84. Duress-alarm devices may be fixed or portable. Operations and security personnel use them to signal a life-threatening emergency. Activation of a duress device will generate an alarm at the alarm-monitoring station. Because of the nature of the alarm, duress devices should never annunciate at the point of threat. These devices are customarily manually operated.

6-85. Fixed duress devices are mechanical switches permanently mounted in an inconspicuous location, such as under a counter or desk. They can be simple push-button switches activated by the touch of a finger or hand or foot-operated switches attached to the floor.

6-86. Portable duress devices are wireless units consisting of a transmitter and a receiver. The transmitter is portable and small enough to be conveniently carried by a person. The receiver is mounted in a fixed location within the facility. Either ultrasonic or RF energy can be used as the communication medium. When activated, the transmitter generates an alarm that is detected (within range) by the receiver. The receiver then activates a relay that is hardwired to the alarm-monitoring system.

## EXTERIOR INTRUSION-DETECTION SENSORS

6-87. Exterior intrusion-detection sensors are customarily used to detect an intruder crossing the boundary of a protected area. They can also be used in clear zones between fences or around buildings, for protecting materials and equipment stored outdoors within a protected boundary, or in estimating the PD for buildings and other facilities.

6-88. Exterior sensors are designed to operate in outdoor environmental conditions. The detection function must be performed with a minimum of unwanted alarms such as those caused by wind, rain, ice, standing water, blowing debris, animals, and other sources. Important criteria for selecting an exterior sensor are the PD, the sensor's susceptibility to unwanted alarms, and the sensor's vulnerability to defeat.

6-89. The PD of an exterior sensor is much more vulnerable to the physical and environmental conditions of a site than that of an interior sensor. Many uncontrollable forces (such as wind, rain, ice, frozen soil, standing or running water, falling and accumulated snow, and blowing dust and debris) may affect an exterior sensor's performance. Although attention generally is directed to circumstances that cause a dramatic drop in the PD, environmental factors can also cause short-term increases in the PD. If controlled intrusions (intrusions by security personnel to verify the current detection capability of an IDS) are done while an IDS temporarily has a higher than usual PD as the result of current site conditions, the results may give a false indication of the general effectiveness of that IDS.

6-90. Because of the nature of the outdoor environment, exterior sensors are also more susceptible to nuisance and environmental alarms than interior sensors. Inclement weather conditions (heavy rain, hail, and high wind), vegetation, the natural variation of the temperature of objects in the detection zone, blowing debris, and animals are major sources of unwanted alarms.

6-91. As with interior sensors, tamper protection, signal-line supervision, self-test capability, and proper installation make exterior sensors less vulnerable to defeat. Because signal-processing circuitry for exterior sensors is generally more vulnerable to tampering and defeat than that for interior sensors, it is extremely important that enclosures are located and installed properly and that adequate physical protection is provided. Several different types of exterior intrusion-detection sensors are available. They can be categorized as—

- Fence sensors.
- Buried line sensors.
- LOS sensors.
- Video motion sensors.

## FENCE SENSORS

6-92. Fence sensors detect attempts to penetrate a fence around a protected area. Penetration attempts (such as climbing, cutting, or lifting) generate mechanical vibrations and stresses in fence fabric and posts that are usually different than those caused by natural phenomena like wind and rain. The basic types of sensors used to detect these vibrations and stresses are strain-sensitive cable, taut wire, and fiber optics. Other types of fence sensors detect penetration attempts by sensing changes in an electric field or in capacitance. Mechanical and electromechanical fence sensors are seldom used and will not be discussed here.

### Strain-Sensitive Cable

6-93. Strain-sensitive cables are transducers that are uniformly sensitive along their entire length. They generate an analog voltage when subject to mechanical distortions or stress resulting from fence motion. Strain-sensitive cables are sensitive to both low and high frequencies. The signal processor usually has a band-pass filter that passes only those signals characteristic of fence-penetration actions. An alarm is initiated when the signal's frequency, amplitude, and duration characteristics satisfy the processor's criteria. Because the cable acts like a microphone, some manufacturers offer an option that allows the operator to listen to fence noises causing the alarm. Operators can then determine whether the noises are naturally occurring sounds from wind or rain or are from an actual intrusion attempt. This feature is relatively costly to implement because it requires additional cable from each signal processor to the security center and, if CCTV is being used, it may be of limited benefit. Strain-sensitive cable is attached to a chain-link fence about halfway between the bottom and top of the fence fabric with plastic ties. One end of the cable is terminated at the signal processor and the other end with a resistive load. The DC through the cable provides line supervision against cutting or electrically shorting the cable or disconnecting it from the processor. A typical installation is shown in Figure 6-11.

### Taut-Wire Sensor

6-94. A taut-wire sensor combines a physically taut-wire barrier with an intrusion-detection sensor network. The taut-wire sensor consists of a column of uniformly spaced horizontal wires up to several hundred feet in length and securely anchored at each end. Typically, the wires are spaced 4 to 8 inches apart. Each is individually tensioned and attached to a detector located in a sensor post. Two types of detectors are commonly used—mechanical switches and strain gauges.

- The mechanical switch consists of a specially designed switch mechanism that is normally open. The tensioned wires are mechanically attached to the switch, and movement of the wire beyond a preset limit causes the switch to close. To counteract small gradual movements of a wire (such as that caused by settling of the fence or by freezing or thawing of soil) switches are usually supported in their housing by a soft plastic material. This material allows the switch to self-adjust when acted upon by gradual external forces and wire effects

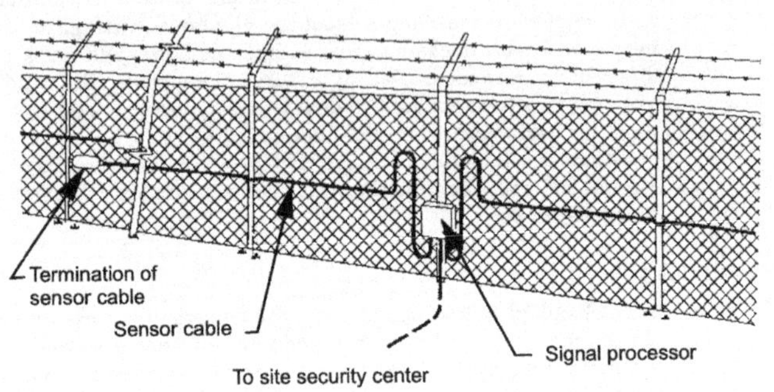


**Figure 6-11. Typical Strain-Sensor Cable Installation**

such as the relaxation of the wire with time and its thermal expansion or contraction.

- Strain-gauge detectors are attached to the taut wire with a nut on a threaded stud. When a force is applied to the taut wire, the resulting deflection is converted by the strain gauge into a change in electrical output that is monitored by a signal processor.

6-95. With sensors that use mechanical switches as detectors, the switches in a single sensor-post assembly are wired in parallel and are connected directly to the alarm-annunciation system. Pulse-count circuitry is not used because a single switch closure, such as that caused by an intruder moving or cutting one wire, is indicative of an intrusion attempt. Strain-gauge detectors in a sensor post are monitored by a signal processor. When the signal from one or more strain gauges satisfies the processor's criteria, an alarm is initiated.

6-96. The taut-wire sensor can be installed as a freestanding fence or can be mounted on an existing fence or wall. Figure 6-12, page 6-32, shows a freestanding configuration.

### Fiber-Optic Cable Sensors

6-97. Fiber-optic cable sensors are functionally equivalent to the strain-sensitive cable sensors previously discussed. However, rather than electrical signals, modulated light is transmitted down the cable and the resulting received signals are processed to determine whether an alarm should be initiated. Since the cable contains no metal and no electrical signal is present, fiber-optic sensors are generally less susceptible to electrical interference from lightning or other sources.

### Electric-Field Sensors

6-98. Electric-field sensors consist of an alternating-current (AC) field generator, one or more field wires, one or more sense wires, and a signal

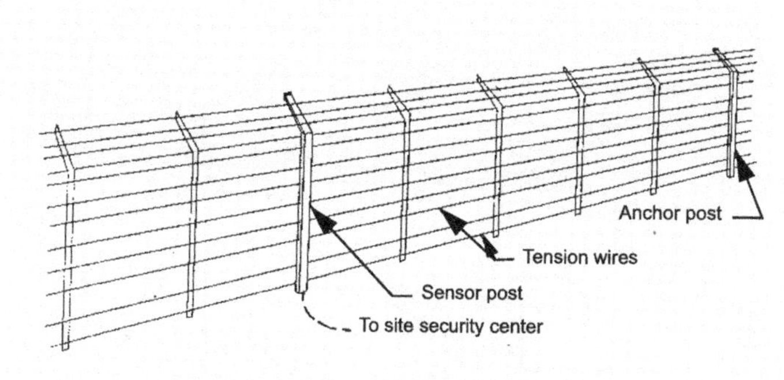


**Figure 6-12. Typical Taut-Wire Installation**

processor. The generator excites the field wires around which an electrostatic-field pattern is created. The electrostatic field induces electrical signals in the sense wires, which are monitored by the signal processor. Under normal operating conditions, the induced signals are constant. However, when an intruder approaches the sensor, the induced electrical signals are altered, causing the signal processor to generate an alarm.

6-99. Several different field- and sense-wire configurations are available. They range from one field wire and one sense wire to as many as four field wires and one sense wire or four field wires and four sense wires. Figure 6-13 shows the detection pattern produced by vertical three-wire (one field and two sense wires) configurations. The three-wire system has a wider detection envelope and is less costly (one less field wire and associated hardware). However, because of the tighter coupling between wires, the four-wire system is less susceptible to nuisance alarms caused by extraneous noise along the length of the zone.

6-100. A signal processor monitors the signals produced by the sense wires. The processor usually contains a band-pass filter that rejects high-frequency signals such as those caused by objects striking the wires. Additional criteria that must be satisfied before the processor initiates an alarm include signal amplitude and signal duration. By requiring the signal to be present for a preset amount of time, false alarms (such as those caused by birds flying through the detection pattern) can be minimized.

6-101. As with taut-wire sensors, electric-field sensors can be freestanding (mounted on their own posts) or attached by standoffs to an existing fence. They can also be configured to follow contours of the ground. The area under the sensor must be clear of vegetation, since vegetation near or touching sense wires can cause false alarms. These sensors can also be installed on the walls and roof of a building.

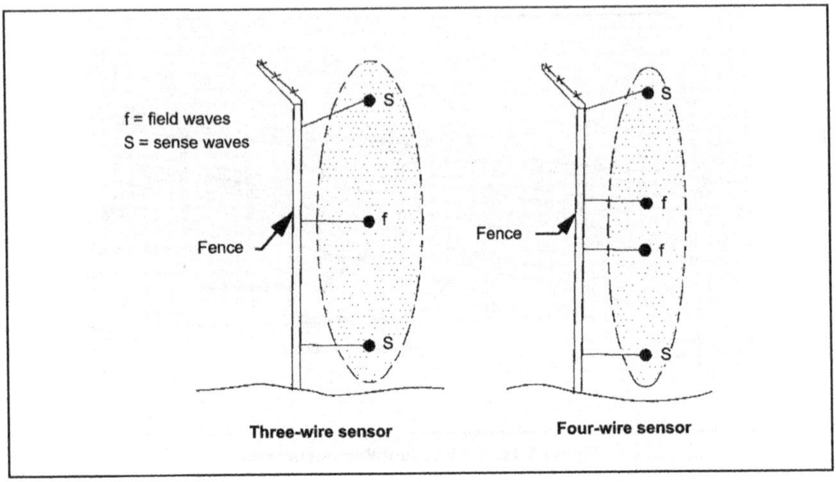


**Figure 6-13. Typical Electric-Field-Sensor Detection Patterns**

### Capacitance Proximity Sensors

6-102. Capacitance proximity sensors measure the electrical capacitance between the ground and an array of sense wires. Any variations in capacitance, such as that caused by an intruder approaching or touching one of the sense wires, initiates an alarm. These sensors usually consist of two or three wires attached to outriggers along the top of an existing fence, wall, or roof edge. Figure 6-14, page 6-34, shows a typical capacitance sensor consisting of three sensor wires attached to the outrigger of a fence. To minimize environmental alarms, the capacitance sensor is divided into two arrays of equal length. The signal processor monitors the capacitance of each array. Changes in capacitance common to both arrays (such as produced by wind, rain, ice, fog, and lightning) are canceled within the processor. However, when changes occur in one array and not the other because of an intruder, the processor initiates an alarm.

## BURIED-LINE SENSORS

6-103. A buried-line sensor system consists of detection probes or cable buried in the ground, typically between two fences that form an isolation zone. These devices are wired to an electronic processing unit. The processing unit generates an alarm if an intruder passes through the detection field. Buried-line sensors have several significant features:

- They are hidden, making them difficult to detect and circumvent.
- They follow the terrain's natural contour.
- They do not physically interfere with human activity, such as grass mowing or snow removal.

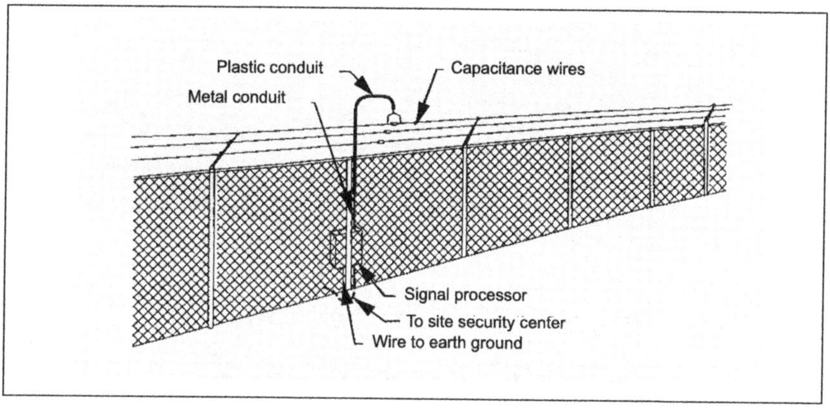


**Figure 6-14. Typical Capacitance-Sensor Configuration**

- They are affected by certain environmental conditions, such as running water and ground freeze/thaw cycles. (Seismic, seismic/magnetic, magnetic, and balanced pressure sensors are seldom used and will not be discussed here.)

6-104. The ported-coax cable sensor consists of two coax cables buried in the ground parallel to each other. An RF transmitter is connected to one cable and a receiver to the other. The outer conductor of each cable is ported (fabricated with small holes or gaps in the shield). The transmitter cable radiates RF energy into the medium surrounding the cables. A portion of this energy is coupled into the receiver cable through its ported shield. (Because of the ported shields, these cables are frequently referred to as leaky cables.) When an intruder enters the RF field, the coupling is disturbed, resulting in a change of signal monitored by the receiver, which then generates an alarm. Two basic types of ported-coax sensors are available—pulse and continuous wave.

- Pulse-type sensors transmit a pulse of RF energy down one cable and monitors the received signal on the other. The cables can be up to 10,000 feet long. The signal processor initiates an alarm when the electromagnetic field created by the pulse is disturbed and identifies the disturbance's approximate location.
- Continuous-wave sensors apply continuous RF energy to one cable. The signal received on the other cable is monitored for electromagnetic-field disturbances that indicate an intruder's presence. Cable lengths are limited to 300 to 500 feet. Additionally, the sensor is available in a single-cable configuration as well as two separate cables. The pattern typically extends 2 to 4 feet above the ground and can be 5 to 13 feet wide, depending on cable spacing and soil composition. Figure 6-15 represents a typical cross-section of a detection pattern created by a ported-cable sensor.

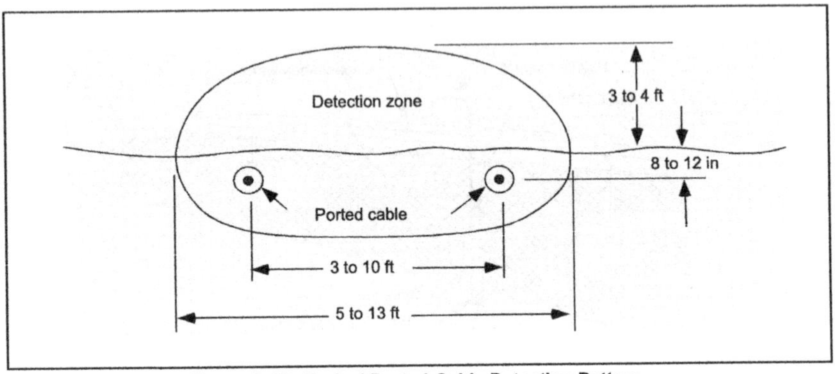


**Figure 6-15. Typical Ported-Cable Detection Pattern**

6-105. Sensor performance depends on properties of the medium surrounding the cables. Velocity and attenuation of the RF wave that propagates along the cables and the coupling between the cables are functions of the dielectric constant of the soil and its conductivity which, in turn, depends on its moisture content. For example, the velocity is greater and the attenuation is less for cables buried in dry, low-loss soil than in wet, conductive soil. Freeze/thaw cycles in the soil also affect the sensor's performance. When wet soil freezes, the wave velocity and the cable coupling increase and the attenuation decreases, resulting in greater detection sensitivity. Seasonal sensitivity adjustments may be necessary to compensate for changing ground conditions.

6-106. Although usually buried in soil, ported cables can also be used with asphalt and concrete. If the asphalt or concrete pavement area is relatively small and only a few inches thick (such as a pedestrian pavement crossing the perimeter), the ported cables can be routed under the pavement. However, for the large and deep pavements, slots must be cut into the asphalt or concrete to accept the cable.

6-107. A portable ported-coax sensor is available that can be rapidly deployed and removed. The cables are placed on the surface of the ground rather than buried. This sensor is useful for temporary perimeter detection coverage for small areas or objects (such as vehicles or aircraft).

## LOS SENSORS

6-108. The LOS sensors, which are mounted above ground, can be either active or passive. Active sensors generate a beam of energy and detect changes in the received energy that an intruder causes by penetrating the beam. Each sensor consists of a transmitter and a receiver and can be in a monostatic or bistatic configuration. Passive sensors generate no beam of energy; they simply look for changes in the thermal characteristics of their field of view. For effective detection, the terrain within the detection zone must be flat and free of obstacles and vegetation.

## Microwave Sensors

6-109. Microwave intrusion-detection sensors are categorized as bistatic or monostatic. Bistatic sensors use transmitting and receiving antennas located at opposite ends of the microwave link, whereas monostatic sensors use the same antenna.

- A bistatic system uses a transmitter and a receiver that are typically separated by 100 to 1,200 feet and that are within direct LOS with each other. The signal picked up by the receiver is the vector sum of the directly transmitted signal and signals that are reflected from the ground and nearby structures. Detection occurs when an object (intruder) moving within the beam pattern causes a change in net-vector summation of the received signals, resulting in variations of signal strength.
  - The same frequency bands allocated by the Federal Communications Commission (FCC) for interior microwave sensors are also used for exterior sensors. Because high-frequency microwave beams are more directive than low-frequency beams and the beam pattern is less affected by blowing grass in the area between the transmitter and the receiver, most exterior sensors operate at the next to highest allowable frequency, 10.525 gigahertz (GHz).
  - The shape of the microwave beam and the maximum separation between the transmitter and the receiver are functions of antenna size and configuration. Various antenna configurations are available, including parabolic-dish arrays, strip-line arrays, and slotted arrays. The parabolic antenna uses a microwave-feed assembly located at the focal point of a metallic parabolic reflector. A conical beam pattern is produced (see Figure 6-16). A strip-line antenna configuration produces a nonsymmetrical beam that is higher than its height. Larger antenna configurations generally produce narrower beam patterns.

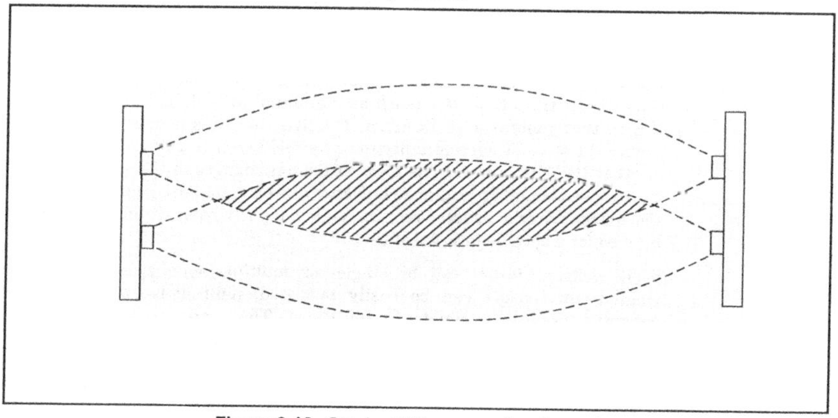

Figure 6-16. Stacked Microwave Configuration

- Monostatic microwave sensors use the same antenna or virtually coincident antenna arrays for the transmitter and receiver, which are usually combined into a single package. Two types of monostatic sensors are available. Amplitude-modulated (AM) sensors detect changes in the net-vector summation of reflected signals similar to bistatic sensors. Frequency-modulated (FM) sensors operate on the Doppler principle similar to interior microwave sensors. The detection pattern is typically shaped like a teardrop (see Figure 6-17). Monostatic sensors can provide volumetric coverage of localized areas, such as in corners or around the base of critical equipment.

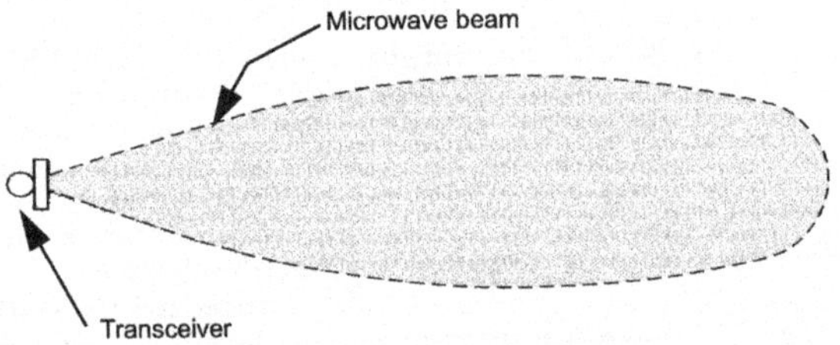


**Figure 6-17. Typical Monostatic-Microwave-Sensor Detection Pattern**

## IR Sensors

6-110. The IR sensors are available in both active and passive models. An active sensor generates one or more near-IR beams that generate an alarm when interrupted. A passive sensor detects changes in thermal IR radiation from objects located within its field of view.

6-111. Active sensors consist of transmitter/receiver pairs. The transmitter contains an IR light source (such as a gallium arsenide light-emitting diode [LED]) that generates an IR beam. The light source is usually modulated to reduce the sensor's susceptibility to unwanted alarms resulting from sunlight or other IR light sources. The receiver detects changes in the signal power of the received beam. To minimize nuisance alarms from birds or blowing debris, the alarm criteria usually require that a high percentage of the beam be blocked for a specific interval of time.

6-112. Active sensors can be single- or multiple-beam systems. Because single-beam sensors can be easily bypassed, multiple-beam systems are generally used in perimeter applications. There are two basic types of multiple-beam configurations—one type uses all transmitters on one post and all receivers on the other post; the second type uses one transmitter and several receivers on each post. Both types are illustrated in Figure 6-18, page 6-38.

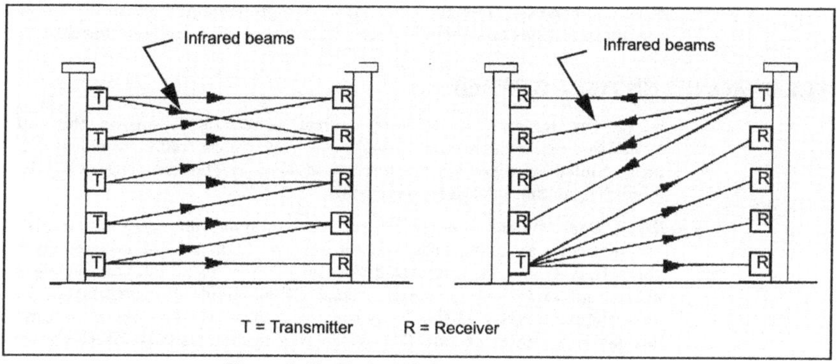


**Figure 6-18. Typical IR-Sensor Beam Patterns**

6-113. The spacing between transmitters and receivers can be as great as 1,000 feet when operation is under good weather conditions. However, conditions such as heavy rain, fog, snow, or blowing dust particles attenuate the IR energy, reducing its effective range to 100 to 200 feet or less.

## VIDEO MOTION SENSORS

6-114. A video motion sensor generates an alarm whenever an intruder enters a selected portion of a CCTV camera's field of view. The sensor processes and compares successive images from the camera and generates an alarm if differences between the images satisfy predefined criteria. Digital devices convert selected portions of the analog video signal into digital data that are compared with data converted previously; if differences exceed preset limits, an alarm is generated.

6-115. The signal processor usually provides an adjustable window that can be positioned anywhere on the video image. Available adjustments permit changing the window's horizontal and vertical sizes, its position, and its sensitivity. More sophisticated units provide several adjustable windows that can be individually sized and positioned. Multiple windows permit concentrating on several specific areas of an image while ignoring others. For example, in a scene that contains several critical assets and multiple sources of nuisance alarms (such as large bushes or trees), the sensor can be adjusted to monitor only the assets and ignore the areas that contain the nuisance-alarm sources.

6-116. The use of video motion-detection systems for exterior applications has been limited, primarily because of difficulties with uncontrolled exterior environments. Lighting variations caused by cloud movement and shadows of slow-moving objects, birds and animals moving within the camera's field of view, camera motion and moving vegetation during windy conditions, and severe weather conditions have traditionally caused a multitude of unwanted alarms in this type of system. Systems using more advanced signal-processing algorithms have improved motion-detection capability and nuisance-alarm rejection; however, they are still subject to high unwanted-alarm rates under certain conditions and should be used with due caution and extreme care.

## ELECTRONIC ENTRY CONTROL

6-117. The function of an entry-control system is to ensure that only authorized personnel are permitted into or out of a controlled area. Entry can be controlled by locked fence gates, locked doors to a building or rooms within a building, or specially designed portals.

6-118. These means of entry control can be applied manually by guards or automatically by using entry-control devices. In a manual system, guards verify that a person is authorized to enter an area, usually by comparing the photograph and personal characteristics of the individual requesting entry. In an automated system, the entry-control device verifies that a person is authorized to enter or exit. The automated system usually interfaces with locking mechanisms on doors or gates that open momentarily to permit passage. Mechanical hardware (such as locking mechanisms, electric door strikes, and specially designed portal hardware) and equipment used to detect contraband material (such as metal detectors, X-ray baggage-search systems, explosives detectors, and special nuclear-material monitors) are described in other documentation. Refer to TM 5-853-1 for additional information on determining entry-control requirements and integrating manual electronic-entry control into a cohesive system.

6-119. All entry-control systems control passage by using one or more of three basic techniques—something a person knows, something a person has, or something a person is or does. Automated entry-control devices based on these techniques are grouped into three categories—coded, credential, and biometric devices.

### CODED DEVICES

6-120. Coded devices operate on the principle that a person has been issued a code to enter into an entry-control device. This code will match the code stored in the device and permit entry. Depending on the application, a single code can be used by all persons authorized to enter the controlled area or each authorized person can be assigned a unique code. Group codes are useful when the group is small and controls are primarily for keeping out the general public. Individual codes are usually required for control of entry to more critical areas. Coded devices verify the entered code's authenticity, and any person entering a correct code is authorized to enter the controlled area. Electronically coded devices include electronic and computer-controlled keypads.

#### Electronic Keypad Devices

6-121. The common telephone keypad (12 keys) is an example of an electronic keypad. This type of keypad consists of simple push-button switches that, when depressed, are decoded by digital logic circuits. When the correct sequence of buttons is pushed, an electric signal unlocks the door for a few seconds.

### Computer-Controlled Keypad Devices

6-122. These devices are similar to electronic keypad devices, except they are equipped with a microprocessor in the keypad or in a separate enclosure at a different location. The microprocessor monitors the sequence in which the keys are depressed and may provide additional functions such as personal ID and digit scrambling. When the correct code is entered and all conditions are satisfied, an electric signal unlocks the door.

## CREDENTIAL DEVICES

6-123. A credential device identifies a person having legitimate authority to enter a controlled area. A coded credential (plastic card or key) contains a prerecorded, machine-readable code. An electric signal unlocks the door if the prerecorded code matches the code stored in the system when the card is read. Like coded devices, credential devices only authenticate the credential; it assumes a user with an acceptable credential is authorized to enter. Various technologies are used to store the code upon or within a card. Hollerith, optically coded, magnetic-spot, capacitance, and electric-circuit cards are seldom used and will not be discussed here. The most commonly used types of cards are described as follows:

### Magnetic-Stripe Card

6-124. A strip of magnetic material located along one edge of the card is encoded with data (sometimes encrypted). The data is read by moving the card past a magnetic read head.

### Wiegand-Effect Card

6-125. The Wiegand-effect card contains a series of small-diameter, parallel wires about one-half inch long, embedded in the bottom half of the card. The wires are manufactured from ferromagnetic materials that produce a sharp change in magnetic flux when exposed to a slowly changing magnetic field. This type of card is impervious to accidental erasure. The card reader contains a small read head and a tiny magnet to supply the applied magnetic field. It usually does not require external power.

### Proximity Card

6-126. A proximity card is not physically inserted into a reader; the coded pattern on the card is sensed when it is brought within several inches of the reader. Several techniques are used to code cards. One technique uses a number of electrically tuned circuits embedded in the card. Data are encoded by varying resonant frequencies of the tuned circuits. The reader contains a transmitter that continually sweeps through a specified range of frequencies and a receiver that senses the pattern of resonant frequencies contained in the card. Another technique uses an integrated circuit embedded in the card to generate a code that can be magnetically or electrostatically coupled to the reader. The power required to activate embedded circuitry can be provided by a small battery embedded in the card or by magnetically coupling power from the reader.

### Laser Card

6-127. The optical memory card, commonly called the laser card, uses the same technology developed for recording video and audio disks for entertainment purposes. Data is recorded on the card by burning a microscopic hole (using a laser) in a thin film covering the card. Data is read by using a laser to sense the hole locations. The typical laser card can hold several megabytes of user data.

### Smart Card

6-128. A smart card is embedded with a microprocessor, memory, communication circuitry, and a battery. The card contains edge contacts that enable a reader to communicate with the microprocessor. Entry-control information and other data may be stored in the microprocessor's memory.

### Bar Code

6-129. A bar code consists of black bars printed on white paper or tape that can be easily read with an optical scanner. This type of coding is not widely used for entry-control applications because it can be easily duplicated. It is possible to conceal the code by applying an opaque mask over it. In this approach, an IR scanner is used to interpret the printed code. For low-level security areas, the use of bar codes can provide a cost-effective solution for entry control. Coded strips and opaque masks can be attached to existing ID badges, alleviating the need for complete badge replacement.

## BIOMETRIC DEVICES

6-130. The third basic technique used to control entry is based on the measurement of one or more physical or personal characteristics of an individual. Because most entry-control devices based on this technique rely on measurements of biological characteristics, they have become commonly known as biometric devices. Characteristics such as fingerprints, hand geometry, voiceprints, handwriting, and retinal blood-vessel patterns have been used for controlling entry. Typically, in enrolling individuals, several reference measurements are made of the selected characteristic and then stored in the device's memory or on a card. From then on, when that person attempts entry, a scan of the characteristic is compared with the reference data template. If a match is found, entry is granted. Rather then verifying an artifact, such as a code or a credential, biometric devices verify a person's physical characteristic, thus providing a form of identity verification. Because of this, biometric devices are sometimes referred to as personnel identity-verification devices. The most common biometric devices are discussed below.

### Fingerprints

6-131. Fingerprint-verification devices use one of two approaches. One is pattern recognition of the whorls, loops, and tilts of the referenced fingerprint, which is stored in a digitized representation of the image and compared with the fingerprint of the prospective entrant. The second approach is minutiae comparison, which means that the endings and branching points of ridges and valleys of the referenced fingerprint are compared with the fingerprint of the prospective entrant.

### Hand Geometry

6-132. Several devices are available that use hand geometry for personnel verification. These devices use a variety of physical measurements of the hand, such as finger length, finger curvature, hand width, webbing between fingers, and light transmissivity through the skin to verify identity. Both two- and three-dimensional units are available.

### Retinal Patterns

6-133. This type of technique is based on the premise that the pattern of blood vessels on the human eye's retina is unique to an individual. While the eye is focused on a visual target, a low-intensity IR light beam scans a circular area of the retina. The amount of light reflected from the eye is recorded as the beam progresses around the circular path. Reflected light is modulated by the difference in reflectivity between blood-vessel pattern and adjacent tissue. This information is processed and converted to a digital template that is stored as the eye's signature. Users are allowed to wear contact lenses; however, glasses should be removed.

### Device Combinations

6-134. Frequently, an automated entry-control system uses combinations of the three types of entry-control devices. Combining two different devices can significantly enhance the system's security level. In some cases, combining devices results in reduced verification times.

## APPLICATION GUIDELINES

6-135. The primary function of an automated entry-control system is to permit authorized personnel to enter or exit a controlled area. Features available to the designer are described below.

- **Enrollment.** All entry-control systems must provide a means of entering, updating, and deleting information about authorized individuals into the system's database files. This is usually accomplished with a dedicated enrollment station for enrolling and disenrolling purposes that is directly connected to the central-processing unit. When credential devices are used, all authorized users must be provided with an appropriate credential. A means should also be provided to disenroll a person quickly without having to retrieve a credential. When using biometric devices, additional enrollment equipment will be required.
- **Entry-control techniques.** Some entry-control functions require additional hardware, while others are accomplished with software. Those features accomplished with software require that the appropriate database be available for every portal affected by them. Typically, these techniques include—
  - Area zones.
  - Time zones.
  - Team zones.
  - Anti-pass back.

  - Antitailgate.
  - Guard tour.
  - Elevator control.
- **Alarms.** Several types of alarms can be used with an entry-control system. These alarms must annunciate audibly and visually in the security center.
- **Entry denial.** Most entry-control devices are configured to permit the user three entry attempts. If more than three unsuccessful entry attempts are made within a specified period, the device generates an alarm. An alarm is also generated if an invalid credential is used or attempted entries are detected that violate specified area, time, or team zoning requirements.
- **Communication failure.** This alarm is generated when a loss of communication between the central processor and the local equipment is detected.
- **Portal open.** If a portal door remains open longer than a predefined time, an alarm is generated.
- **Duress.** This alarm is generated when a special duress code is entered at a keypad.
- **Guard overdue.** This duress alarm is generated when a security guard is determined to be overdue at a checkpoint during a predefined guard tour.
- **Software tamper.** This type of alarm is generated when unauthorized persons are detected attempting to invoke certain system commands or modify database files.

## PERFORMANCE CRITERIA

6-136. The overall performance of an entry-control system can be evaluated by examining the verification error rate and the throughput rate. An entry-control system can produce two types of errors—denial of admission of a person who should be admitted or admission of a person who should not be admitted. These are commonly referred to as false-reject errors (type I errors) and false-accept errors (type II errors). Although a false-reject error does not constitute a breach of security, it does create an operational problem that must be handled by an alternative method. False-accept errors constitute a breach of security. Ideally, both false-reject and false-accept error rates should be zero; in practice, however, they are not. In fact, they tend to act in opposition to each other. When the system is adjusted to minimize the false-accept error rate, the false-reject error rate usually increases. Verification error rates are typically measured in percent (number of errors/number of attempts x 100 percent). These error rates are typically very low for coded and credential devices, but many become significant if biometric devices are used.

6-137. The throughput rate is the number of persons that can pass through an entry point in a given unit of time and is usually expressed in persons per minute. It is the time required to approach the entry-control device and for the device to verify information (verification time) and the time required passing through the entry point. Typically, an individual can approach the device and pass through in 3 to 5 seconds. Verification time depends on the type of device and may vary from 3 to 15 seconds. Table 6-5 provides a list of typical verification times for different types of entry-control devices.

**Table 6-5. Typical Verification Times of Entry-Control Devices**

| Device | Verification Time |
|---|---|
| Keypad | 3 seconds |
| Card reader | 3 seconds |
| Keypad/card reader | 6 seconds |
| Biometric/keypad | 6 to 15 seconds |
| Biometric/card reader | 6 to 15 seconds |
| Biometric | 2 minutes |

## DATA TRANSMISSION

6-138. A critical element in an integrated ESS is the DTM that transmits information from sensors, entry-control devices, and video components to display and assessment equipment. A DTM link is a path for transmitting data between two or more components (such as a sensor and alarm reporting system, a card reader and controller, a CCTV camera and monitor, or a transmitter and receiver). The DTM links connect remote ESS components to the security center. An effective DTM link ensures rapid and reliable transmission media, transmission technique, associated transmission hardware, and degree of security to be provided for the communication system.

6-139. A number of different media are used in transmitting data between elements of an IDS, an EECS, and a CCTV system. These include wire lines, coaxial cable, fiber-optic cable, and RF transmission.

- **Wire line.** Wire lines are twisted pairs that consist of two insulated conductors twisted together to minimize interference by unwanted signals.
- **Coaxial cable.** Coaxial cable consists of a center conductor surrounded by a shield. The center conductor is separated from the shield by a dielectric. The shield protects against electromagnetic interference.
- **Fiber optics.** Fiber optics uses the wide bandwidth properties of light traveling through transparent fibers. Fiber optics is a reliable communication medium best suited for point-to-point, high-speed data transmission. Fiber optics is immune to RF electromagnetic interference and does not produce electromagnetic radiation emission. The preferred DTM for an ESS is fiber-optic cables unless there are justifiable economic or technical reasons for using other types of media.
- **RF transmission.** Modulated RF can be used as a DTM with the installation of radio receivers and transmitters. An RF transmission system does not require a direct physical link between the points of communication, and it is useful for communicating over barriers such as bodies of water and heavily forested terrain. A disadvantage is that the signal power received depends on many factors (including transmission power, antenna pattern, path length, physical

obstructions, and climatic conditions). Also, RF transmission is susceptible to jamming and an adversary with an appropriately tuned receiver has access to it. The use of RF will be coordinated with the communications officer to avoid interference with other existing or planned facility RF systems.

6-140. There are two basic types of communication links—point-to-point and multiplex lines. A point-to-point link is characterized by a separate path for each pair of components. This approach is cost effective for several component pairs or when a number of scattered remote areas communicate with a single central location. The multiplex link, commonly referred to as a multidrop or multipoint link, is a path shared by a number of components. Depending on the number and location of components, this type of configuration can reduce the amount of cabling required. However, the cost reduction from reduced cabling is somewhat offset by costs of equipment required to multiplex and demultiplex data.

6-141. Data links used to communicate the status of ESS devices or other sensitive information to the security center must be protected from possible compromise. Attempts to defeat the security system may range from simple efforts to cut or short the transmission line to more sophisticated undertakings, such as tapping and substituting bogus signals. Data links can be made more secure by physical protection, tamper protection, line supervision, and encryption.

## CCTV FOR ALARM ASSESSMENT AND SURVEILLANCE

6-142. A properly integrated CCTV assessment system provides a rapid and cost-effective method for determining the cause of intrusion alarms. For surveillance, a properly designed CCTV system provides a cost-effective supplement to guard patrols. For large facilities, the cost of a CCTV system is more easily justified. It is important to recognize that CCTV alarm-assessment systems and CCTV surveillance systems perform separate and distinct functions. The alarm-assessment system is designed to respond rapidly, automatically, and predictably to the receipt of ESS alarms at the security center. The surveillance system is designed to be used at the discretion of and under the control of the security center's console operator. When the primary function of the CCTV system is to provide real-time alarm assessment, the design should incorporate a video-processing system that can communicate with the alarm-processing system.

6-143. A candidate site for a CCTV assessment system will typically have the following characteristics:

- Assets requiring ESS protection.
- A need for real-time alarm assessment.
- Protected assets spaced some distance apart.

6-144. Figure 6-19, page 6-46, shows a typical CCTV system configuration. A typical site will locate CCTV cameras—

- Outdoors, along site-perimeter isolation zones.
- Outdoors, at controlled access points (sally ports).

- Outdoors, within the protected area, and at viewing approaches to selected assets.
- Indoors, at selected assets within the protected area.

6-145. The security console is centrally located in the security center. The CCTV monitors and the ancillary video equipment will be located at this console, as will the ESS alarm-processing and -annunciation equipment.

## CCTV CAMERA COMPONENTS

6-146. An optical-lens system that captures and focuses reflected light from the scene being viewed onto an image target is common to all CCTV cameras. The image target converts reflected light energy into electrical impulses in a two-dimensional array of height and width. An electronic scanning system (reading these impulses in a predetermined order) creates a time-sensitive voltage signal that is a replica of optical information captured by the lens and focused on the target. This voltage signal is then transmitted to a location where it is viewed and possibly recorded. For components and technical information regarding CCTV cameras, see the appropriate TMs.

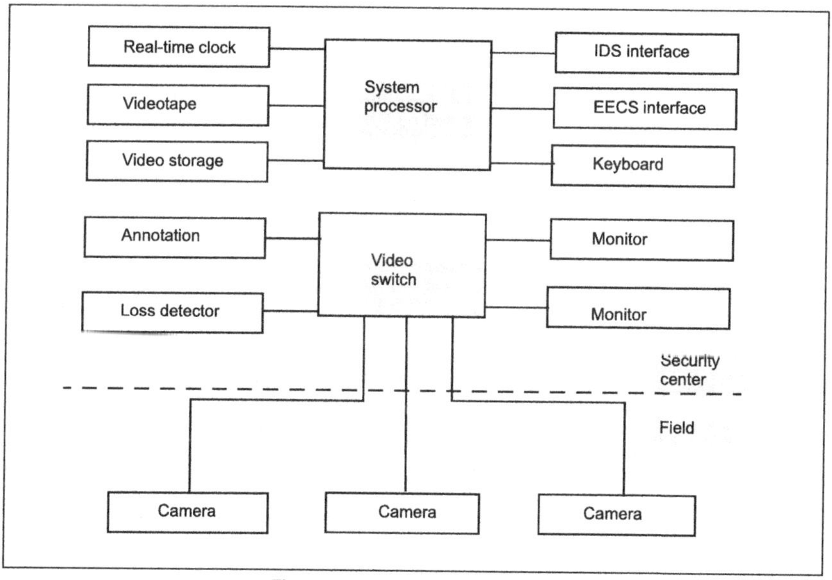


**Figure 6-19. Typical CCTV System**

## VIDEO SIGNAL AND CONTROL LINKS

6-147. A CCTV transmission system is needed to convey video signals from various facility cameras to the security center and to carry commands from the security center to the cameras. Information may be sent via metallic cable, RF, or optical transmission.

### Metallic Cable

6-148. Metallic video cables are electrical conductors manufactured specifically for the transmission of frequencies associated with video components. Coaxial cable is a primary example of this type of transmission media. Devices such as video-equalization amplifiers, ground loop correctors, and video-distribution amplifiers may be required.

### RF Transmission

6-149. For a system that has widely separated nodes, RF transmission may be a good alternative to metallic cable and associated amplifiers. The information can be transmitted over a microwave link. A microwave link can be used for distances of about 50 miles, as long as the receiver and the transmitter are in the LOS.

### Fiber-Optic Cable

6-150. In fiber-optic cable systems, electrical video signals are converted to optical light signals that are transmitted down the optical fiber. The signal is received and reconverted into electrical energy. An optic driver and a receiver are required per fiber. The fiber-optic transmission method provides a low-loss, high-resolution transmission system with usable length three to ten times that of traditional metallic in cable systems. Fiber-optic cable is the transmission media favored by DA.

## CCTV-SYSTEM SYNCHRONIZATION

6-151. Timing signals are processed within the image-scan section of the CCTV camera. These signals may be generated internally from a crystal clock, derived from the camera's AC power source, or supplied by an external signal source. The camera should be capable of automatic switchover to its internal clock in case of external signal loss. When CCTV cameras are supplied by a common external (master) signal source or are all powered from the same AC power source, all cameras scan in synchronism. In this case, a console CCTV monitor will display a smooth transition when switched from one video source to another. Without this feature, the monitor display breaks up or rolls when switched between video sources. The rolling occurs for as long as it takes the monitor to synchronize its scan with that of the new video source, typically one second. The resynchronization delay will be experienced by all system components that receive video information, including recorders. To avoid this delay, the designer must specify that all cameras are powered from the AC power phase or must specify master synchronization for the design.

## VIDEO PROCESSING AND DISPLAY COMPONENTS

6-152. As shown in Figure 6-19, page 6-46, CCTV camera signals propagate through the video transmission system and through coverage at the security center. In very simple configurations with only a few cameras and monitors, a hardwired connection between each camera and console monitor is adequate. As the number of cameras increases, the need to manage and add supplemental information to camera signals also increases. Psychological testing has demonstrated that the efficiency of console-operator assessment improves as the number of console monitors is reduced, with the optimum number being four to six monitors. Effectiveness is also enhanced by the use of alarm-correlated video. Major components of the video-processor system are the video switcher, the video-loss detector, the alarm-processor communication path, the master video-sync generator, video recorders, and monitors.

- **Video switchers.** Video switchers are required when the number of cameras exceeds the number of console monitors or when a monitor must be capable of selecting video from one of many sources. Video switchers are capable of presenting any of multiple video images to various monitors, recorders, and so forth.
- **Video-loss detector.** Video-loss detectors sense the continued integrity of incoming camera signals.
- **ESS interface and communication path.** There must be a means of rapid communication between the ESS alarm-annunciation and video-processor systems. The alarm processor must send commands that cause the video switcher to select the camera appropriate for the sensor reporting an alarm. The video-processor system must report system tampering or failures (such as loss of video) to the alarm processor. The path should also pass date-and-time synchronizing information between processors so that recorded video scenes and printed alarm logs are properly correlated.
- **Master video-sync generation and distribution.** Master video sync includes a crystal-controlled timing generator, distribution amplifiers, and a transmission link to each camera.
- **Video recorders.** Video recorders provide the means to record alarm-event scenes in real time for later analysis. A recorder typically receives its input through dedicated video-switcher outputs. To support recorder playback, the recorder output is connected to a dedicated switcher input and must be compatible with the switcher-signal format. In addition, the recorder receives start commands from the switcher, and compatibility must exist at this interface. Videocassette recorders should be used when alarm events are to be recorded for later playback and analysis. The cassettes can record in time lapse for up to 240 hours (depending on the user-selected speed) and will change to real-time recording on command. The cassettes can be erased and reused or archived if required.
- **Monitors.** Monitors are required to display the individual scenes transmitted from the cameras or from the video switcher. In alarm-assessment applications, the monitors are driven by dedicated outputs of the video switcher and the monitors display video sources selected by the switcher. For security-console operations, the 9-inch monitor is the

smallest screen that should be used for operator recognition of small objects in a camera's field of view. Two 9-inch monitors can be housed side by side in a standard 19-inch console. If the monitors are to be mounted in freestanding racks behind the security console, larger units will be used.

6-153. Video-processor equipment will be specified to append the following alphanumeric information so that it appears on both monitors and recordings. The equipment must allow the operator to program the annotated information and dictate its position on the screen. This information includes—

- Time and date information.
- Video-source or alarm-zone identification.
- Programmable titles.

## CCTV APPLICATION GUIDELINES

6-154. Site-specific factors must be taken into consideration in selecting components that comprise a particular CCTV system. The first is the system's size in terms of the number of cameras fielded, which is the minimum number needed to view all ESS sensor-detection fields and surveillance cameras. Another factor is that some CCTV cameras may require artificial light sources. Finally, there are CCTV-system performance criteria and physical, environmental, and economic considerations. Each is discussed in detail in TM 5-853-4.

### Scene Resolution

6-155. The level to which video details can be determined in a CCTV scene is referred to as resolving ability or resolution. It is generally accepted that for assessment purposes, three resolution requirements can be defined. In order of increasing resolution requirements, they are detection, recognition, and identification.

- Detection is the ability to detect the presence of an object in a CCTV scene.
- Recognition is the ability to determine the type of object in a CCTV scene (animal, blowing debris, or crawling human).
- Identification is the ability to determine object details (a particular person, a large rabbit, a small deer, or tumbleweed).

6-156. A CCTV assessment system should provide sufficient resolution to recognize human presence and to detect small animals or blowing debris. Given an alarmed intrusion sensor, it is crucial that the console operator be able to determine if the sensor detected an intruder or if it is simply responding to a nuisance condition. (Refer to TM 5-853-4 for detailed design applications.)

### Illumination Levels

6-157. For interior applications where the same camera type is used in several different areas and the scene illumination in each area is constant, specify the manually adjustable iris. This allows a manual iris adjustment appropriate for each particular area's illumination level at the time of installation. If the camera must operate in an area subject to a wide dynamic range of illumination levels (such as would be found outdoors), specify the automatically adjusted iris feature.

### Cost Considerations

6-158. The cost of a CCTV system is usually quoted as cost-per-assessment zone. When estimating the total system cost, video-processor equipment costs and the video-transmission system's costs must be included. Other potentially significant costs are outdoor lighting system upgrades and the site preparation required to support the CCTV cameras. The CCTV systems are expensive compared to other electronic security subsystems and should be specified with discretion.

## DESIGN GUIDELINES

6-159. The design and application of CCTV systems are quite complex and should be left to professionals who are abreast of the current state-of-the-art systems. Some of the general design guidelines include the following:

- **System familiarity.** Before designing an effective CCTV assessment system, the designer must be familiar with the ESS's sensor placement and the detection field's shape.
- **CCTV camera placement and lighting.** The placement of exterior cameras requires more attention than that of interior cameras because of weather and illumination extremes. The field-of-view alignment, illumination range, and balanced lighting are major design factors. Exterior CCTV design considerations include environmental housings, camera mounting heights, system types, and so forth. Indoor design considerations include the mounting location and tamper detection. The layout for indoor alarm-assessment cameras is subject to three constraints—
  - The camera's location should enclose the complete sensor detection field in the camera's field of view.
  - Lighting that is adequate to support alarm assessment will be provided.
  - Protection from tampering and inadvertent damage by collision during normal area operations will be provided.

# Chapter 7

# Access Control

Perimeter barriers, intrusion-detection devices, and protective lighting provide physical-security safeguards; however, they alone are not enough. An access-control system must be established and maintained to preclude unauthorized entry. Effective access-control procedures prevent the introduction of harmful devices, materiel, and components. They minimize the misappropriation, pilferage, or compromise of materiel or recorded information by controlling packages, materiel, and property movement. Access-control rosters, personal recognition, ID cards, badge-exchange procedures, and personnel escorts all contribute to an effective access-control system.

## DESIGNATED RESTRICTED AREAS

7-1. The installation commander is responsible for designating and establishing restricted areas. A restricted area is any area that is subject to special restrictions or controls for security reasons. This does not include areas over which aircraft flight is restricted. Restricted areas may be established for the following:

- The enforcement of security measures and the exclusion of unauthorized personnel.
- Intensified controls in areas requiring special protection.
- The protection of classified information or critical equipment or materials.

### DEGREE OF SECURITY

7-2. The degree of security and control required depends on the nature, sensitivity, or importance of the security interest. Restricted areas are classified as controlled, limited, or exclusion areas.

- A controlled area is that portion of a restricted area usually near or surrounding a limited or exclusion area. Entry to the controlled area is restricted to personnel with a need for access. Movement of authorized personnel within this area is not necessarily controlled since mere entry to the area does not provide access to the security interest. The controlled area is provided for administrative control, for safety, or as a buffer zone for in-depth security for the limited or exclusion area. The commander establishes the control of movement.
- A limited area is a restricted area within close proximity of a security interest. Uncontrolled movement may permit access to the item. Escorts and other internal restrictions may prevent access within limited areas.

- An exclusion area is a restricted area containing a security interest. Uncontrolled movement permits direct access to the item.

7-3. The security protection afforded by a restricted area pertains particularly to subversive-activity control; that is, protection against espionage, sabotage, or any such action adversely affecting national defense. Within this context, the designation "restricted area" is not applicable to an area solely for protection against common pilferage or misappropriation of property or material that is not classified or not essential to national defense. For example, an area devoted to the storage or use of classified documents, equipment, or materials should be designated as a restricted area to safeguard against espionage. An installation communications center should also be so designated to safeguard against sabotage. On the other hand, a cashier's cage or an ordinary mechanic's tool room should not be so designated, although the commander may impose controls to access. This may be a simple matter of posting an "off limits to unauthorized personnel" sign. The PM or the physical-security manager acts as an advisor to the commander. In his recommendations, he must consider evaluating the purpose of designating a restricted area and coordinating with the intelligence officer and the staff judge advocate (SJA).

7-4. A restricted area must be designated in writing by the commander and must be posted with warning signs according to AR 190-13. In areas where English is one of two or more languages commonly spoken, warning signs will be posted in English and in the local language (see Figure 7-1).

7-5. An installation may have varying degrees of security. It may be designated in its entirety as a restricted area, with no further restrictions; or it may be subdivided into controlled, limited, or exclusion areas with restrictions of movement and specific clear zones. Figure 7-2 depicts a simplified restricted area and the degrees of security.

## CONSIDERATIONS

7-6. There are other important considerations concerning restricted areas and their lines of division. These considerations include the following:

- A survey and analysis of the installation, its missions, and its security interests. This can determine immediate and anticipated needs that require protection. Anticipated needs are determined from plans for the future.
- The size and nature of the security interest being protected. Safes may provide adequate protection for classified documents and small items; however, large items may have to be placed within guarded enclosures.
- Some security interests are more sensitive to compromise than others. Brief observation or a simple act by an untrained person may constitute a compromise in some cases. In others, detailed study and planned action by an expert may be required.
- All security interests should be evaluated according to their importance. This may be indicated by a security classification such as confidential, secret, or top secret.

**RESTRICTED AREA**

THIS (INSTALLATION OR ACTIVITY) HAS BEEN DECLARED A RESTRICTED AREA BY AUTHORITY OF (TITLE, COMMANDING GENERAL OR COMMANDING OFFICER) IN ACCORDANCE WITH THE PROVISIONS OF THE DIRECTIVE ISSUED BY THE SECRETARY OF DEFENSE ON 20 AUGUST 1954, PURSUANT TO THE PROVISIONS OF SECTION 21, INTERNAL SECURITY ACT OF 1950. UNATHORIZED ENTRY IS PROHIBITED.

ALL PERSONS AND VEHICLES ENTERING HEREIN ARE LIABLE TO SEARCH. PHOTOGRAPHING OR MAKING NOTES, DRAWINGS, MAPS, OR GRAPHIC REPRESENTATIVES OF THIS AREA OR ITS ACTIVITIES ARE PROHIBITED UNLESS SPECIFICALLY AUTHORIZED BY THE COMMANDER. ANY SUCH MATERIAL FOUND IN THE POSSESSION OF UNAUTHORIZED PERSONS WILL BE CONFISCATED.

**Figure 7-1. Sample Restricted-Area Warning**

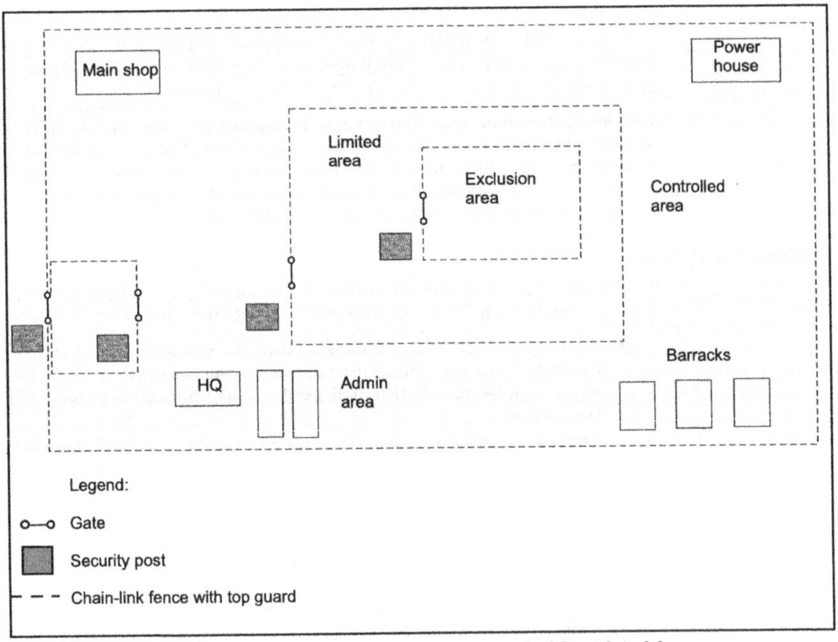


**Figure 7-2. Schematic Diagram of a Simplified Restricted Area and the Degrees of Security**

- Parking areas for privately owned vehicles (POVs) are established outside of restricted areas. Vehicle entrances must be kept at a minimum for safe and efficient control.
- Physical protective measures (such as fences, gates, and window bars) must be installed.

## EMPLOYEE SCREENING

7-7. Screening job applicants to eliminate potential acts of espionage and sabotage and other security risks is important in peacetime and is critical during a national emergency. Personnel screenings must be incorporated into standard personnel policies.

7-8. An applicant should be required to complete a personnel security questionnaire, which is then screened for completeness and used to eliminate undesirable applicants. A careful investigation should be conducted to ensure that the applicant's character, associations, and suitability for employment are satisfactory. The following sources may be helpful in securing employment investigative data:

- State and local police (including national and local police in overseas areas).
- Former employers.
- Public records.
- Credit agencies.
- Schools (all levels).
- References. (These references should include those names not furnished by the applicant. These are known as throw offs, and they are obtained during interviews of references furnished by applicants.)
- Others as appropriate. (These may include the FBI, the US Army Criminal Records Repository, and the Defense Investigative Agency).

7-9. Medical screening considerations should be made (based on an applicant's position [such as a guard]) to evaluate physical and mental stamina. Once an applicant has been identified for employment, he is placed on an access-control roster.

## IDENTIFICATION SYSTEM

7-10. An ID system is established at each installation or facility to provide a method of identifying personnel. The system provides for personal recognition and the use of security ID cards or badges to aid in the control and movement of personnel activities.

7-11. Standard ID cards are generally acceptable for access into areas that are unrestricted and have no security interest. Personnel requiring access to restricted areas should be issued a security ID card or badge as prescribed in AR 600-8-14. The card's/badge's design must be simple and provide for adequate control of personnel.

7-12. A security ID card/badge system must be established for restricted areas with 30 or more employees per shift. Commanders may (at their discretion) authorize a card/badge system in restricted areas for less than 30 people.

## ID METHODS

7-13. Four of the most commonly used access-control ID methods are the personal-recognition system, the single-card or -badge system, the card- or badge-exchange system, and the multiple-card or -badge system.

### Personal-Recognition System

7-14. The personal-recognition system is the simplest of all systems. A member of the security force providing access control visually checks the person requesting entry. Entry is granted based on—

- The individual being recognized.
- The need to enter has been established.
- The person is on an access-control roster.

### Single-Card or -Badge System

7-15. This system reflects permission to enter specific areas by the badge depicting specific letters, numbers, or particular colors. This system lends to comparatively loose control and is not recommended for high-security areas. Permission to enter specific areas does not always go with the need to know. Because the ID cards/badges frequently remain in the bearer's possession while off duty, it affords the opportunity for alteration or duplication.

### Card- or Badge-Exchange System

7-16. In this system, two cards/badges contain identical photographs. Each card/badge has a different background color, or one card/badge has an overprint. One card/badge is presented at the entrance to a specific area and exchanged for the second card/badge, which is worn or carried while in that area. Individual possession of the second card/badge occurs only while the bearer is in the area for which it was issued. When leaving the area, the second card/badge is returned and maintained in the security area. This method provides a greater degree of security and decreases the possibility of forgery, alteration, or duplication of the card/badge. The levels of protection described in TM 5-853-1 require multiple access-control elements as the levels of protection increase. In the case of the badge exchange, this system counts as two access-control elements.

### Multiple-Card or -Badge System

7-17. This system provides the greatest degree of security. Instead of having specific markings on the cards/badges denoting permission to enter various restricted areas, the multiple card/badge system makes an exchange at the entrance to each security area. The card/badge information is identical and allows for comparisons. Exchange cards/badges are maintained at each area only for individuals who have access to the specific area.

## MECHANIZED/AUTOMATED SYSTEMS

7-18. An alternative to using guards or military police (MP) to visually check cards/badges and access rosters is to use building card-access systems or biometric-access readers. These systems can control the flow of personnel entering and exiting a complex. Included in these systems are—

- Coded devices such as mechanical or electronic keypads or combination locks.
- Credential devices such as magnetic-strip or proximity card readers.
- Biometric devices such as fingerprint readers or retina scanners.

7-19. Access-control and ID systems base their judgment factor on a remote capability through a routine discriminating device for positive ID. These systems do not require guards at entry points; they identify an individual in the following manner:

- The system receives physical ID data from an individual.
- The data is encoded and compared to stored information.
- The system determines whether access is authorized.
- The information is translated into readable results.

7-20. Specialized mechanical systems are ideal for highly sensitive situations because they use a controlled process in a controlled environment to establish the required database and accuracy. One innovative technique applied to ID and admittance procedures involves dimension comparisons. The dimension of a person's full hand is compared to previously stored data to determine entry authorization. Other specialized machine readers can scan a single fingerprint or an eye retina and provide positive ID of anyone attempting entry.

7-21. An all-inclusive automated ID and access-control system reinforces the security in-depth ring through its easy and rapid change capability. The computer is able to do this through its memory. Changes can be made quickly by the system's administrator.

7-22. The commercial security market has a wide range of mechanized and automated hardware and software systems. Automated equipment is chosen only after considering the security needs and the environment in which it operates. These considerations include whether the equipment is outdoors or indoors, the temperature range, and weather conditions. Assessment of security needs and the use of planning, programming, and budgeting procedures greatly assist a security manager in improving the security posture.

## CARD/BADGE SPECIFICATIONS

7-23. Security cards/badges should be designed and constructed to meet the requirements of AR 600-8-14. Upon issuing a card/badge, security personnel must explain to the bearer the wear required and the authorizations allowed with the card/badge. This includes—

- Designation of the areas where an ID card/badge is required.

- A description of the type of card/badge in use and the authorizations and limitations placed on the bearer.
- The required presentation of the card/badge when entering or leaving each area during all hours of the day.
- Details of when, where, and how the card/badge should be worn, displayed, or carried.
- Procedures to follow in case of loss or damage of the card.
- The disposition of the card/badge upon termination of employment, investigations, or personnel actions.
- Prerequisites for reissuing the card/badge.

## VISITOR IDENTIFICATION AND CONTROL

7-24. Procedures must be implemented to properly identify and control personnel. This includes visitors presenting their cards/badges to guards at entrances of restricted areas. Visitors are required to stay with their assigned escort. Guards must ensure that visitors stay in areas relating to their visit; an uncontrolled visitor, although conspicuously identified, could acquire information for which he is not authorized. Foreign-national visitors should be escorted at all times.

7-25. Approval for visitors should be obtained at least 24 hours in advance (if possible). Where appropriate, the installation should prepare an agenda for the visitor and designate an escort officer. Measures must be in place to recover visitor cards/badges on the visit's expiration or when they are no longer required.

7-26. Physical-security precautions against pilferage, espionage, and sabotage require the screening, ID, and control of visitors. Further information about visiting requirements and procedures are found in ARs 12-15 and 381-20. Visitors are generally classed in the following categories:

- Persons with whom every installation or facility has business (such as suppliers, customers, insurance inspectors, and government inspectors).
- Individuals or groups who desire to visit an installation or facility for personal or educational reasons. Such visits may be desired by educational, technical, or scientific organizations.
- Individuals or groups specifically sponsored by the government (such as foreign nationals visiting under technical cooperation programs and similar visits by US nationals). Requests for visits by foreign nationals must be processed according to AR 380-10.
- Guided tours to selected portions of the installation in the interest of public relations.

7-27. The ID and control mechanisms for visitors must be in place. They may include the following:

- Methods of establishing the authority for admitting visitors and any limitations relative to access.

- Positive ID of visitors by personal recognition, visitor permit, or other identifying credentials. Contact the employer, supervisor, or officer in charge to validate the visit.
- The use of visitor registration forms. These forms provide a record of the visitor and the time, location, and duration of his visit.
- The use of visitor ID cards/badges. The cards/badges bear serial numbers, the area or areas to which access is authorized, the bearer's name, and escort requirements.

7-28. Individual groups entering a restricted area must meet specific prerequisites before being granted access. The following guidance is for group access into a restricted area:

### Visitors

7-29. Before allowing visitors into a restricted area, contact the person or activity being visited. After verifying the visitor's identity, issue a badge, complete the registration forms, and assign an escort (if required). Visitors may include public-utility and commercial-service representatives.

### Very Important Persons

7-30. The procedures for admitting very important persons (VIPs) and foreign nationals into restricted areas are contained in AR 12-15. Special considerations and coordination with the protocol office are necessary. A 24-hour advance notice is desirable for these requests, along with an agenda for the visit and the designation of an escort, if appropriate.

### Civilians Working on Jobs Under Government Contract

7-31. To allow these personnel to conduct business in restricted areas, the security manager must coordinate with the procurement office. The security manager must also identify movement-control procedures for these employees.

### Cleaning Teams

7-32. Supervisors using cleaning teams must seek technical advice from the physical-security office on internal controls for each specific building. This may include providing escorts.

### DOD Employees in Work Areas After Normal Operating Hours

7-33. Supervisors establish internal controls based on coordination with the security manager. They also notify security personnel of the workers' presence, type, and duration of work.

## ENFORCEMENT MEASURES

7-34. The most vulnerable link in any ID system is its enforcement. Security forces must be proactive in performing their duties. A routine performance of duty will adversely effect even the most elaborate system. Positive enforcement measures must be prescribed to enhance security. Some of these measures may include—

- Designating alert and tactful security personnel at entry control points.
- Ensuring that personnel possess quick perception and good judgment.
- Requiring entry-control personnel to conduct frequent irregular checks of their assigned areas.
- Formalizing standard procedures for conducting guard mounts and posting and relieving security personnel. These measures will prevent posting of unqualified personnel and a routine performance of duty.
- Prescribing a uniform method of handling or wearing security ID cards/badges. If carried on the person, the card must be removed from the wallet (or other holder) and handed to security personnel. When worn, the badge will be worn in a conspicuous position to expedite inspection and recognition from a distance.
- Designing entry and exit control points of restricted areas to force personnel to pass in a single file in front of security personnel. In some instances, the use of turnstiles may be advisable to assist in maintaining positive control.
- Providing lighting at control points. The lighting must illuminate the area to enable security personnel to compare the bearer with the ID card/badge.
- Enforcing access-control measures by educating security forces and employees. Enforcement of access-control systems rests primarily with the security forces; however, it is essential that they have the full cooperation of the employees. Employees must be instructed to consider each unidentified or improperly identified individual as a trespasser. In restricted areas where access is limited to a particular zone, employees must report unauthorized individuals to the security force.
- Positioning ID card/badge racks or containers at entry control points so that they are accessible only to guard-force personnel.
- Appointing a responsible custodian to accomplish control procedures of cards/badges according to AR 600-8-14. The custodian is responsible for the issue, turn in, recovery, and renewal of security ID cards/badges.

7-35. The degree of compromise tolerable in the ID system is in direct proportion to the degree of security required. The following control procedures are recommended for preserving the integrity of a card/badge system:

- Maintenance of an accurate written record or log listing (by serial number) all cards and badges and showing those on hand, to whom they are issued, and their disposition (lost, mutilated, or destroyed).
- Authentication of records and logs by the custodian.
- A periodic inventory of records by a commissioned officer.
- The prompt invalidation of lost cards/badges.
- The conspicuous posting at security control points of current lists of lost or invalidated cards/badges.
- The establishment of controls within restricted areas to enable security personnel to determine the number of persons within the area.
- The establishment of the two-person rule (when required).

- The establishment of procedures to control the movement of visitors. A visitor-control record will be maintained and located at entry control points.

### SIGN/COUNTERSIGN AND CODE WORD

7-36. This method of verifying identity is primarily used in a tactical environment. According to the local SOP, the sign/countersign or code-word procedures must be changed immediately if compromised.

## DURESS CODE

7-37. The duress code is a simple word or phrase used during normal conversation to alert other security personnel that an authorized person is under duress. A duress code requires planning and rehearsal to ensure an appropriate response. This code is changed frequently to minimize compromise.

## ACCESS-CONTROL ROSTERS

7-38. Admission of personnel to a restricted area is granted to those identified and listed on an access-control roster. Pen-and-ink changes may be made to the roster. Changes are published in the same manner as the original roster.

7-39. Rosters are maintained at access control points. They are kept current, verified, and accounted for by an individual designated by the commander. Commanders or their designated representatives authenticate the rosters. Admission of persons other than those on the rosters is subject to specific approval by the security manager. These personnel may require an escort according to the local SOP.

## METHODS OF CONTROL

7-40. There are a number of methods available to assist in the movement and control of personnel in limited, controlled, and restricted areas. The following paragraphs discuss the use of escorts and the two-person rule:

### ESCORTS

7-41. Escorts are chosen because of their ability to accomplish tasks effectively and properly. They possess knowledge of the area being visited. Escorts may be guard-force personnel, but they are normally personnel from the area being visited. Local regulations and SOPs determine if a visitor requires an escort while in the restricted area. Personnel on the access list may be admitted to restricted areas without an escort.

### TWO-PERSON RULE

7-42. The two-person rule is designed to prohibit access to sensitive areas or equipment by a lone individual. Two authorized persons are considered present when they are in a physical position from which they can positively detect incorrect or unauthorized procedures with respect to the task or operation being performed. The team is familiar with applicable safety and security requirements, and they are present during any operation that affords access to sensitive areas or equipment that requires the two-person rule. When application of the two-person rule is required, it is enforced constantly by the personnel who constitute the team.

7-43. The two-person rule is applied in many other aspects of physical-security operations, such as the following:

- When uncontrolled access to vital machinery, equipment, or materiel might provide opportunity for intentional or unintentional damage that could affect the installation's mission or operation.
- When uncontrolled access to funds could provide opportunity for diversion by falsification of accounts.
- When uncontrolled delivery or receipt for materials could provide opportunity for pilferage through "short" deliveries and false receipts.
- When access to an arms or ammunition storage room could provide an opportunity for theft. Keys should be issued so that at least two people must be present to unlock the locks required under the provisions of AR 190-11.

7-44. The two-person rule is limited to the creativity of the PM and the physical-security manager. They should explore every aspect of physical-security operations in which the two-person rule would provide additional security and assurance and include all appropriate recommendations and provisions of the physical-security plan. An electronic-entry control system may be used to enforce the two-person rule. The system can be programmed to deny access until two authorized people have successfully entered codes or swiped cards.

## SECURITY CONTROLS OF PACKAGES, PERSONAL PROPERTY, AND VEHICLES

7-45. A good package-control system helps prevent or minimize pilferage, sabotage, and espionage. The local SOP may allow the entry of packages with proper authorization into restricted areas without inspection. A package-checking system is used at the entrance gate. When practical, inspect all outgoing packages except those properly authorized for removal. When a 100 percent inspection is impractical, conduct frequent unannounced spot checks. A good package-control system assists in the movement of authorized packages, material, and property.

7-46. Property controls are not limited to packages carried openly, but they include the control of anything that could be used to conceal property or material. Personnel should not be routinely searched except in unusual situations. Searches must be performed according to the local SOP.

7-47. All POVs on the installation should be registered with the PM or the installation's physical-security office. Security personnel should assign a temporary decal or other temporary ID tag to visitors' vehicles to permit ready recognition. The decal or the tag should be distinctly different from that of permanent-party personnel.

7-48. When authorized vehicles enter or exit a restricted area, they undergo a systematic search, including (but not limited to) the—

- Vehicle's interior.
- Engine compartment.
- External air breathers.
- Top of the vehicle.
- Battery compartment.
- Cargo compartment.
- Undercarriage.

7-49. The movement of trucks and railroad cars into and out of restricted areas should be supervised and inspected. Truck and railroad entrances are controlled by locked gates when not in use and are manned by security personnel when unlocked. The ID cards/badges are issued to operators to ensure proper ID and registration for access to specific loading and unloading areas.

7-50. All conveyances entering or leaving a protected area are required to pass through a service gate manned by security forces. Drivers, helpers, passengers, and vehicle contents must be carefully examined. The examination may include—

- Appropriate entries in the security log (including the date, operator's name, load description, and time entered and departed).
- A check of the operator's license.
- Verification of the seal number with the shipping document and examination of the seal for tampering.

7-51. Incoming trucks and railroad cars must be assigned escorts before they are permitted to enter designated limited or exclusion areas. Commanders should establish published procedures to control the movement of trucks and railroad cars that enter designated restricted areas to discharge or pick up cargo (escorts will be provided when necessary).

7-52. The best control is provided when all of these elements are incorporated into access-control procedures. Simple, understandable, and workable access-control procedures are used to achieve security objectives without impeding operations. When properly organized and administered, access-control procedures provide a method of positively identifying personnel who have the need to enter or leave an area.

# TACTICAL-ENVIRONMENT CONSIDERATIONS

7-53. Access-control procedures during tactical operations may establish additional challenges for the commander. In some instances, the commander cannot provide a perimeter barrier (such as a fence) based on METT-TC. Commanders are still required to provide security measures for restricted areas, although they may not always have the necessary assets. Early-warning systems and the use of guards become crucial. A restricted area may become a requirement without prior notice during an operation. Figure 7-3 and Figure 7-4, page 7-14, are examples of temporary tactical restricted and exclusion areas.

7-54. Commanders must plan for these considerations when developing their budget. Funding must be requested and set aside to support physical-security requirements during tactical operations. Resources will not always be available; therefore, commanders must implement procedures that support access-control measures. Improvising will become common practice to overcome shortfalls concerning access-control equipment in the field.

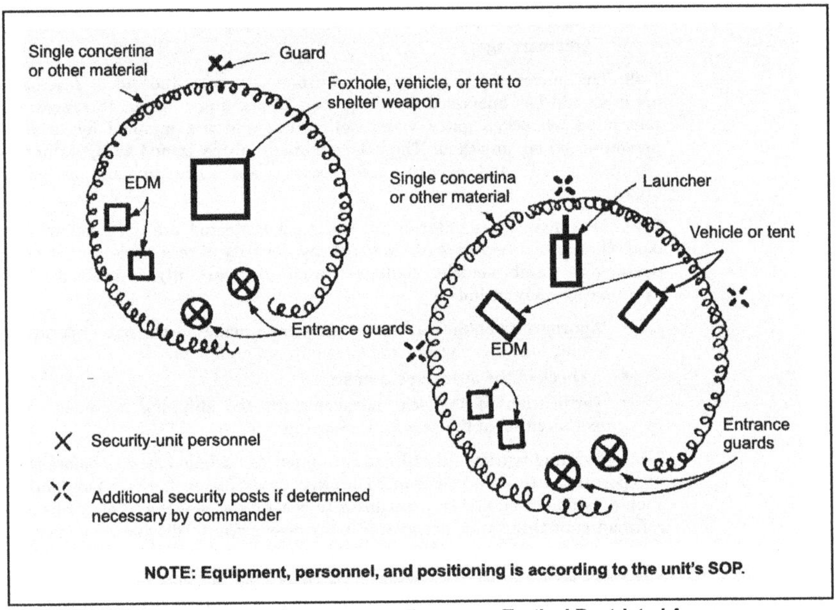


**Figure 7-3. Sample Layout of Temporary Tactical Restricted Areas**

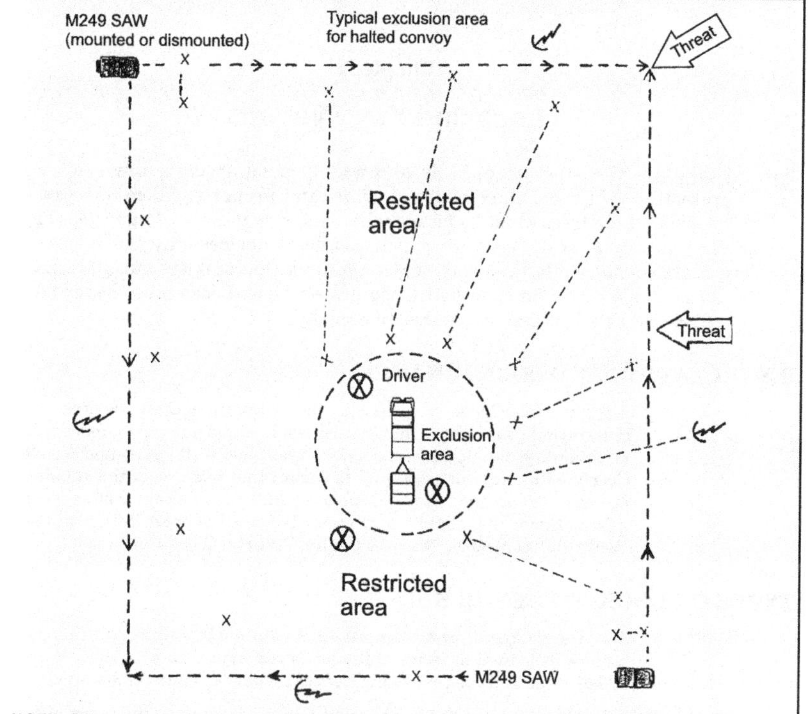


**NOTE: Other vehicles and personnel, to include the LPs, form a defensive perimeter inside the primary directions of fire of the M60 machine guns. During attack, alternate machine-gun crews should be placed in the perimeter away from primary machine guns and be spaced so that they can be used as replacements and so that large gaps do not appear in the perimeter.**

Legend:

X — MP rifleman

◄ - - Principal direction of fire

( ) Temporary exclusion area (size depends on terrain and threat force)

Listening post

**Figure 7-4. Sample Layout for a Temporary Tactical Exclusion Area**

# Chapter 8

# Lock and Key Systems

Locks are the most acceptable and widely used security devices for protecting facilities, classified materials, and property. All containers, rooms, and facilities must be locked when not in actual use. Regardless of their quality or cost, locks are considered delay devices only. Some locks require considerable time and expert manipulation to open, but all locks can be defeated by force and with the proper tools. Locks must never be considered as a stand-alone method of security.

## INSTALLATION AND MAINTENANCE

8-1. The USACE is responsible for installing locking devices in newly constructed facilities. Installation-level engineers are responsible for maintaining the locking devices. Physical-security personnel must work closely with engineer personnel to ensure that locks meet the standards and are installed according to applicable regulations. One source of assistance and information is the DOD Lock Program Technical Support Hotline at the Naval Facilities Engineering Services Center, Port Hueneme, California.

## TYPES OF LOCKING DEVICES

8-2. The degree of protection afforded by a vault, a safe, or a filing cabinet may be measured in terms of the lock's resistance. Locking devices are listed in TM 5-805-8. Types of locking devices include key and combination locks.

8-3. ARs 190-11, 190-51, 50-5, and 50-6 prescribe specific types of locks for specific types of facilities. AR 380-5 prescribes standard facilities for storing classified material and contains guidance for different storage requirements.

### KEY LOCKS

8-4. Key locks consist of, but are not limited to, the following:

- Cylindrical locksets are often called key-in-knob or key-in-lever locks. They are normally used to secure offices and storerooms. The locking cylinder located in the center of the doorknob distinguishes these locks. Some cylindrical locksets have keyways in each of the opposing knobs that require a key on either side to lock and unlock them. Others unlock with a key, but may be locked by pushing or rotating a button on the inside knob. These locks are suitable only for very low-security applications. Using these locks may require compensatory measures in the form of additional locks on containers within the room.
- Dead-bolt locks are sometimes called tubular dead bolts. They are mounted on the door in a manner similar to cylindrical locksets. The primary difference is in the bolt. When the bolt is extended (locked),

the dead bolt projects into the doorframe at least one inch, and it cannot be forced back (unlocked) by applying pressure to the end of the bolt. The dead-bolt lock has the potential for providing acceptable levels of protection for storerooms and other areas where more security is desired. It is recommended for use in military housing as an effective security measure in the installation's crime-prevention program. In situations where there is a window in or adjacent to the door, a double-cylinder dead-bolt lock (one that requires a key to open from either side) should be used.

- Mortise locks are so named because the lock case is mortised or recessed into the edge of the door. The most common variety of mortise locks has a doorknob on each side of the door. Entrance doors often have an exterior thumb latch rather than a doorknob. The mortise lock can be locked from inside by means of a thumb turn or by a button. Mortise locks are considered low-security devices since they weaken the door in the mortised area.
- Drop-bolt locks (often referred to as jimmy-proof locks) are normally used as auxiliary locks similar to dead bolts. Both the drop-bolt lock body and the strike have interlocking leaves similar to a door hinge. When closed, locking pins in the lock body drop down into the holes provided in the strike and secure the locking system. Since the lock body and the strike are interconnected with locking pins when closed, the lock essentially becomes a single unit and is extremely difficult to separate.
- Rim-cylinder locks are mounted to the door's inside surface and are secured by screws in the door face. These locks are generally used with drop-bolt and other surface-mounted locks and latches. They consist of an outer barrel, a cylinder and ring, a tailpiece, a back mounting plate, and two mounting screws. The tailpiece screws are usually scored so that the lock can be tailored to fit varying door thicknesses.
- Unit locks are ideal in heavily traveled facilities (such as hospitals or institutional buildings). These locks are a complete, one-piece unit that slides into a notch cut into the door's latch edge. The one-size cutout of the door edge simplifies the door preparation for the lock.
- Mechanical, push-button combination locks are digital (push buttons numbered 1 through 9) combination door-locking devices used to deny area access to any individual not authorized or cleared for a specific area. These locks are normally used for access control and should be backed up by door locking devices when the facility is unoccupied.
- Padlocks are detachable locks that are typically used with a hasp. Low-security padlocks, sometimes called secondary padlocks, are used to deter unauthorized access, and they provide only minimal resistance to force. Low-security locks are made with hardened steel shackles. Precautions must be taken to avoid confusing these locks with similar brass or bronze locks. The brass or bronze locks are commonly used but do not meet the security requirements of the hardened shackled locks. High-security padlocks may be used to secure AA&E. They provide the maximum resistance to unauthorized entry when used with a high-security hasp.

8-5. Some locks have interchangeable cores, which allow the same key system to include a variety of locks. Padlocks, door locks, cabinet locks, and electrical key switches can all be operated by the same key system. Because these cores are removable by a special key, this system allows for rapid rekeying of locks in the event that the key is compromised.

8-6. Locks are keyed in several different ways. When several locks are keyed differently, each is operated by its own key. When they are keyed alike, one key will open them all. Locks that are master-keyed are keyed differently, yet have one key that will open them all. Master-keying is done for convenience and represents the controlled loss of security. Master-keying is not used unless permitted by regulation.

## COMBINATION LOCKS

8-7. Combination locks are available as padlocks or as mounted locks. They are low-security padlocks with combinations that are either fixed or changeable. Combination locks may be either mechanical or electronic. They are operated by entering a particular sequence of numbers. When the correct combination is entered, the lock's bolt is retracted. Combination locks used for securing classified material must meet Federal Specification FF-L-2740.

8-8. Although the lock is the most accepted and widely used security device, it is only a delay device and should never be considered as a positive bar to entry. A lock can (and will) be defeated. The best defense for locking devices is a good key-control program. Refer to AR 190-51, Appendix D, for standard key and lock procedures. Additional key and lock procedures for AA&E can be found in AR 190-11, Chapter 3.

# Chapter 9

# Security Forces

The security force for an installation or a facility provides the enforcement element in the physical-security program. This force consists of personnel specifically organized, trained, and equipped to protect the command's physical-security interests. It is a commander's most effective tool in a comprehensive, integrated, physical-security program. Vulnerability tests are periodically conducted to determine and ensure the state of readiness of security forces (see Appendix K).

## TYPES OF SECURITY FORCES

9-1. On installations, security forces may be MP forces, security police, DOD civil-service security guards, or contract guards. Interior guard duties are performed by installation unit troops on a roster basis. MP forces normally perform security duties that require higher degrees of training and experience. These include—

- Security of restricted areas.
- Security of specific sensitive gates.
- Supervisory or coordinated roles with other military or DOD civil-service security guards.
- Responsibility for monitoring and responding to intrusion alarms.

9-2. An MP unit may perform the entire physical-security function alone based on METT-TC, the area, and the facilities. When an MP unit cannot assume responsibilities for all of the physical-security requirements in the command, other forces may be required. Additional forces may consist of the following:

- Personnel furnished by units of the installation's command on a daily or weekly basis. While this method has the single advantage of providing additional manpower, it has the disadvantages of rapid turnover and the lack of training. If this manpower is used, personnel should be assigned the least sensitive posts or patrols. For extended augmentation, units may be attached to MP units. The MP unit may also be augmented by reserve units or units in rotation.
- The combat-arms branches (especially the infantry) may attach their forces to MP units and may be designated as security guards assisting in the required operations.
- Military or paramilitary units of the host country may also be attached to or operate in coordination with MP forces. They may also be supplemented with national police of their own country.

- The installation's band may be a source of military force during wartime. (The band is assigned enemy-prisoner-of-war [EPW] duty as a wartime duty.) The band is doctrinally capable of providing security at the division tactical operations center (DTOC) and ASPs, assisting in the perimeter defense of the command post (CP), and operating the dismount point for the CP. It is capable of providing access control at the DTOC and the ASPs and augmenting or relieving security personnel on the defensive perimeter.

9-3. Civil-service security guards are uniformed civilian employees from a government agency. They are customarily trained and organized along military lines. The organization may be completely civil service or may be composed of civil-service personnel under military supervision. In either case, they are under the operational control of the PM or the security officer.

9-4. Labor-service personnel (local civilian personnel) have been organized and used successfully in theaters of operation. These types of units were organized after World War II and since that time have established enviable records in the physical-security field. They are distinctively uniformed, organized, and equipped. They have set and maintained the highest security standards, resulting in a minimal loss of property. While not military organizations, these units have successfully developed a high sense of duty and esprit de corps that has been reflected in their outstanding contributions to the physical security of installations in overseas commands.

## AUTHORITY AND JURISDICTION

9-5. It is most important that the PM or the security officer determine (and instruct his security force in) the extent and limitations of the commander's jurisdiction in the field of law enforcement and investigations. Those jurisdictions include—

- Jurisdiction of place.
  - Military installations and facilities. Whether state or federal law or both are applicable on a military installation or facility depends largely on the nature of jurisdiction over the land involved. The amount of federal jurisdiction may vary between different areas of the installation or facility. The legal formalities of acquiring jurisdiction over land under the control of the Secretary of the Army are accomplished at DA level and according to the provisions of AR 405-20. Information and advice relating to jurisdictional questions should be referred to the local SJA.
  - Areas outside of military installations. Areas outside of military installations are generally subject to state and local laws; however, there are exceptions. Information and advice in this regard should be obtained through the local SJA.
  - Overseas areas. In overseas areas, jurisdiction varies according to the military situation and existing international treaties, contracts, and agreements. Guidance should be obtained in each instance from the commander and the SJA and set forth in appropriate command directives.

- Jurisdiction of personnel.
  - Jurisdiction of personnel generally follows the limitations of jurisdiction of the installation.
  - MP forces have jurisdiction and authority over personnel as described in AR 190-14 and related publications.
  - Authority for federal civilian employees assigned to security, police, and guard duties is derived from the installation's commanding officer. These personnel can have no more authority than he possesses and are subject to any limitations imposed thereon.
  - Security-force personnel may enforce all offenses under the Uniform Code of Military Justice (UCMJ), military regulations, federal laws and regulations, and state laws where applicable.
  - Security-force personnel may be given the same authority as MP forces over all personnel subject to military jurisdiction, including apprehension, detention, and search.
  - Civilian security-force personnel have no specific grant of authority over civilians other than the right of citizen's arrest.
  - The commander is the source of jurisdiction and authority for all other personnel assigned to security-force duties.

## PERSONNEL SELECTION

9-6. Regardless of the use of structural, mechanical, or electronic equipment, the human element in security operations makes the difference between success or failure. Commanders and supervisors have a responsibility to ensure that security personnel who control access to restricted areas and classified activities are qualified based on criteria in AR 380-67. Personnel who perform physical-security duties must be disciplined and alert, have sound judgment, be confident and physically fit, and possess good interpersonal communication skills.

## SECURITY CLEARANCE

9-7. Security-clearance criteria for security positions must be based on the security classifications of the information to which access will be granted. Security positions are normally designated as sensitive and require a secret security clearance. ARs 381-20 and 380-67 describe criteria and procedures governing security clearances. Appropriate civilian-personnel regulations should also be consulted when civilians are involved.

9-8. Positive evaluation of the reliability of all personnel must be made before they are entrusted with classified or sensitive information. (The Individual Reliability Program is prescribed in AR 190-56.) Follow-up action must be made on all personnel who are granted a security clearance to ensure that they continue to meet the criteria for their clearance. Personnel not meeting or adhering to the prescribed standards must have their security clearances revoked and thereby lose their access to areas containing classified information or material (see AR 380-67).

## ORGANIZATION AND EMPLOYMENT OF FORCES

9-9. The organization of a security force will vary, depending on circumstances and the forces available. Forces consist of—

- Mobile patrols. A mobile detachment of ground, sea, or air forces dispatched to gather information or carry out a security mission.
- The response force. A mobile force with appropriate fire support (usually designated by the area commander) to deal with Level II threats in the rear area (Army). This is normally an MP function.
- Reserves. That portion of a force withheld from action or uncommitted to a specific course of action so as to be available for commitment at the decisive moment. Its primary purpose is to retain flexibility throughout an offensive action.
- Any combination of these three.

9-10. Instructions to the security force should be issued in writing. These instructions are normally in the form of general, special, or temporary orders. They should be carefully and clearly worded and include all phases of each assignment. They should be reviewed at least monthly to ensure that they are current. Categories of instructions of each are as follows:

- General orders are those orders that concern the security force as a whole and are applicable at all posts and patrols.
- Special orders pertain to a permanent post or patrol. Each permanent post or patrol should have special orders issued concerning the location, duties, hours manned, arms, ammunition, and other equipment required and the instructions for using force in enforcement and apprehension activities.
- Temporary orders are issued for a short period and cover a special or temporary situation. If it can be predetermined, such orders should indicate the period of time for which they are valid.

9-11. A security-force SOP that outlines policies, organization, authority, functions, and other required information should be prepared for required reading. Each security-force member should be held responsible for full knowledge and understanding of the contents of the SOP. Each installation PM, physical-security officer, or chief of a guard force should conduct periodic inspections and tests to determine each individual's degree of understanding of these instructions. Instructions should be provided in writing regarding the safeguarding and control of the SOP. Its contents may not be classified; however, the information could assist an intruder in breaching security.

## HEADQUARTERS AND SHELTERS

9-12. The location of the security force's headquarters will depend on the size and layout of the installation or facility. The objectives are the efficient control of the security force and the adequate security of vital activities. On a small installation, there is frequently only one full-time entrance that may be supplemented by several part-time entrances. At these installations, the logical location of the headquarters would be at or near the main entrance. On larger installations, it might be better to locate the headquarters near the center of the cantonment area.

9-13. The security force's headquarters should be the control point for all physical-security matters for the installation and the monitoring point for protective alarm and communication systems. This office should have a reliable and independent means to contact nearby civil authorities. A list of key telephone numbers should be available for use in emergency operations.

9-14. Personnel shelters should be available to protect the guards from the elements. The design can be temporary or hardened and include adequate space for guard-force personnel only. The facility should have heat, ventilation, storage space for essential accessories, lighting that will not expose the occupant, and good visibility in all directions.

## EXECUTION OF SECURITY ACTIVITIES

9-15. Security personnel must exercise good interpersonal communication skills when carrying out their duties with other employees. Bad employee relations can result if security personnel become impertinent and assume powers not rightfully theirs. Security personnel must understand the methods and techniques that will detect security hazards and assist in identifying violators and intruders.

9-16. Written reports or journals are recommended for security activities. These should be prepared by either the security force's supervisor or the personnel at the security post. These reports should record all activities, actions, and visits at the security post.

9-17. It must be strongly emphasized that security personnel will be used for security duties only and should not be given other routine functions except as directed by the commander or his representative. Security personnel should have no fire-fighting or similar duties regularly assigned. Such emergencies offer an excellent diversion to cover an intruder's entrance. Consequently, during such times, security personnel must be exceptionally alert when performing their duties. However, the security force may be cross-trained in other areas (such as fire fighting) so that they may be used when required and when circumstances permit (such as when they are off duty).

9-18. Personnel who are assigned to fixed posts should have a designated method of relief. The security force's shift supervisor should establish a relief schedule (about every two hours) according to local policies and the SOP. A simple but effective plan of operation should be worked out for the security force to meet every foreseeable emergency. Practice alarms should be conducted frequently to test the plan's effectiveness. Such plans should be designed to prevent a diversion at one point on the installation, drawing off the guards or distracting their attention from another section of the installation where unauthorized entry may be made. Routes and times for security patrols should also be varied at frequent intervals to preclude establishing a routine that may be observed by potential intruders.

## TRAINING REQUIREMENTS

9-19. The extent and type of training required for security forces will vary according to the importance, vulnerability, size, and other factors affecting a particular installation or facility. The training program's objective is to ensure that all personnel are able to perform routine and emergency duties competently and efficiently.

### BENEFITS OF PROPER TRAINING

9-20. Efficient and continuing training is the most effective means of obtaining and maintaining maximum proficiency of security-force personnel. Regardless of the selection process, new personnel seldom have all of the qualifications and experience necessary to do the job. In addition, new or revised job requirements frequently mean that personnel must be retrained. Training can bridge the void between ability and job requirement.

9-21. Supervisors need to remember that all personnel do not have the same training needs. It is a waste of valuable time to train an individual in a subject that he has already mastered. Past experience, training, acquired skills, and duty assignments should be evaluated for each person as an aid in planning an effective training program.

9-22. A good training program benefits both the installation and the security force. The task of supervising the security force is made easier, there is much less wasted time, fewer mistakes are made, and there is less friction with other agencies. A good training program helps to instill confidence through developing increased skill proficiency. The training program provides for more flexibility and better physical protection, fewer required personnel, and less time to learn duties. Training establishes systematic and uniform work habits.

### BASIC TRAINING

9-23. As a minimum, personnel (including civil-service security personnel) who have not had security training should receive training in their security duties. This training includes—

- The care and use of weapons, if required. No person should be placed on security duty unless weapons training has occurred within the past 12 months. Weapons training must be according to AR 190-14.
- Areas of responsibility and authority of security personnel, particularly on apprehension, search and seizure, and the use of force.
- The location and use of first aid and fire-control equipment and electrical switches.
- Duties in case of emergencies such as alerts, fires, explosions, and civil disturbances.
- Common forms of sabotage and espionage activity.
- The location of hazardous and vulnerable equipment and material.

### IN-SERVICE TRAINING

9-24. All newly assigned individuals are given special instructions for each post. When possible, their first assignment should be with an experienced person. Additional in-service training and periodic retraining to review basic material and procedures are continuous requirements.

9-25. Scheduling in-service training and classes to enable all of the security force or a complete shift to participate is often difficult. Therefore, the supervisor must exercise good judgment when scheduling training to ensure that each person has the opportunity to receive the training.

### EVALUATION OF TRAINING

9-26. Testing designed to evaluate performance is a necessary step in the training program. These tests may be oral or written or may be a type of performance test. They should be administered annually to ensure that the entire force maintains high standards of proficiency. A testing program also helps to improve training by—

- Discovering gaps in learning.
- Emphasizing main points.
- Evaluating instructional methods.

9-27. Security training received by personnel at their units must be entered in unit training charts or records. The record serves to—

- Indicate individual degrees of skill.
- Establish priorities of instruction.
- Present a consolidated picture of the security force's training status.
- Help certify guard personnel.

## SUPERVISION

9-28. A security supervisor is tasked with overseeing and directing the work and behavior of other members of the security force. To obtain maximum performance from each member of his force, the supervisor must have a complete understanding of leadership principles and be capable of applying them.

9-29. The supervisor is responsible for understanding the operations of all posts. Additionally, he is often responsible for selecting, inducting, training, and ensuring the productivity, safety, morale, and advancement of guard force members.

9-30. To ensure an alert, presentable, and efficient security force, the leadership must provide consistent and intelligent supervision. To earn the respect and cooperation of the guard force, supervisors must be professional in their conduct. The security force's morale and efficiency is a direct reflection of the quality of its supervision.

9-31. The ratio of supervisory personnel to security personnel should be determined by the individual characteristics of each installation. At small installations, the ratio may be higher than at large installations.

9-32. There must be sufficient supervision to enable the inspection of each post and patrol. It is also essential that supervisors be in contact with security headquarters to control emergencies that may arise. Specific duties of a supervisor include the inspection and briefing of the relief shift and the inspection of posts, vehicles, and equipment during visits to posts and patrols.

### SUPPLEMENTS TO SUPERVISION

9-33. Various means and devices may be used as supplements to personnel supervision. These include the following:

- **Recorded tour systems.** Personnel record their presence at strategic points throughout an installation by using portable watch clocks or similar devices. These are effective means of ensuring that such points are regularly covered. This system provides an after-the-fact type of supervision.
- **Supervisory tour systems.** A signal is transmitted to a manned central headquarters at the time the post is visited. These systems provide instantaneous supervision and a means of detecting interference with normal security activities and initiating an investigation or other appropriate action.

9-34. All personnel on security duty should be required to report periodically to headquarters by the usual means of communication. The frequency of such reports will vary, depending on a number of factors. Regularity should be avoided to preclude setting a pattern by which an intruder can gauge an appropriate time for entrance.

### MANAGEMENT

9-35. The physical-security supervisor is responsible for managing and developing the security organization. A physical-security program is greatly enhanced by a well-developed educational program.

9-36. The physical-security supervisor acts as an advisor and assists in formulating policies for the installation's physical-security measures. The goal should be the best security within the restrictions of the commander's budget guidance. Physical-security planners must remember that anyone can provide adequate security with unlimited funds; however, this is not a realistic approach. There must be a constant endeavor to effect justifiable economy where possible without jeopardizing the physical-security program.

## UNIFORMS

9-37. All security-force personnel are required to wear the complete prescribed uniform as outlined in their special orders. Deviations from the prescribed uniform should not be made except for items to protect the guard force's health, comfort, and safety. The duty uniform will be worn during all tours of duty and may be worn during off-duty hours only between the place of residence and the place of duty. Each member of the security force is required to maintain high standards of appearance.

## VEHICLES

9-38. The security force should be furnished with sufficient and reliable vehicles to maintain patrol standards established by the installation commander. Vehicles assigned to the force should be equipped with two-way radios to obtain the greatest possible use of all personnel and vehicles.

## FIREARMS

9-39. Before issuing weapons, the security force will be briefed on the use of force. Security-force personnel will be issued weapons as prescribed by AR 190-11 and the unit's SOP. The commander may prescribe other weapons for the security force based on needs and requirements. Weapons normally are loaded with live ammunition, except where prohibited for safety reasons. The use of privately owned weapons while on duty is not authorized. Weapons and ammunition issued to security-force personnel will not be removed from the installation except in the course of official duty. When not in use, weapons are secured in arm racks in storage rooms as prescribed by AR 190-11.

9-40. Weapons are inspected as necessary to ensure proper maintenance. A written report is prepared and filed on the discharge of any weapon except for authorized and supervised training. The patrol supervisor or an MP investigator prepares the report (DA Form 3975).

9-41. Ammunition supplies for the security force's use must be maintained in secured storage containers according to AR 190-11. Ammunition must be issued only under proper supervision for authorized purposes. Ammunition issued to members of the security force must be accounted for by individual members immediately upon completion of duty. Any ammunition unaccounted for will be the subject of a report of its disposition by the individual.

## COMMUNICATIONS

9-42. The security force should be equipped with two-way radios. These may be vehicle-mounted and portable, or they may be telephones. A secure-voice capability should be used where possible. This equipment is considered essential for the efficient operation of the security force and the accomplishment of its assigned mission. Proper use and care by security personnel will enhance the equipment's usefulness and capability.

## MISCELLANEOUS EQUIPMENT

9-43. Security managers or supervisors should obtain other equipment necessary to accomplish their security mission. Items in this category may include (but are not limited to) warning lights; sirens; spotlights; portable lights; flashlights; first aid kits; traffic-control devices; and items of wear for the health, comfort, and safety of security personnel. Some of this equipment may require local purchase.

## MILITARY WORKING DOGS

9-44. The requirements for physical protection of installations or facilities within the US and overseas theaters of operation continue to increase. Manpower available for this purpose has always been (and probably will continue to be) limited. The MWD, properly trained and properly used, can enhance a physical-security program. See AR 190-12 and DA Pam 190-12 for information regarding the use of MWDs.

## SUMMARY

9-45. A security force is the critical element of a successful physical-security program. It is as strong as its weakest member. A comprehensive training program is essential to a knowledgeable, disciplined, and alert security force. A well-trained security force will be prepared to respond to a security breach.

## Chapter 10

# In-Transit Security

In-transit security subjects the movement of cargo to different, and frequently, more demanding aspects of physical security. Cargoes may be moved via port, rail, pipeline, or convoy. Regardless of the mode of movement, commanders must aggressively apply the principles of physical security to their protection. Security forces must be provided at the most vulnerable areas of each cargo movement.

## IN-PORT CARGO

10-1. Ports and harbors are prime targets for enemy and criminal activities. Perimeter areas of these facilities are more vulnerable because of the extensive distance and exposed beach or pier areas. Terminal areas may include fully developed piers and warehouses or may be an unimproved beach where logistics-over-the-shore (LOTS) or roll-on/roll-off (RORO) operations are conducted.

10-2. If a Theater Army Area Command (TAACOM) MP unit must provide security for cargo in a port, the main effort is to provide security from the perimeter of the port outward. Security measures focus on aggressive patrolling to detect, report and, if need be, combat enemy threats. Measures may include—

- Conducting route and area reconnaissance patrols.
- Developing police intelligence in the area of operations (AO).
- Controlling traffic in the area surrounding the port.
- Conducting mounted or dismounted patrols (with MWDs, if available) around the port's perimeter.
- Establishing an access-control/ID section.
- Watching for diversions of supplies out of the port.
- Providing a response force to react to incidents inside the port's perimeter.
- Providing observation and early warning of threat ground and air attacks.

10-3. When providing security for cargo, the focus is on providing a security overwatch for the cargo as it moves from the port to the combat area. Inside a port's perimeter, access to cargo is limited by—

- Operating random mounted or dismounted patrols (with MWDs, if available).
- Using combined patrols as a response force for incidents inside the perimeter.

- Controlling access to the most restricted areas.

10-4. On occasion, the MP may have to safeguard highly critical cargo inside a port's perimeter. The type and degree of security provided is based on logistical security information. Some examples are the—

- Types and values of the cargo stored.
- Vulnerability of the cargo to a land threat.
- Likelihood of theft, diversion, pilferage, or sabotage by military personnel, local workers, black marketers, or enemy agents.
- Location and nature of the port facilities.
- HN agreements.
- Degree of entrance and exit controls.

10-5. Safeguarding the most critical cargo waiting to be transferred to land transport is the priority. The following measures help to safeguard stored cargo:

- Establishing access-control procedures.
- Searching bundles and packages being taken from the area.
- Examining trip tickets and documentation of cargo vehicles.

10-6. If the restricted area is a pier or other maritime environment, access from the water must be controlled as well as from the land. Entry on the landward side of a pier can be limited with fencing, pass control, and aggressive patrolling; but the part of the pier that protrudes over the water is accessible from the sides and from below. Methods for securing the pier along its water boundaries include—

- Patrols (both walking on the pier and in small boats).
- Protective lighting.
- Log booms.
- Nets.
- Buoys or floats.
- Anchored or pile-mounted navigational aids and signaling devices.
- Barges.

10-7. While most of the barriers described above will stop or impede access to facilities from boats or swimmers, nets are among the most effective. Well-marked, partially submerged objects are also effective; however, there may be legal prohibitions against placing barriers that may constitute a hazard to navigation. These barriers should be placed only after coordination with and approval by the appropriate legal and HN authorities. Sometimes it is best to close off the waterside of a pier. A floating boom will keep small boats out. Suspending a cable or a chain-link net from the bottom of the boom will deny access underwater.

10-8. At least two security zones must be established on a facility's waterside—the reaction zone and the keep-out (exclusion) zone. Security forces in these zones notify vessels, craft, and swimmers that they are entering restricted waters and should alter their course. Security forces may stop and search intruders if necessary. Security zones should extend at least 1,000 meters from the nearest protected asset; however, in some port areas this large security zone is not possible. In such cases, other measures (such as boat patrols) must be increased to mitigate the possibility of attack.

10-9. A reaction zone extends from the high-water mark to a distance beyond the maximum range of anticipated waterborne threats. Security forces will stop and challenge intruders inside the reaction zone.

10-10. The keep-out zone is the zone closest to the protected assets. It extends from the asset to the maximum range of anticipated threat weapons. Security forces should prevent the entry of all unauthorized craft or vessels into this zone. The tactical response force (in this case, a boat) may be used. In addition to organic security, forces may be provided by HN or contracted personnel.

10-11. To keep the cargo secured while transferring from one transport method to another, the traffic moving in and out of cargo-handling areas must be controlled. MP forces can—

- Set up a single access-control point.
- Erect field-expedient barriers. Truck trailers or other large vehicles can be used to constrict the traffic flow if permanent barriers are not in place.
- Limit entry to mission-essential personnel, vehicles, and equipment (as designated by the port authority).

10-12. A holding area should be provided if gates are used by vehicles other than cargo vehicles. Cargo vehicles can pull into the holding area while they are being checked. The holding area should be large enough to handle the volume and size of traffic. A wooden deck or platform at, or slightly higher than, the level of the truck bed can be used to facilitate checking. The platform must be at least as long as the vehicle (such as an empty flatbed trailer). Such a platform makes it quicker and easier to observe and check cargo.

10-13. Cargo is less likely to be diverted if a close watch is kept on cargo documentation and container safety. Containerized cargo is less likely to be stolen or sabotaged. However, containers must be watched closely as they are filled and sealed. Cargo can be pilfered before the seal is applied. An unsealed container can be moved to a stacking area; or someone may apply a false seal, break the seal later, remove the cargo, and then apply a legitimate seal.

10-14. At access-control points—

- Inbound and outbound containers should be inspected. Signs of damage or unserviceability should be observed.
- Containers must be inspected for the presence of seals or locks and hinges. Their serviceability should also be checked.
- The document's transport number, container number, and seal number should be checked to ensure that they match those numbers on the transportation control-and-movement document. (Check the seals by handling them, not simply by a visual check.)
- Containers with valid documents only should be allowed to pass inbound or outbound through the control point.

## RAIL CARGO

10-15. Because a train's movement is determined directly by the condition of the tracks, cargo moving by rail is particularly vulnerable to attack. The destruction of switches, signals, or the track may be a delaying harassment; or it could trigger a major catastrophe. Since railroads can be such high-value targets, the commander may task MP or other US forces to provide on-board security for critical cargo.

10-16. Most train crews consist of four or five people who control the train—the engineer, a conductor, a fireman, a senior brakeman, and a brakeman or a flagman. The conductor is the train commander unless a transportation railway service officer is assigned to the train. The train commander is responsible for the train's operation and security. He makes all decisions affecting the train. The security force's commander is responsible for the cargo's security. The train crew and the security force watch for and report any discrepancies or interruptions to normal procedures at any time during the movement. Information about the movement is usually sent along the movement route by the chief dispatcher through a telephone circuit.

10-17. A four- to six-person security force is usually enough to secure railway shipments of sensitive freight, but additional security forces may be needed for moving critical cargo. In addition to a military security force, the shipper or loading agency may send specially trained personnel with highly sensitive cargo. The number of MP in a train security force depends on the—

- Sensitivity of the freight.
- Priority of need for the freight.
- Terrain over which the train will pass.
- Length of the train.
- Duration of the trip.
- Degree of threat.

10-18. Security forces prepare and maintain a record (by car number) of guarded cars in the train. Security forces can ride in—

- A specific car that requires protection.
- The caboose.
- A security-force car. (If only one security car is used, it should be near the center of the train; if more than one is used, cars should be spaced to provide the best protection for the train.)

10-19. The security force on a train must keep a constant check on car doors, seals, wires, and locks to detect tampering. The following instances must be noted and reported immediately:

- Irregularities in procedures.
- The presence or actions of unauthorized persons.
- Deficiencies or incidents that occur.

10-20. When planning rail-cargo security, the time schedule for the rail movement must be obtained. A map reconnaissance of the route should be provided, detailing bridges and tunnels that are especially vulnerable.

10-21. Security-force actions should be planned at scheduled stops or relief points, and forces should be deployed according to these plans. Locations of MP units and other friendly forces should be plotted along the route, and their radio frequencies and call signs should be noted. An intelligence report covering the route should also be obtained. This report should indicate sites where sabotage may occur, attacks may be expected, or thefts and pilferage are likely.

10-22. The shipper is responsible for the security of all carload freight until it is turned over to the Transportation Railway Service and the loaded cars are coupled to a locomotive for movement. The shipper or field transportation officer should complete the freight waybill or the government bill of lading. This report shows the car number, a brief description of contents, the weight of the load, the consignor, the consignee, the origin, and the destination. In addition, it may show special instructions for the movement or security of the car and its contents. Careful documentation is essential for—

- Securing the shipment.
- Locating cars with critical cargo.
- Ensuring that priority movement is authorized.

10-23. Transportation officers are responsible for the completeness, correctness, and proper handling of waybills. Each car must have a waybill; this allows cars to be detached or left behind should they become defective en route. If this occurs, a team from the security force must remain with the cargo until they are relieved.

10-24. Railway cars are sealed after loading. A seal shows that a car has been inventoried and inspected. The standard method of sealing a railway boxcar door (in addition to padlocks or wires) is with a soft metal strap or a cable seal that contains a serial number. Maintaining rigid accountability of all seals is necessary to prevent the undetected replacement of an original seal with another. While sealing does not prevent pilferage, a broken seal is a good indicator that the car and its contents have been tampered with. Train security forces or operating crews can easily check the seals on cars when the train stops. Broken seals should be reported immediately to help pinpoint the time and place of a possible theft. When vehicles are shipped by railcar, sensitive and high-value items must not be secured in the vehicles. Container-express (CONEX) and military-van (MILVAN) containers are ideal for shipping these and other small items on flatcars since they greatly reduce the chance of pilferage. These containers must be locked and sealed and, if possible, placed door to door for additional security.

10-25. When operations permit, cars containing highly pilferable freight, high-priority cargo, or special shipments are grouped in the train to permit the most economical use of security forces. When flatcars or gondolas are used to transport sensitive or easily pilfered freight, security forces should be placed where they can continuously observe and protect these cars.

10-26. When the train is stopped, security forces should dismount and check both sides of the train, verifying that seals, locks, and wires are intact. They must report a broken seal immediately to help pinpoint the time and place of the theft.

10-27. If the security force is relieved by another security force while en route, a joint inspection of the cars is conducted. The relief force signs the record being kept on the guarded cars. Consignees assume responsibility for the security of loaded freight cars at the time they arrive at their destination. When the trip is complete, the receiver or his agent will inspect the cars. The security force obtains a receipt for the cars, which is then attached to the trip report. The trip report should include—

- Dates and times the trip started and ended.
- Any additional information required by the local SOP or command directive.
- Recommendations for correcting deficiencies or for improving future security on trains.

10-28. Because unloading points are highly vulnerable to pilferage and sabotage, cars should be unloaded as soon as possible to reduce the opportunity for loss. MP forces are normally not available for the security of freight in railway yards. For more information regarding rail cargo, see FM 55-20.

## PIPELINE CARGO

10-29. Pipeline systems are widely used in a theater of operation to transport bulk petroleum products or other liquids. Such systems are open to a number of security threats from the point of entry to the point of final delivery. Pipeline systems are composed of storage and dispersing facilities, pump stations, and extended pipelines. They also include discharging facilities for tankers at ports or other water terminals.

10-30. The type and extent of risk to a pipeline varies with the level of conflict in the AO. In a communications zone, the chief hazard is likely to be pilferage. Pipelines can be tapped by loosening the flange bolts that join sections of pipe or by cutting holes in the hose line. The risk rises if gasoline is scarce and expensive on the civilian market. Sabotage is a security hazard during all levels of conflict. It is committed by any method such as simply opening pipe flanges, cutting a hose line, or setting fires and causing explosions to destroy portions of the line.

10-31. In areas of conflict, the likelihood of sabotage and interdiction increases. Pipeline systems are vulnerable to air attacks, especially aboveground sections of the pipeline, pump stations, and storage facilities.

10-32. Security forces should be deployed in the best manner to provide coverage to the most vulnerable portions of the pipeline that are at the greatest risk to enemy, terrorist, partisan, and ground attack. Patrols should be set up to screen isolated areas and remote pumping stations. Sensors should also be considered, along with aerial security. Security patrols will—

- Detect, report, and respond to attacks on or sabotages of the pipeline.
- Monitor critical parts of the pipeline on a routine but random basis.
- Monitor ground sensors and other intrusion-detection devices. These are often used at pump stations and elsewhere along the pipeline to detect and identify threats to the system.

- Check line-pressure devices in pipeline and pumping facilities. These devices monitor the flow and detect breaks in the line, which may indicate pilferage of gasoline (or other petroleum products).

10-33. Dedicated security forces are rarely sufficient in number for the surveillance of an entire pipeline system. All available supporting forces (in the course of their normal duties) should observe and report items of intelligence for further investigation. Examples of suspicious activities in the pipeline area might include the unusual presence of commercial tanker trucks, the appearance of gasoline drums or cans, or an increased use of motor vehicles in fuel-scarce areas. Other resources available to the commander for coordination and support include HN and MP elements responsible for the AO, as well as the security officer of the petroleum group or battalion.

## CONVOY MOVEMENT

10-34. As convoy movements are tactical in nature and are discussed in detail in FM 19-4, they will be briefly discussed here. When moving by convoy, consideration should be made for the following:

- Congested traffic areas.
- Travel during night hours when traffic is reduced rather than travel during daylight hours when traffic congestion is heaviest.
- National holidays. Traffic may be three times heavier than on a normal day. Also, if you are moving a convoy overseas on a national holiday, the HN people may not be receptive to your action and the result may be unwanted reactions on their part.
- The use of a marked HN police vehicle in conjunction with the convoy. The HN people are more receptive to an activity when it is represented by one of their own. Additionally, the HN police may be able to diffuse a potential crisis.
- Security of the convoy. Security of the convoy is foremost important both during movement and stops. During extended or overnight stops, special consideration must be given to securing the loaded vehicles.

# Chapter 11

# Inspections and Surveys

Inspections and surveys are valuable tools to a commander's physical-security program. These tools collectively measure and identify the readiness of a commander's physical-security program. The survey provides the installation commander with an overall security posture of the installation.

## INSPECTIONS

11-1. Physical-security inspections are conducted at DA installations, activities, and facilities by trained physical-security inspectors. Some facilities on an installation may be exempt from inspection due to their mission. These facilities are inspected under the guidance of regulations and directives unique to those activities. Inspection personnel will be trained and will conduct inspections according to AR 190-13.

11-2. A physical-security inspection is a recorded assessment of physical-security procedures and measures implemented by a unit or an activity to protect its assets. The inspection is recorded on DA Form 2806-1-R (see AR 190-13 for using the form).

### COORDINATION

11-3. Liaison and coordination should be established with other agencies on the installation before an inspection. The director of facility engineers can provide information to benefit the overall security program. Other agencies, such as MI (threat analysis) and local law-enforcement agencies, may have input essential to the security program.

### SECURITY LIBRARY

11-4. A security library is necessary to help personnel prepare for and conduct an inspection. This library may include—

- The mission and history of each activity to be inspected.
- Previous inspection reports.
- A copy of the most current risk analysis.
- The SOPs and ARs specific to physical security.

### ENTRANCE INTERVIEWS

11-5. Entrance interviews are usually required before the actual inspection. During the interview, the inspector establishes a rapport with the unit representative. The inspector identifies the following during the interview:

- All members of the inspection team.

- An overview of the last inspection.
- Areas to be inspected and the order of inspection.
- A review of waivers, work orders, and exceptions.
- Changes to the unit's mission (if any).

## CONDUCTING INSPECTIONS

11-6. The inspection should be conducted from the outside to the inside of the facility, activity, or area with regard to the following:

- Observation of the facility will be conducted during all hours of the day.
- Interviews of managerial and operational personnel will be performed.
- Security forces should be inspected so as not to disrupt the mission (if possible).
- An assessment should be made of security-force training, especially if security-force knowledge proves inadequate.
- Inspection of entry and movement control by the guard force should not hinder operations.
- All communications systems used by the guard force should be thoroughly inspected. The guard force should have two reliable and efficient means of communication, one of which is a radio.
- Inspections should be conducted according to regulations appropriate for the facility.

## EXIT INTERVIEWS

11-7. Exit interviews should be conducted as soon as possible after the inspection. The commander should be informed of any deficiencies or compliments noted. A rating on the inspection's results will not be provided during the exit interview. The approving authority, not the inspector, will determine the inspection rating. The rating will be forwarded to the unit along with the final report.

11-8. Recommendations will be made according to regulations. Written reports should be forwarded through channels in a timely manner according to the PM's SOP. The commander's report of actions taken will be required and reviewed by the PM's staff.

# SURVEYS

11-9. A physical-security survey differs from an inspection in that a survey covers a formal assessment of an installation's physical-security program. Each survey includes a complete reconnaissance, study, and analysis of installation property and operations. The survey provides the commander with an assessment of the installation's overall security posture. It consists of the threat and the mission, and it advises the commander on the installation's physical-security program's strengths and weaknesses.

## PHYSICAL-SECURITY SURVEY

11-10. The physical-security survey is a formal recorded assessment of an installation's physical-security program. See AR 190-13 for further information on this type of survey.

## SECURITY-ENGINEERING SURVEYS

11-11. While a security-engineering survey is largely an engineer function, it must be coordinated with physical-security personnel to be successful. A security-engineering survey is the process of identifying (by means of an on-site survey) engineering requirements associated with facility enhancements for physical security and antiterrorism, including an IDS installation. This type of survey should be conducted when planning new construction, renovations, or upgrades to existing facilities where there are likely to be physical-security requirements. A security-engineering survey may also be requested by the PM or an equivalent security officer to evaluate existing security. This survey—

- Identifies assets to be protected.
- Identifies threats to these assets and the level of protection required to protect them.
- Identifies the protective measures.
- Determines the cost of the protective measures.

## Appendix A

# Metric Conversion Chart

This appendix complies with current Army directives which state that the metric system will be incorporated into all new publications. Table A-1 is a conversion chart.

**Table A-1. Metric Conversion Chart**

| Metric to English | | | English to Metric | | |
|---|---|---|---|---|---|
| **Multiply** | **By** | **To Obtain** | **Multiply** | **By** | **To Obtain** |
| **Length** | | | | | |
| Centimeters | 0.0394 | Inches | Inches | 2.54 | Centimeters |
| Meters | 3.28 | Feet | Feet | 0.0305 | Meters |
| Meters | 1.094 | Yards | Yards | 0.9144 | Meters |
| Kilometers | 0.621 | Miles (stat) | Miles (stat) | 1.5609 | Kilometers |
| Kilometers | 0.540 | Miles (naut) | Miles (naut) | 1.853 | Kilometers |
| Millimeters | 0.039 | Inches | Inches | 25.40 | Millimeters |
| **Area** | | | | | |
| Square centimeters | 0.1550 | Square inches | Square inches | 6.45 | Square centimeters |
| Square meters | 10.76 | Square feet | Square feet | 0.0929 | Square meters |
| Square meters | 1.196 | Square yards | Square yards | 0.836 | Square meters |
| **Volume** | | | | | |
| Cubic centimeters | 0.610 | Cubic inches | Cubic inches | 16.39 | Cubic centimeters |
| Cubic meters | 35.3 | Cubic feet | Cubic feet | 0.0283 | Cubic meters |
| Cubic meters | 1.308 | Cubic yards | Cubic yards | 0.765 | Cubic meters |
| Milliliters | 0.0338 | US liq ounces | US liq ounces | 29.6 | Milliliters |
| Liters | 1.057 | US liq quarts | US liq quarts | 0.946 | Liters |
| Liters | 0.264 | US liq gallons | US liq gallons | 3.79 | Liters |
| **Weight** | | | | | |
| Grams | 0.0353 | Ounces | Ounces | 28.4 | Grams |
| Kilograms | 2.20 | Pounds | Pounds | 0.454 | Kilograms |
| Metric tons | 1.102 | Short tons | Short tons | 0.907 | Metric tons |
| Metric tons | 0.984 | Long tons | Long tons | 1.016 | Metric tons |

## Appendix B

# Sample Installation Crime-Prevention Handbook

This appendix provides guidance on planning, organizing, directing, and controlling installation crime-prevention programs. It provides guidance on developing an installation program, criminal analyses to identify crimes, guidance on which crimes to address, command and individual countermeasures for particular crimes, and program-evaluation procedures.

## SECTION I — INSTALLATION CRIME-PREVENTION PROGRAMS

B-1. In the past few years, the Army has shifted an increasingly larger percentage of its manpower from combat-service-support activities to combat organizations. This change means that fewer MP personnel are available to support a larger number of units. To meet this challenge, it is necessary to reevaluate the way we do business and to emphasize those programs or procedures that have the greatest impact on our installation crime rates. Crime prevention is one program that can have a major impact on installation crime rates at a relatively minor cost in both dollars and manpower. It takes less effort to discourage a criminal from perpetrating a crime or to teach a soldier to avoid becoming a victim than it does to investigate a crime, identify the offender, prosecute him, and punish him. In addition, a proactive approach to law enforcement can help maintain the high quality of service life that can improve the retention of first-term soldiers.

B-2. The Army is a large organization that performs a variety of activities in many different environments. Crimes that are major problems on one installation may be totally absent from others. For example, most military installations have a significant number of robberies while most depots have none. Because of this, any rigid, centrally controlled program—no matter how carefully thought out—is bound to be inappropriate in many locations. Therefore, DA has elected to provide only the most general guidance and to allow commanders to develop crime-prevention programs that address their local problems.

### CRIME-PREVENTION WORKING GROUPS

B-3. The installation is the smallest practical level for implementing crime-prevention programs. If these programs are developed and implemented at a lower level, then crime is often not eliminated but is merely displaced from units with good programs to units with less effective programs. Also, crime does not affect personnel only when they are in their place of duty. In many cases, a company commander's troops are victimized in areas over which he has little control. Unit commanders are responsible for implementing many anticrime measures; however, the selection of overall program goals, the ID of appropriate countermeasures, and quality control should be done at the installation level.

B-4. Crime prevention must always be recognized as a commander's program rather than as an MP program. MP personnel have the expertise to analyze data, identify major problems, and develop lists of possible countermeasures. They should perform these functions in support of an installation crime-prevention council appointed by the installation commander and composed of representatives of all of the installation's major organizations and activities. The advantages of using this type of system are—

- It provides representatives of all major segments of the post population with a forum where they can identify criminal problems that are of the greatest concern to them.
- It allows the representatives of all major commands to review the available options to counter a crime and to select the level of resource commitment that is compatible with their missions and internal priorities.
- It helps ensure that the resources of the entire community, rather than only those of the MP force, are mobilized to attack the problem.
- It is easier to obtain the support of the whole population if its representatives are instrumental in the development of the program.

## CRIME-PREVENTION OFFICERS

B-5. The installation's crime-prevention officer is normally a senior NCO or an officer who has a solid background as an MP investigator or a physical-security inspector (PSI). He supports the installation council by performing a crime-data analysis to identify problem areas, drafting programs for the council's consideration, inspecting the implementation of council-mandated measures, and coordinating the efforts of unit crime-prevention officers in the implementation of the crime-prevention program.

B-6. As a member of the PM's staff, the crime-prevention officer develops the law-enforcement section of the crime-prevention program, develops and maintains the written crime-prevention plan, and coordinates crime-prevention programs with civilian police agencies and community groups.

B-7. Crime-prevention officers are also appointed in each organization down to the company level. At this level, written crime-prevention plans are not required; however, SOPs are established. The crime-prevention officers serve as their organizations' focal points for coordinating installation crime-prevention plans; they supervise the implementation of the installation's program within their organizations.

## CRIME-PREVENTION PROGRAM DEVELOPMENT

B-8. The starting point for developing a crime-prevention program must be a thorough analysis of criminal activity on the installation. This identifies significant criminal problems that are susceptible to crime-prevention efforts. Crimes that are most susceptible to crime-prevention measures are those for which a high probability of reoccurrence exists. Crimes such as murder normally are not repetitive and are poor candidates for inclusion in the crime-prevention program. Since it is seldom practical to attack all criminal problems simultaneously, they should be prioritized based on their impact on the command's ability to perform its mission and their impact on installation personnel. Next, the whole range of countermeasures that can be used to combat each problem must be identified (see Figure B-1). Table B-1, page B-4, identifies (by offense) programs that have been successful in countering specific criminal problems. Sections III and V of this appendix contain discussions of the strengths, weaknesses, and applicability of the countermeasures listed in Table B-1. Once developed and prioritized, the list of criminal problems and possible countermeasures must be presented to the installation crime-prevention council for action. The council should decide which crimes will be addressed and which countermeasures will be used for each crime. The council must then identify specific objectives for its anticrime campaigns.

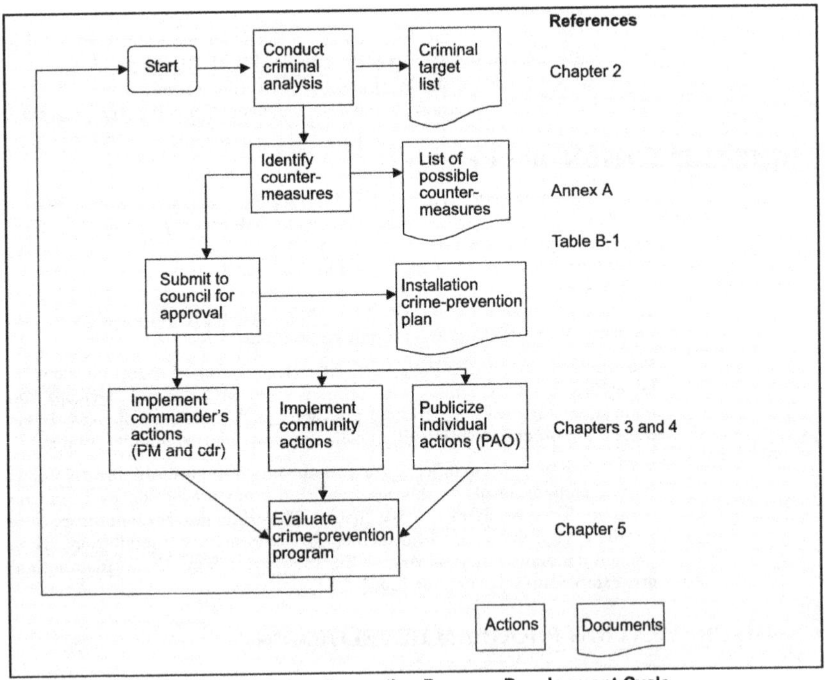


**Figure B-1. Crime-Prevention Program-Development Cycle**

**Table B-1. Offenses Countermeasure Matrix**

| Crime | Command/Law-Enforcement Countermeasures | | | | | | | | | | | | Community Programs | | | | | |
|---|---|---|---|---|---|---|---|---|---|---|---|---|---|---|---|---|---|---|
| | Crime Hot line | Lighting | Environmental Changes | Patrol/Surveillance | Publicity Campaigns | Residential Security Survey | Juvenile Programs | Fraud Programs | Employee Theft | Arson Programs | CCTV | Warning Signs | Neighborhood Watch | Operation ID | Neighborhood Walks/Escorts | Mobile Patrols | Project Lock | Education |
| Arson | | | | X | X | | | | | X | | | X | | | X | | X |
| Auto theft | X | X | | | X | | | | | | | | X | | X | X | X | X |
| Burglary/ housebreaking | X | X | X | X | X | X | | | | | | | X | X | X | X | | X |
| Employee theft | X | | | | X | | | | X | | X | X | | X | | | | X |
| Fraud | X | | | | X | | | X | | | | X | | | | | | X |
| Larceny | X | | X | | | X | | | X | | | | | X | | | | X |
| Rape | X | X | X | X | X | X | | | | | | | X | | X | X | | X |
| Robbery | X | X | X | X | X | | | | | | X | | X | | X | X | | X |
| Juvenile delinquency | X | | | X | X | | X | | | | | | X | | X | X | | X |
| Vandalism | X | X | | X | X | | X | | | | | | X | | X | X | | X |

B-9. Objectives must identify—

- What crime will be reduced.
- What target population will be addressed.
- What specific changes and behaviors on the part of the victims or perpetrators will be encouraged.
- What actions the command must take to reduce the opportunity for the crime to occur.

B-10. Once objectives have been clearly defined, specific areas of responsibility should be assigned to each council member (based on their organization's primary area of responsibility) and major milestones should be identified for developing the campaign against each targeted crime.

## TRAINING

B-11. The prerequisite skills for successful performance as an installation crime-prevention officer are best developed through on-the-job experience as the supervisor of MP investigations or physical-security inspections. More important than any technical skill is the cultivation of a frame of mind that instinctively examines each case to determine not only what occurred, but also how the crime could have been prevented. Technical skills (such as criminal-data analysis) that may not have been developed as an MP investigator or physical-security supervisor are presented in courses taught by several civilian agencies. These classes should be used to the fullest extent possible.

## CIVILIAN CRIME-PREVENTION ORGANIZATIONS

B-12. There are many civilian crime-prevention organizations at the national, state, and local levels. Many of these organizations have produced crime-prevention material (including posters, radio spots, and leaflets). Material and programs sponsored by civilian agencies should be used to support Army crime-prevention efforts. However, when material from a source outside of DOD is used, a copyright release must be obtained. Normally, it is necessary to get a release for each separate item that is used. If there is any doubt as to the necessity of securing a copyright release, the crime-prevention officer should refer the matter to the local SJA.

# SECTION II — CRIMINAL ANALYSIS

B-13. Criminal analysis is a system for identifying trends and patterns where they may exist. It is a routine, ongoing function for the PM and battalion- and brigade-level staffs. Criminal analysis is the foundation upon which the installation force-protection program is based. Moreover, criminal analysis is an integral component of the police intelligence-operations function and is applicable across the operational continuum. An effective criminal analysis establishes the following:

- Crimes having a significant impact on the installation.
- The segments of the population being victimized.
- The ID of criminals/perpetrators.
- The most common time of occurrence.
- The areas that experience the highest number of incidents.
- Offense information (such as types of weapons or victims' actions that contribute to the offense).
- Information critical to an installation's VA.
- Information essential in formulating a successful patrol-distribution plan.
- Police information and criminal intelligence fused with tactical intelligence.

B-14. With this type of information, specific countermeasures are developed to reduce the opportunity for a crime to occur or to remove the incentives for perpetrators. Without an effective criminal analysis, the overall security effort is unfocused. Moreover, the installation/base patrol-distribution plan may be skewed. When this occurs, broad countermeasures are implemented and meaningful results are lost.

B-15. A professional analysis of criminal data is essential for protecting soldiers, units, and installations. The Military Police Management Information System (MPMIS) provided the initial software capable of assisting in the development of criminal-analysis information. As this system is modernized and replaced by the Military Police Automated Control System (MPACS), the MP Corps and the CID will have interface with the Army Battle Command System (ABCS). The analysis of police, criminal, and tactical information results in the development of police intelligence (PI). The PI, which is coordinated with the Assistant Chief of Staff, G2 (Intelligence) (G2) or the Intelligence Directorate (Joint Command) (J2), provides relevant information and intelligence (RII) that greatly contribute to the commander's critical information requirements (CCIR). The RII provided vertically and horizontally ensures that the common operational picture is properly shared.

## SOURCES OF INFORMATION

B-16. The basic source for identifying crimes that warrant examination is the installation's Law Enforcement and Discipline Report (DA Form 2819). The information in this report can be used to plot the amount of and seasonal variations for each major type of crime. This will tell little about conditions that produce the crime, but it is useful in identifying crimes that can be eliminated from detailed analyses due to low frequency rates.

B-17. The Law Enforcement and Discipline Report identifies which major category of crime should be targeted, but it does not tell which type of crime is causing the most problems. For example, It may indicate that robbery is a problem, but it will not discriminate between robbery of commercial establishments (such as the post exchange [PX], the Class VI store, or banks) and muggings of individuals. Since countermeasures for these two types of robbery are different, it is necessary to collect additional information. The best sources for this information are the military-police report (MPR) and the CID report of investigation (ROI). If there are fewer than 200 cases in the past year for a particular type of crime, they should all be examined. With a larger annual caseload, a random sample that is large enough to give results of plus or minus 5 percent accuracy should be examined.

B-18. General factors must be identified for any type of crime. These are the types of victims, the perpetrators, geographical data, and chronological data.

### VICTIMS

B-19. It is essential to determine (as precisely as possible) the segment of a post population that is being victimized. Junior soldiers live in different areas on post than officers and senior NCOs. They patronize different clubs and, for the most part, work in different areas. Information programs must use multiple sources to ensure that the same message reaches the different portions of the population. If the specific population segment that is being victimized is not identified, it is possible to spend a large amount of resources and have little impact on the targeted crime because the message is not getting to the people who need it.

## PERPETRATORS

B-20. As with the victim profile, data on perpetrators is essential to ensure that countermeasures are targeted against the correct population. When on-post auto thefts are committed by civilians with legitimate reasons for entering the installation, it may not make sense to increase security (as a countermeasure) at post entrances. Countermeasures are developed based on specific profiles identified by case type and the information described above. Successful countermeasures depend on correct criminal analyses. An analysis and profile of perpetrators will help to focus MP/CID countermeasures.

## GEOGRAPHICAL DATA

B-21. The MPR and the ROI should contain a street address or a description of the incident's location (such as "in a parking lot on the east side of building 1409"). This type of description fulfills requirements for identifying the location of the incident in court. It is further enhanced by identifying the normal duty hours and the type of activity that takes place in the location (for example, "in a parking lot on the east side of the NCO club (building 1401), normal operating hours 1800-0400" or "in a parking lot on the east side of building 1408, a troop billets for Company A, 15th Cav"). This type of information may help to develop a list of specific types of areas where a particular crime is occurring, making it easier for the PM to provide increased MP patrols in these areas. For example, if there have been several incidents in which vehicles have been broken into and stereos removed, it would help to know that 90 percent of them occurred in the parking lot by troop billets.

## CHRONOLOGICAL DATA

B-22. As with geographical data, the more specific the time-of-occurrence pattern is for a crime, the easier it is to apply sufficient resources to affect the crime rate. For each crime having a significant impact, determine the following:

- Major seasonal variations.
- Monthly variations. Is there a concentration of crimes immediately before or after payday?
- Weekend occurrences. Is there a concentration of incidents on weekends? Each day represents about 14.25 percent of the week. Concentrations of crime higher than that for any particular day may be significant.
- Time of day. Is there a particular period that accounts for a disproportionate share of the incidents?

## CRIME-SPECIFIC FACTORS

B-23. In addition to the factors that should be examined for all crimes, the following crime-specific factors are useful in analyzing specific offenses:

- Housebreaking and burglary.

  - The type of building that was attacked (family-housing unit, troop barracks, PX, and so forth).
  - Whether the facility was occupied or unoccupied.
  - The point of entry (door, window, and so forth).
  - The method of entry (unsecured door, forced door, forced window, and so forth).
  - The property that was stolen. (Was it marked?)
- Robbery.
  - The number of perpetrators.
  - The perpetrator's method of operation.
  - The type of weapon used.
  - The type of robbery (street mugging, residential robbery, or commercial robbery).
  - The victim's injuries.
  - Actions by the victim that contributed to his being targeted.
- Larceny.
  - The type of property taken.
  - Whether the property was secured or unsecured.
  - The perpetrator's method of operation.
- Auto theft.
  - The type of vehicle stolen (POV, motorcycle, and so forth).
  - Whether the vehicle was secured or unsecured.
  - Whether the vehicle was recovered and where it was recovered.
  - Whether the vehicle was stripped of parts.
  - The perpetrator's method of operation.
- Forgery.
  - The type of document that was forged.
  - How the document was obtained.
  - The type of ID used in passing the forged document.
  - The con games or techniques used.
- Rape and sex offenses.
  - The perpetrator's method of operation.
  - The relationship between the victim and the perpetrator (blood relatives, acquaintances, or strangers).
  - The degree of force used.
- Aggravated assault or murder.
  - The relationship between the victim and the perpetrator.
  - The motivation.
  - The weapon used.

B-24. Personnel who have tried to use MPRs and ROIs as a sole source of data know that in many cases the required information is not contained in sufficient detail to be useful. To correct this situation, investigators and inspectors must be trained to recognize conditions that contributed to the crime as well as the information needed to identify and prosecute the offender. As a minimum, the PM ensures that all law-enforcement personnel know general and specific crime factors that are required to analyze each type of crime. He must also provide feedback through the law-enforcement operations staff when incomplete reports are received. If this is done, high-quality data can be collected without generating additional reports to collect criminal information for analysis.

B-25. The MPRs and ROIs are not the only sources of information on which to base criminal analyses. Physical-security inspection results, summaries of common deficiencies noted by the inspector general, data from the SJA claims section, and information summaries from reports of survey on lost government material can all produce worthwhile information on conditions that lead to the commission of crimes. Additionally, a review of civilian criminal statistics and civilian police information may contribute to the pattern analysis. The reviewing authority must understand which information is relevant. The MP commanders ensure that the CCIR are published in operations orders (OPORDs), operation plans (OPLANs), and SOPs. This effort ensures that MP forces at every level recognize relevant information.

## INDIVIDUAL CRIMINAL ANALYSIS

B-26. A criminal analysis is most effective when it is applied to the class of criminal offenses with a high probability of recurrence. Single-incident crimes do not lend themselves to analyses. Most crimes against persons do not usually benefit from analyses, with the notable exceptions of rape, robbery, and related combinations of offenses (such as kidnapping and rape; robbery and attempted murder; burglary and rape; and burglary, robbery, and kidnapping). Analyzing isolated criminal offenses has some value (such as gaining knowledge of where these offenses are most likely to reoccur). However, this knowledge is usually difficult to use effectively for preventive purposes.

B-27. There are universal factors available for analysis for most crimes. The availability of these factors varies greatly between criminal types and specific reported offenses.

B-28. In addition to the universal factors, there are an almost infinite number of factors that may be considered specific to a particular criminal class or type. These crime-specific factors are data elements that are usually recorded during the reporting of a particular type of offense and are used for analysis purposes.

B-29. Crime-specific factors provide information that can be used by the analyst to connect crimes with similar characteristics. Information regarding physical evidence may have considerable value in the analysis of several criminal types (such as burglary and auto theft). Therefore, the suitability of different crimes to analyses depends on the general factors, specific modus operandi (MO) factors, and physical evidence. With these considerations in mind, different criminal types and classifications are examined to determine their applicability.

## RESIDENTIAL HOUSEBREAKING AND BURGLARY

B-30. Although housebreaking and burglary are two of the most difficult crimes to prevent, they are the most suitable for analyses. The typical housebreaker or burglar establishes an MO pattern based on successful past offenses. Usually, a burglar will continue to commit similar crimes until he is apprehended.

B-31. The available information for evaluation must consider the specificity, accuracy, and value of the information received. Informational factors are presented in the order of relative importance. It should be emphasized that the factors listed in Figure B-2 are for analysis purposes; the order for investigative solutions may vary.

## COMMERCIAL HOUSEBREAKING

B-32. The analysis of commercial housebreaking is in many respects easier than that of residential housebreaking or burglaries. More pertinent information is available to the analyst for analyzing commercial housebreaking (which is more specialized and will exhibit more specific MO characteristics). The analyst normally has specific information regarding the point and method of entry, the victim/target description, and the property-loss description. The time factors for commercial housebreaking may be of less value in a commercial-housebreaking analysis than in a residential-burglary analysis since most commercial housebreakings occur at night or during weekends.

B-33. The commercial housebreaker is generally more mobile than the residential burglar. An analysis of commercial housebreaking is not as

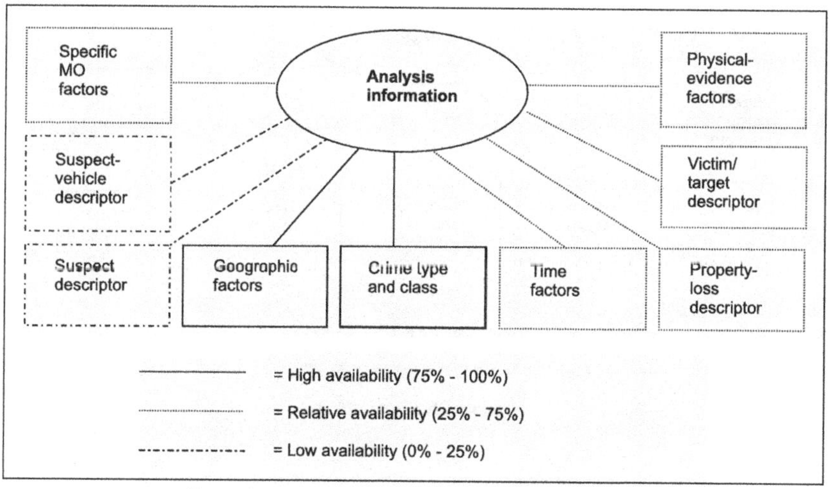


**Figure B-2. Residential Housebreaking and Burglary**

restricted in geographical area as an analysis of residential burglaries. The commercial burglar may travel a considerable distance to attack a particular type of business. The residential burglar is less discriminating because he has a greater number of potential targets.

B-34. The analysis of commercial housebreaking should be systematically directed to examining the information factors shown in Figure B-3. A commercial housebreaker may be a juvenile criminal, an addict, a professional thief, a soldier, or an enemy agent. Each of these types provides a different set of factors to analyze.

## ROBBERY

B-35. Robberies are well suited for analysis. This class of criminal offender usually operates in a given geographical area. The commercial robber generally seeks a sizeable amount of cash; while targets of the street robber frequently include credit cards, checks, and other valuables in addition to cash.

B-36. The presence of physical evidence is more probable in street robberies than in commercial robberies because the criminal offender frequently discards evidence (such as purses and wallets) after completing the offense. The mugger frequently uses surprise as part of his MO in order to reduce his chance of apprehension by physical ID.

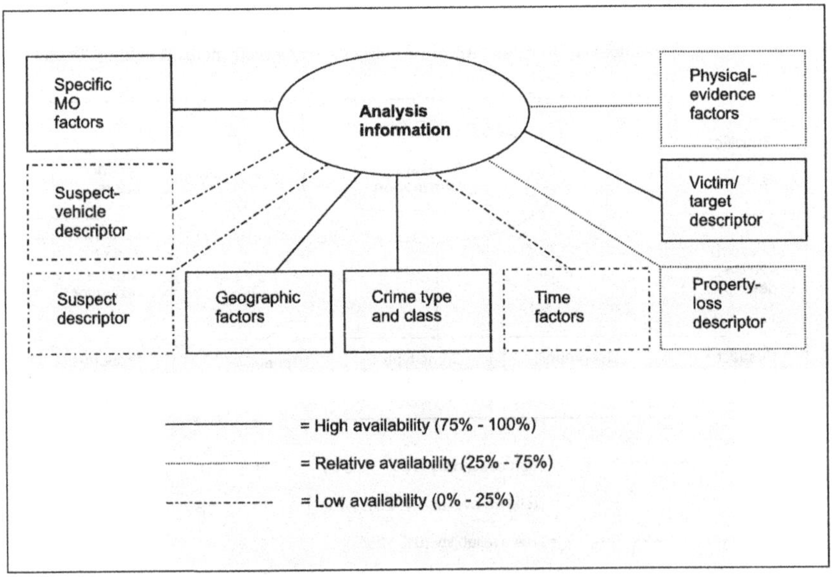


**Figure B-3. Commercial Housebreaking**

B-37. The probability of universal and crime-specific informational factors for a typical robbery is a systematic method for examining the case. See Figure B-4 for the analysis of a robbery case.

## AUTO THEFT

B-38. Auto theft is probably the best-suited crime for analysis. The analyst frequently has more information available on any given auto theft than any other crime. Suspect descriptive information is usually lacking in auto larceny cases. However, the greater availability of crime-specific information makes auto theft extremely receptive to an analysis.

B-39. Auto-theft offenders include joyriders, professional car dismantlers, and wholesalers of stolen vehicles. In many cases, these offenders will establish and maintain a particular MO until apprehended.

B-40. Helpful elements in analyzing auto-theft cases are the availability of information, the presence of physical evidence, and the ability of the stolen property to be traced. Of special importance are the geographic factors, suspect-vehicle descriptors, and property-loss descriptors. In analyzing an auto theft, the analyst usually has information concerning two geographic locations for analysis—where the vehicle was stolen and where it was recovered. The suspect-vehicle descriptors, property-loss descriptors, and victim/target descriptors are the same (see Figure B-5). This greatly augments the analysis of auto thefts. In an examination of specific MO factors in an auto

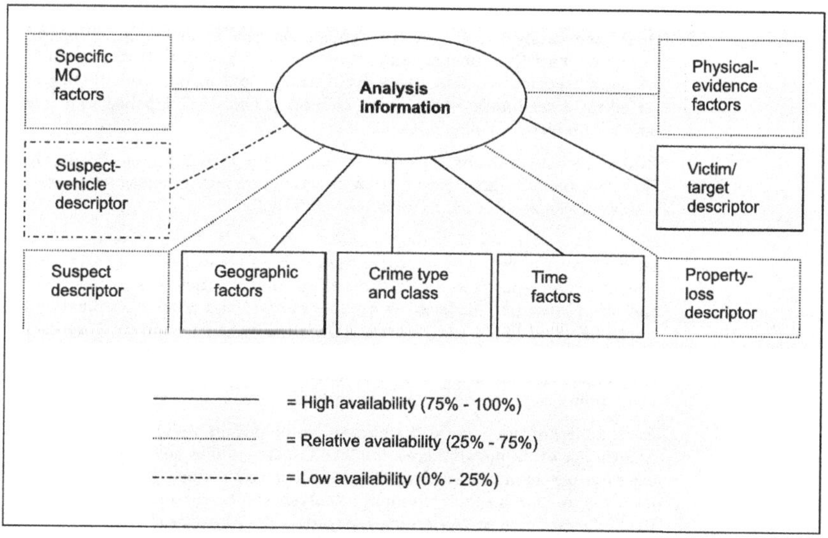


Figure B-4. Robbery

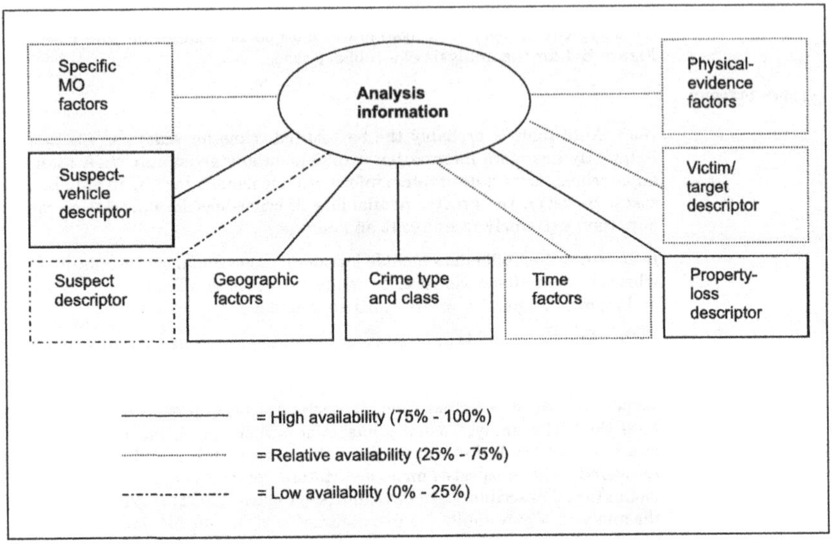


**Figure B-5. Auto Theft**

theft, the analyst will place emphasis on the vehicle's condition when recovered. Factors of primary importance in analyzing auto-theft cases are the vehicle's recovery location, the vehicle's make and model, and the degree of stripping. Coordination with local civilian police or HN police may reveal potential buyers of stolen automobiles.

B-41. Larceny and theft cases are not usually well suited for analysis. However, benefits are derived from analyzing larceny offenses when specific factors are selected, such as auto accession thefts.

B-42. The problem with trying to analyze general theft cases results from the large volume of reported offenses, the large number of possible crime types, and the many possible MO patterns. To analyze theft cases, the analysis operation must first restrict the number of cases and crime classification to a workable level. This may be done by classifying general theft cases into special categories and analyzing them by considering only special MO factors. Logical classifications include thefts of autos, auto accessories, bicycles, items on shipping docks, and tools and equipment.

B-43. Many of the general theft cases are helpful in analyzing other types of crimes. For example, an increasing trend in stolen automobile parts may have common perpetrators involved in auto theft and stripping. When particular patterns appear, a specific in-depth analysis can be conducted. It is essential that information developed during the analysis process be shared with the G2/ J2, the SJA, the CID, and others (as authorized).

## RAPE AND SEX CRIMES

B-44. Rape and sex crimes fall into two distinct categories—those in which the offender is known to the victim and those in which he is unknown (stranger to stranger). Those cases in which the victim knows the offender are of limited informational value to the analyst. The in-depth analysis of rape and sex crimes is restricted to those cases in which there is no apparent relationship between the suspect and the victim. Rape and sex offenses in which the victim knows the suspect are often crimes of opportunity. Usually, these offenders do not establish a particular MO. On the other hand, the rapist or the sex offender committing stranger-to-stranger crimes will usually provide definite MO patterns, making this crime well suited for analysis. For example, the burglary/rape motive is particularly well-suited for analysis, as is the kidnapper/child molester. The advantages to analyzing stranger-to-stranger rape and sex offenses lies in the seriousness of the crime, its relative rareness, and the availability of information for a particular offense or group of offenses.

B-45. As with robbery, the suspect descriptors are important to the analyst for examining rape and sex offenses (see Figure B-6). The victim descriptors are also important in analyzing rape and sex offenses. In many cases, the perpetrator will restrict his attacks to victims of a certain age, a particular race, or a particular occupation grouping.

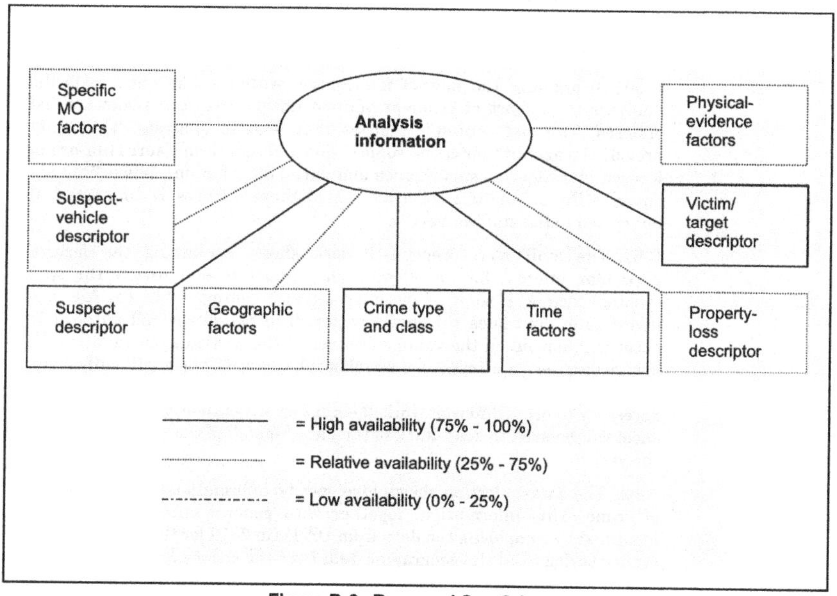


Figure B-6. Rape and Sex Crimes

## FORGERY AND FRAUD

B-46. Forgery and fraud cases can provide correlative information for other reported criminal offenses such as burglary and strong-armed robbery. The analysis of forgeries is also aided by the fact that most forgers are repeaters who have established definite MO patterns.

B-47. An analysis of forgery and fraud cases can generate information that is disseminated to area merchants in the form of check-warning bulletins. The use of these bulletins can enhance community relations.

B-48. The primary problem in analyzing forgery and fraud cases is the immediate availability of information. In some cases, a delay of between three or four days to several months occurs between the time the offense was committed and the time it was reported.

## ASSAULT AND MURDER

B-49. Most criminal offenses constituting assault and murder do not lend themselves to analysis. The comparative rarity of these crimes involving complete strangers makes it difficult to plot and predict assaults and murders. A tremendous amount of time and research is necessary to predict assaults and murders effectively. This data may identify specific areas with high rates of violent crime. This allows the commander to identify unsafe areas and warn soldiers to stay away from them.

# CRIMINAL-ANALYSIS PROCEDURES

B-50. To organize the flow of information within a PMO and to facilitate analyses in support of crime-prevention efforts, the staff reviews a list of offenses. This list contains offenses that lend to analyses. The analyses provide the azimuth for crime suppression and countermeasures (for example, focused investigation/surveillance and patrol distribution). When final action on an MPR or an ROI covering one of these offenses is completed, it is forwarded to the staff for review.

B-51. The staff uses preprinted work sheets containing the factors of particular interest for an offense. They annotate and review the specific criminal factors. If some of the information is not available, the MP or CID investigator continues the investigation. The period covered by each work sheet will depend on the volume of cases at the particular installation. For a low-incidence-rate crime (such as robbery), one work sheet will suffice for the entire calendar year. For a more frequent crime (such as larceny), it may be necessary to use a different work sheet for each month or quarter. A summary sheet can be used to keep track of the data from all of the reporting periods in the year.

B-52. The data-collection sheets give specific information on individual types of crime. This information, together with general crime trends that are identified by comparing the data from DA Form 2819 for the current period to earlier periods and the geographic data from the crime-occurrence map, gives most of the information that is required to develop tightly targeted crime-prevention programs. While this system requires additional forms (the data-collection sheets and the crime-occurrence map) to be developed for use by the

MP Intelligence Officer (US Army) (S2), it does not increase the patrol's workload.

B-53. In addition to the information recorded in the forms, handwritten notes should be kept on items or trends that become apparent to the analyst during the review. For example, as a result of reading the cases on robbery to extract information for the collection form, the analyst may notice that a high proportion of the victims were attacked while crossing an unlighted athletic field near a troop billet area. This should be recorded so that it may be used in publicity campaigns and so that it may be passed along to patrols who routinely patrol the troop billet areas.

B-54. A geographic analysis is performed to determine information not available in the original data-element source (for example, a crime report). When a pin is used to indicate a crime location on a map, the relationship to other reported crimes of the same type becomes apparent to the analyst. Mapping is usually the source for a geographic analysis.

B-55. Mapping-analysis techniques involve the use of a map to depict the actual geographical relationships between particular criminal events according to prescribed data elements. A pin map displaying the actual locations of burglaries over a period of time is an example of such a technique.

B-56. In selecting a mapping technique, several things must be considered. These include the number of data elements that are to be recorded in the mapping, the retrievability of stored data, and the number of maps to be maintained.

B-57. The number of data elements to be recorded on a map can dictate the technique to be used. For example, if the case number is to be recorded in conjunction with the location of the offense, a dot (color-coded, self-adhesive paper disk) map or a flag-pin map can be used, whereas a typical color-coded pin map cannot. If particular MO factors are recorded, various color-coded and marked pins can be applied to identify specific MO patterns.

B-58. In addition, the length of time a particular map or set of maps is maintained should be considered. No set rules are established regarding a map's maintenance time; however, a number of agencies develop annual statistics based upon maps. Generally, maps are maintained for short-term (quarterly) and long-term (annually) periods for each type of crime. This decision should be based on the volume of criminal activity, physical-space limitations, the specific mapping technique adopted, and the number of maps to be maintained. As geospatial-terrain data becomes available, mapping techniques can be fully integrated into the MPACS.

B-59. Stored map data can be easily retrieved for comparing data. Pin maps can be photographically recorded and digitized for software applications. Computers and scanning equipment can be used to store the data for later retrieval. Color-coded paper dots placed on acetate overlays covering area maps facilitate the retrieval of recorded data. Digitized photographs of the maps are excellent sources of evidence for trials and court-martials. Moreover, the developed maps may contribute to the CCIR.

B-60. The number of maps to be maintained must be based on physical-space limitations, staffing, the types of crimes selected for analysis, and the maintenance period for each map. There are five types of crime suited to geographical analysis (burglary, robbery, housebreaking, auto theft, and sex crimes). Other crimes (such as theft of government property) may be depicted as a geographical analysis and contribute to RII. Housebreaking and burglaries can be placed on the same map by using different colored pins for each type of crime. The incorporation of numbered pins of different colors can be used to designate the type of premise attacked. Auto-theft and recovery maps may be kept on the same map with recorded information on thefts of vehicle parts.

B-61. The number of maps maintained should reflect the needs of the agency according to the volume of reported crimes and the different data elements or informational factors to be recorded. Therefore, the more detailed the analysis, the more maps are required. The actual number of maps maintained must be left to the discretion of the PM or the commander.

B-62. Periodically, a summary of the information that is collected, countermeasures, and operational/tactical procedures are briefed to the installation commander. Whether this is done monthly, quarterly, or semiannually will depend on the volume of reported crime and the commander's preference.

## CRIMINAL-ANALYSIS SUMMARY

B-63. Criminal analysis is an important element to the MP function of PI. The techniques mentioned in this appendix are not all-inclusive. The PM/CID commander establishes formal procedures for integrating data and developing PI. It may require the formation of an ad hoc team, or it may be accomplished with computers and the MPACS. Criminal analysis requires close coordination with the G2/J2, the public affairs office (PAO), civilian police, and others based on METT-TC.

# SECTION III — COMMAND AND LAW-ENFORCEMENT COUNTERMEASURES

B-64. This section discusses actions that commanders or PMs can take to reduce crime on Army installations. In many cases, the actions described have applications other than strictly crime prevention. However, only the crime-prevention aspects are discussed.

B-65. Most of the data available on the effectiveness of law-enforcement measures in reducing crime comes from studies conducted by civilians. While most of the findings of these studies are applicable to crime-prevention programs on Army installations, differences in population, the degree of control that can be exercised by authorities, and other environmental factors may dictate that the civilian recommendation be modified before implementation on military installations.

## CRIME HOT LINES

B-66. A crime hot line is a dedicated crime-reporting telephone number located at the MP or security desk. This hot line allows anyone in the community to make an immediate report of an observed crime or suspicious activity. An effective crime-reporting program is a deterrent to crime and enhances law-enforcement responses to such incidents. Considerations in implementing a reporting program include the following:

- It should be publicized that personnel reporting incidents are allowed to remain anonymous if they desire. Some individuals will report their observations only if they know they can remain anonymous. If deemed feasible, neighborhood-watch blocks could be provided designated block numbers that could be used when reporting crimes or other suspicious activities. This system would allow neighborhoods to receive feedback on the disposition of the reported incident.
- The hot line's phone number should be easy to remember. This could include a number where extension digits are all the same, are in ascending or descending order, or spell out a word (such as 4357 [HELP]). Sticker labels listing the number could be placed on the phone with other emergency numbers for quick reference.
- The program should be well-publicized.
- Members on the installation should be educated on the desired procedures for reporting incidents.
- Each call should be documented. Records should be maintained concerning the results of these calls to evaluate the program's effectiveness.

## CRIME PREVENTION THROUGH ENVIRONMENTAL DESIGN

B-67. Crime-prevention practitioners are recognizing the importance of considering design and physical planning in crime reduction. The crime-prevention officer has an opportunity to influence the design of facilities through the installation planning board. However, to be effective, he must understand a number of concepts about the relationship between the physical design of buildings and crime occurrences. These include the concepts of territoriality, natural surveillance, and defensible space.

### TERRITORIALITY

B-68. Historically, a single-family home on its own piece of land and somewhat isolated from its neighbors (but often by as little as a few feet) has been considered to be the family's territory. The single-family home sits on a piece of land buffered from neighbors and the public street by intervening grounds. At times, symbolic shrubs or fences reinforce a boundary. The positioning of lights in windows that look out on the grounds also act to reinforce the claim.

B-69. Unfortunately, as the population has grown and the need for housing has increased, the trend toward developing single-family units has been paralleled, if not surpassed, by the development of row houses, apartment buildings, and various high-rise structures. Architects, planners, and designers involved in developing structures have not paid a great deal of attention to crime control or the need for an individual or a family group to identify with its home in a manner that might affect crime. Therefore, most families living in apartment buildings consider the space outside their apartment door to be distinctly public. In effect, they relegate responsibility for all activity outside the immediate confines of their apartment to public authorities. A question is whether environmental design can be used to extend the boundaries of these private realms, subdividing public space outside quarters so that more of the common space comes under the resident's influence and responsibility.

B-70. Through extensive research of efficiently functioning housing developments, a number of mechanisms have been identified that may be used in the design process (or may be added after construction). These mechanisms encourage the residents of multifamily dwellings to identify more with the ground or area around their immediate home site and to assume responsibility for its protection. Presented below is a brief discussion of a number of the mechanisms.

- **Site design.** If the grounds around a set of quarters can be directly identified with a particular building and the residents of that building take a personal interest in the use or upkeep of that area, they will play a role in protecting it. Through proper site design, a recreational area adjoining a building may be used as a buffer zone by providing play equipment for young children and seating areas for adults. The fact that children play and adults sit in these areas serves to increase the residents' concerns with the activities taking place there. Strangers are usually recognized and their activities come under observation and immediate questioning.
- **Street design.** Research has shown that by the placement, enclosure, or rerouting of streets and traffic the nature of a particular area can be changed and the crime rate reduced. For example, a particular portion of a street might be closed to vehicular traffic, and play equipment and seats may be added. In a number of areas where this technique has been used, it has been found that most residents know or at least recognize people up and down the block and strangers on the street are identified. Similar approaches that involve rerouting traffic, using one-way streets, or blocking off streets has lowered the crime rate in some areas.
- **Symbolic barriers.** The types of barriers that planners may use in laying out an area include open gateways, light standards, low walls, and plantings. Both physical and symbolic barriers serve the same purpose—to inform an individual that he is passing from a public to a private space. Symbolic barriers identified by residents as boundary lines serve as defining areas of comparative safety. Many places warrant the use of symbolic barriers, including transition points between a public street and the semipublic grounds of a building; an area between a building's lobby and its corridors; or hallways on particular floors of a building.
- **Internal design.** Although economics may sometimes enter the picture, a building's interior may be designed for specific groupings of apartment units and shared entrances. These factors may cause the residents of these apartments to develop a concern for the space immediately adjacent to their dwelling. For example, on each floor of an apartment building, two to four families might be required to share

a common corridor area. The apartment doors would be grouped around that common corridor, and access to elevators or stairs might be screened by a glazed partition. The net effect would be that the floor's residents would adopt the corridor as a collective extension of their dwelling unit and would take an increased interest in its maintenance and use.

- **Facilities and amenities.** The location of particular facilities (such as play and sitting areas and laundry facilities) will tend to give an area a high intensity of use and support the idea of territoriality. The presence of residents involved in various activities (children at play and people chatting or engaged in other types of activities) allows for casual surveillance by concerned members of the family and screens out possible intruders.

B-71. Reducing the number of apartment units grouped together to share a collectively defined area and limiting the number of buildings that comprise a housing project are important factors for creating an environment that residents will help to protect. Research has documented the fact that housing projects comprised of fewer high-rise buildings (two to four) have lower crime rates than projects containing a larger number of buildings. Based on this finding, it is argued that there appears to be much less freedom of movement in the public spaces of the smaller high-rise projects. Unlike buildings and large developments, every building of a small grouping usually has an entrance directly off a public street. These dwellings more closely resemble middle-income, high-rise developments and look more private.

B-72. As a crime-prevention officer, you may not be in a position to directly use these techniques. However, your familiarity with these approaches and the value of their use in the crime-prevention process are important elements in your arsenal of tools to create public involvement in reducing crime. In particular, the purpose of outlining these tools is not to equip you to be a designer, but rather to equip you to communicate with those who are involved in that profession. The discussion that follows will further enhance your ability to converse with designers.

## NATURAL SURVEILLANCE

B-73. Experience has shown that the ability to observe criminal activity may not be adequate to stimulate an observer to respond with assistance to the person or property being victimized. The decision to act depends on the presence of motivational conditions, including—

- The degree to which the observer has developed a sense of personal and property rights that are being violated by the criminal act.
- The degree to which the observer feels that the event is within his area of influence.
- The observer's ability to clearly identify whether the act is unusual for the particular area.
- The observer's identification with either the victim or the property being vandalized.
- The degree to which the observer believes he can effectively alter the course of events he is observing.

B-74. Based on these conditions, a number of mechanisms have been identified that can be used to design the grounds and internal areas of apartment units, housing developments, and other residential areas to facilitate natural monitoring of activities taking place. By providing opportunities for surveillance through the positioning of windows in relation to stairs, corridors, or outside areas, continual natural observation will be maintained and crime will be deterred. If such steps are taken, the security of observed areas will be understood by the potential criminal, making him think twice before committing a crime.

B-75. The first of these natural surveillance mechanisms involves the positioning of service areas and access paths leading to apartment buildings to facilitate surveillance by residents and authorities. For example, buildings might be designed so that their entries face and are within 50 feet of a street, so that well-lit paths lead to the front door or the lobby, and so that the lobby is arranged to afford good visibility from the street. Other related steps focus on the strategic placement of windows, fire stairwells, lobby lights, and mailboxes so that they can be easily viewed from the street. Elevator waiting areas on each floor can also be designed so that they can be seen from the street level. Research has proven that if steps such as these are taken, residents will be more likely to become involved with protecting the facility, MP patrols will be in a better position to observe what is going on, and criminals will be discouraged from vandalizing the site.

B-76. A second technique that might be used to increase surveillance is to design facilities so that people within them will naturally view commonly used paths, entries, and play and seating areas during their normal household activities. This concept also focuses on the strategic placement of windows, lighting, and open areas so that natural surveillance by residents is improved.

B-77. Another mechanism involves the subdivision of housing areas into small, recognizable, and identifiable groupings that improve visual surveillance possibilities. Research has shown that in housing developments where the surveillance of a neighbor's outside activities was possible, residents were found to be very familiar with everyone's comings and goings. The overall effect was to cement collective identity and responsibility through social pressure.

## DEFENSIBLE SPACE

B-78. Defensible space is a term for a range of combined security measures that bring an environment more under the control of its residents. A defensible space is a residential environment that can be used by inhabitants for the enhancement of their lives while providing security for their families, neighbors, and friends. The physical mechanisms suggested to create safety and improve upkeep (as part of the defensible-space concept) are self-help tools wherein design catalyzes the natural impulses of residents rather than forcing them to surrender their shared social responsibilities to any formal authority.

B-79. Research has revealed investigative techniques that might be used to modify existing housing areas to make them more secure. The following methods may require alteration or adaptation to the particular situation on your installation:

- Widening major pathways and using colored decorative paving.
- Differentiating small private areas (front lawns) outside each dwelling unit from the public path with low, symbolic walls.
- Adding public-seating areas in the center of public paths far enough from private-dwelling units to eliminate conflicts of use but close enough to be under constant surveillance by residents.
- Designing play areas as an integral part of open space.
- Adding new and decorative lighting to highlight various paths and recreation areas at night and extending the residents' surveillance potential and feeling of security.
- Adding seats and path networks to recreational facilities where large, central court areas exist. This increases the interest and usability of the areas.
- Redesigning parking and play areas around buildings to create the illusion that the buildings are grouped where natural opportunities exist.
- Modernizing building entrances to create breezeways into building courts and to accommodate a telephone intercom for opening entry doors to the lobby.
- Providing video surveillance of public grounds and central paths by security of public monitors.
- Installing audio surveillance capabilities in elevators and at the doors of residences.

## CRIME-PREVENTION MODEL

B-80. The model for crime prevention through environmental design is based on the theory that action must be taken to counter crime before it occurs. The critical element in this model is the environmental-engineering component. It provides both direct and indirect controls against criminal activity by reducing the opportunity for crime through science and technology and the use of various urban planning and design techniques. The model explains what environmental engineering is and how it supports crime prevention. With this information, you may be in a better position to understand and respond to questions and discussions on how urban design and planning can have an impact on the installation's criminal element.

## THE ENVIRONMENTAL INFLUENCE ON CRIMINAL BEHAVIOR

B-81. The basic theory that supports crime prevention through environmental design is that urban environments can influence criminal behavior in two ways. First, the physical surroundings in which people live have an effect on each individual. These physical characteristics include noise, pollution, overcrowding, and the existence and unmonitored spreading of refuse and other unsightly waste. The second element that must be dealt with in the environmental-engineering formula concerns the social characteristics of the community that provide individuals with social relationships to which they must respond. Characteristics such as alienation, loneliness, anxiety, and dehumanization are seen as keys to criminal behavior.

B-82. In terms of these environmental characteristics, buildings are all too often constructed to be dangerous, with corridors and passageways hidden from public view. Elevators, basements, and storage and washroom areas are also laden with danger due to their design. Various large-scale housing developments are not secure in that they are often isolated from the main flow of traffic (both human and automobile) and are closed to public use and public view.

B-83. With regard to altering the social characteristics of the community and their relationship to criminal behavior, it should be recognized that behavior is future-oriented, not past-oriented. A man steals so that he can have a car or money in the future, not because in the past he experienced psychic trauma, a broken home, poverty, or delinquent associates. Criminal behavior can be explained directly in terms of the consequences of behavior and in terms of noncriminal variables such as poverty, race, or social class. Criminal behavior is viewed as a problem to be dealt with and not symptomatic of other problems (such as poverty, mental conflict, class conflict, unemployment, or undereducation). To change criminal behavior, it must be dealt with directly by removing the environmental reinforcement that maintains the behavior. The approach advocated is to change the environment to which the individual responds.

## ACTION APPROACHES TO CRIME PREVENTION THROUGH PHYSICAL PLANNING

B-84. The primary focus of crime prevention has been on what architects, planners, and other nonpolice professionals can do in terms of various physical-planning strategies to reduce criminal opportunity. Experienced MP personnel have long recognized that certain physical conditions can contribute to the rate and nature of crime. They have also developed a capability to identify high crime-risk locations by noting such factors as poor lighting and weak points of entry as potential targets. The critical job is to identify specific areas concerning physical planning and design that can be responded to and actions can be taken against on the installation.

B-85. Attempting to reduce crime or the fear of crime by regulating physical environments is easier said than done. In fact, although crime prevention can be built into almost every aspect of community planning, it is often ignored for a number of reasons. For example, fragmentation of responsible agencies is a key problem. In addition, crime has historically been looked upon as the exclusive responsibility of MP forces; not of those in charge of education, housing, or health and welfare. Yet, with an understanding of crime prevention combined with the knowledge that design techniques can change the opportunity for criminal behavior, PSIs will be able to talk the language of the planner and the designer and to be able to advise them from a police perspective.

B-86. It is notable that a number of civilian police agencies have become involved in the physical-planning process and have achieved notable results from their work. For example, the Fremont, California, Police Department has been involved in a planning process and maintains that law enforcement should become an integral part of the master- or comprehensive-plan review to screen all redevelopment plans for safety and crime hazards. Working with other units of municipal government as well as architects and designers, the department drew up a set of model guidelines for the evaluation of projects. The model included evaluation criteria dealing with such subjects as the accessibility of buildings to patrol units; traffic flow and off-street parking provisions; and the location and regulation of cul-de-sacs, playgrounds, common greens, fences, and security entrances. In addition, working with agencies such as the American Institute of Architects, the National Public Works Association, the Association of Public Utilities, and others, the department identified a number of subjects that are of specific concern to police officers and that should be considered in the design and planning stage. As a result of these efforts, the following list of design concerns was developed by the department:

- Building setbacks (front, side, and rear).
- Wall construction, interior and exterior (industrial, commercial, and residential).
- Door construction, setbacks and security (industrial, commercial, and residential) (including carports, garages, and sliding-glass doors).
- Windows and skylights, setbacks, heights (from ground), show-window displays, and the type of frame or pane.
- Stairs (stairwells and staircases).
- Balconies.
- Utility boxes.
- Fences, walls, hedges, screens, setbacks, heights, and louvers.
- Parking (public and private).
- Lighting (industrial, commercial, and residential).
- Streets, sidewalks, and walkways (locations, slopes, curvature, grades, and the length of a block).
- Alleys (blind and through alleys).
- Visibility of valuables (people, safes, cash registers, and personal property).
- Signs (street signs and signals, traffic signs and signals, and advertising signs).
- Accessibility; approach, entrance, and exit (pedestrian, vehicular, services, residential, commercial, and industrial).
- Public utilities and easements (gas, water, telephone, and electrical).
- Public areas and facilities (public restrooms, parks, bus stops and shelters, playgrounds, recreation halls, and so forth).
- Street trees and shrubbery (types, heights, and locations).

B-87. With this information, you can improve the security aspects of the community's physical-planning process. There is a probability that the work and recommendations of the National Advisory Commission on Criminal Justice Goals and Standards may help you in your efforts to get involved in the design process. More specifically, the Commission noted that "every police agency should participate with local planning agencies and organizations, public and private, in community physical planning that affects the rate or nature of crime or the fear of crime."

B-88. The future role of MP forces in crime prevention through physical planning will depend on the PSI's initiative. The perspectives and knowledge of what is happening in this field, combined with a working knowledge of the language, should equip PSIs to sell this approach as part of the overall program. It is important to point out that others who represent professions other than crime prevention are aware of the relationships between urban planning and crime. Remember, they deal in concepts, approaches, and ideas that have not involved the realities that law-enforcement personnel face.

## SPECIALIZED PATROL TACTICS AND SURVEILLANCE

B-89. Specialized patrol operations use a variety of tactics in attempting to control identified crime problems. The most common tactics include uniformed tactical patrols and suspect and area surveillances. The following paragraphs discuss these tactics in terms of their target crimes, operation requirements and characteristics, and established or perceived levels of effectiveness.

B-90. The appropriateness of a given tactic depends on the characteristics of a particular crime problem. The selection of specialized patrol tactics should be made on the basis of a careful and continuous analysis of crimes. Most crimes can be addressed by more than one tactic. Several tactics might be tried in an effort to find the best one, and it is quite possible that the most effective approach to a given crime problem will include the combination of several tactics.

### UNIFORMED TACTICAL PATROLS

B-91. A uniformed tactical patrol is the most traditional and widely used form of specialized patrol. It is a simple, straightforward approach to specialized patrol that involves the same procedures and techniques used by MP officers on routine patrol. These include constant visible movement throughout an area to generate a sense of police presence, careful observation of street activity, vehicle and pedestrian stops, and citizen contacts. The difference between uniformed tactical patrols and routine patrols are that uniformed tactical patrols use these tactics in an intense, concentrated fashion. MP officers are relieved of the responsibility for responding to routine calls for services so that they can devote their full time and attention to patrol, thus intensifying its impact. In addition, uniformed tactical operations typically deploy a number of MP officers in target areas, thereby increasing the level of patrol in these areas.

B-92. Uniformed tactical patrols can be used to control virtually any type of suppressible crime (for example, crimes that can be viewed from locations where the police have a legitimate right to be and those that can be potentially affected by police operations). These suppressible crimes include street robberies, purse snatches, vehicle thefts, burglaries, and housebreakings. Uniformed tactical patrols can also have an impact on other types of crime as officers use observation, field interrogation, and citizen contacts to develop information on the locations, activities, vehicles, and associates of suspects.

B-93. The primary purpose of uniformed tactical patrol is deterrence. This tactic is based on the assumption that highly visible, active patrols will deter potential offenders. By increasing the perceived probability of apprehension, conspicuous patrol is thought to reduce the likelihood that crimes will occur. If the deterrence should fail, heightened patrol coverage is believed to increase the probability of the immediate apprehension of the suspects.

B-94. Uniformed tactical patrols are often used to saturate an area that is experiencing a particularly serious crime problem. Although it has been widely used for years, saturation patrol has never been clearly and adequately defined. Exactly what level and intensity of patrol constitutes saturation has never been determined, nor have the effects of different levels of patrol been clearly established. It is difficult to prescribe the level of uniformed tactical patrols that should be used to disrupt a crime pattern in a particular area. This should be determined through an analysis of the size and characteristics of the area of concentration of each potential target crime pattern coupled with an assessment of manpower availability.

B-95. Some patterns can be effectively handled by a very small increase in the patrol level. For example, in Portland, Oregon, Operation Crime Reduction Involving Many People (CRIMP) had a significant impact on street robberies and assaults in the city's skid-row area by initiating two-officer, uniformed foot beats in the area. This was sufficient to saturate the primary locations of the target crimes during the high-crime hours, and it led to a substantial reduction in these crimes with little apparent spillover into adjacent areas.

B-96. Other departments have used saturation patrols on a much larger scale. For example, in the mid-1950s, the New York City Police Department attempted to saturate an entire precinct by assigning over 200 additional officers to the precinct's patrol force. Foot beats covering extremely small areas were arranged in straight lines so that an officer could keep the entire street area of his beat in view at all times. The four-month experiment led to a dramatic reduction in crime in the precinct. Compared with the same period in the previous year, street muggings fell by 89.9 percent, burglaries where entry was made from the front of the building dropped by 78 percent, and so on. The only crime category that was not affected was the relatively private crime of murder. Since crime displacement was not examined in the experiment, its true impact remains unknown. The experiment strongly suggests that massive additions of police manpower can have a substantial effect on crime. The problem is that most departments do not have the ability to conduct even a limited version of this experiment.

B-97. The amount of resources required for saturation patrols can vary tremendously, and there is no definite way of determining how much additional patrol is needed. This can best be decided on a problem-by-problem basis, using experience and evaluations of past efforts as a guide. As a rule, patrol augmentation should be sufficient to affect rather quickly the perceptions of would-be offenders concerning the level of police activity in a particular area. It appears that for such a strategy to be effective, sufficient resources should be applied in a limited area to ensure a true saturation effect.

B-98. Uniformed tactical patrols can use various modes of transportation. Patrol cars are most often used; however, foot patrols can be effective in congested service districts and bikes have been used to good advantage in high-density residential areas. The mode of transportation should be selected based on visibility, mobility, and access to citizens.

B-99. Some specialized units deploy MP officers in unmarked police cars. This is done in an effort to effect a balance between overt and covert operations, hopefully gaining many of the advantages of both. Unmarked cars may also be readily available since in many departments investigators work primarily during the day, which leaves their vehicles free for specialized patrol in the evening and early morning watches. This approach has serious drawbacks. First, unmarked MP cars are somewhat less visible than marked cars, yet they are still easily recognizable as police vehicles to large segments of the public (especially when the officers in them are in uniform). Second, the use of these cars in uniformed tactical patrols could lead to the sacrifice of some of the deterrent effects of high visibility without realizing the apprehension benefits of truly covert patrols.

B-100. Several patrol techniques have been tried to increase the visibility of uniformed patrol and to enhance the sense of police presence. A tandem patrol uses two marked patrol cars that follow each other at intervals of one-half to several blocks. Two units can also patrol on parallel streets one block over or in an alley. Another approach combines foot and vehicle patrols in an effort to increase visibility. Officers park their marked cars in conspicuous locations in high-crime areas and then are transported to other high-crime areas where they patrol on foot. The frequent repetition of this procedure is seen as a way of multiplying patrol visibility.

B-101. Unless a target area is more or less completely saturated (as in the New York experiment), MP patrols should move in a random, unpredictable pattern. This will make it difficult for potential offenders to plan their crimes on the basis of observation of patrol activities. One department found that its uniformed-tactical-patrol operation was actually helping burglars to commit break-ins. Interviews with confessed burglars indicated that they were aware of a visible patrol passing through the neighborhoods at regular intervals and they planned their crimes accordingly.

B-102. In addition to increasing the level and visibility of patrol in high-crime areas, uniformed tactical patrols often use aggressive patrol tactics involving frequent vehicle and pedestrian stops. MP patrols stop, question, and perhaps search citizens when there are reasonable grounds for suspecting that the citizens may have committed, may be committing, or may be about to commit a crime. Since the concept of reasonable suspicion is vague, MP officers have a considerable amount of discretion in conducting field interviews. Field interviews that do not result in either immediate arrest of the citizen or in alleviation of the officer's suspicions are usually documented in field-interview cards or spot reports. Field interviews serve to generate information about the activities of probable suspects; more importantly, they make the suspects aware that the police know of their presence in a given area, regard them as suspicious, and are watching them closely. This is expected to reduce the likelihood that they will commit crimes, at least in the area in which the tactical force is working.

B-103. The extensive use of field interviews is often highly controversial. Tactical units that emphasize field interviews have been accused of being storm troopers who constantly harass citizens and abuse their rights. There is sometimes a certain amount of truth to the allegations of harassment, especially when the activities of a particular type or group of suspects are being closely monitored. Not surprisingly, individuals who are stopped and questioned frequently are likely to complain, particularly if they have reason to be concerned about close police scrutiny. A recent study of field interviews in San Diego found that when conducted at moderate levels, field-interview activities do not have a major effect on police and community relations. The majority of citizens in all three study areas accept field interviews as legitimate and properly conducted police activity. The majority of all citizens who were the subjects of contacts felt that the contact was justified and properly conducted. The study found that the suspension of field interviews in a particular area was associated with a significant increase in the level of suppressible crime and when reinstated in the area, these crimes declined. The results of the San Diego study provide confirmation for the widely held belief that field interviews can have an important deterring effect on suppressible crime without doing irreparable damage to police and community relations.

B-104. The potentially negative impact of field interviews on certain segments of the community can be held to a minimum if the interviews are conducted in a professional manner. Citizens should be informed of the reasons why they are being stopped. They should be detained for as little time as possible and should not be subjected to verbal or physical abuse. There is also no need to stop everything that moves.

B-105. While experience and a limited amount of research indicate that uniformed tactical patrols can have a positive impact on the level of suppressible crime in areas, the overall effectiveness of this tactic is a controversial and much-debated issue. The principal concern is that rather than reducing crime, uniformed tactical patrols may simply lead to its displacement from one area to another or from one period to another.

B-106. When conducting uniformed tactical patrols, a PM activity should carefully monitor crime trends for indications of possible displacement effects. This is an important aspect of evaluating the impact of tactical operations, and it will provide information to guide future deployments and tactical decisions.

## AREA SURVEILLANCE

B-107. Covert patrol and surveillance of high-crime areas can be used to make apprehensions for crime problems. These problems include those for which there are no suspects who warrant personal surveillance, the suspects are too numerous to permit personal surveillance, and there are too many potential targets to conduct either physical or electronic stakeouts. Examples of these types of problems would be a rash of residential burglaries or auto thefts in a particular area.

B-108. This tactic simply involves the covert patrol of a particular area and the observation of suspicious or unusual activities and occurrences that might indicate the likelihood of a crime. Suspicious individuals are not stopped but are watched until they either commit an offense or the officers' suspicions are removed.

B-109. The list of various techniques that can be used in an area surveillance is virtually endless. The following are some techniques that have been effectively used by specialized patrol units:

- Mingling with citizens at the crime scene to pick up information on possible suspects.
- Maintaining rooftop surveillance of a shopping center's parking lot to locate larcenies from vehicles.
- Surveilling housing areas by posing as maintenance workers.
- Following likely crime victims such as elderly citizens leaving a bank.
- Surveilling rooftops for unusual activity from aircraft or higher buildings. Binoculars are used to facilitate surveillance, and rooftops are marked so that street units can be dispatched to check out suspicious circumstances.

B-110. As in all types of plain or uniformed patrol, care should be taken to ensure that area surveillance is truly covert. Rental vehicles that can be changed frequently provide an excellent, though expensive, means of covert transportation.

B-111. Police on covert patrol should be dressed to blend in with the environment in which they are working, and they should have apparently legitimate, nonpolice-related reasons for being where they are. Several specialized units have found that surveillance teams composed of one male and one female officer can work in many situations without arousing suspicion. An apparently married or romantically involved couple lingering in a park, meandering slowly down the street, or sitting together in a parked car would generally appear less suspicious than two male MP officers doing the same things. Finally, it should be noted that in some small neighborhoods where residents know each other well, covert surveillance may be difficult, if not impossible, since the presence of any stranger arouses immediate curiosity and suspicion.

## SUMMARY

B-112. These tactics represent the basic approaches that specialized patrol operations normally take in trying to control suppressible crime. Some of the tactics (such as uniformed tactical patrols) are directed primarily at crime deterrence, while others (such as suspect surveillance) are used to achieve apprehensions for target crimes. The tactics are most commonly used independently of one another. However, there are some indications that the combined use of several tactics in an integrated operation might be an effective way of coping with particular types of crime. Especially promising is the coordinated use of highly visible and covert patrols. A visible patrol force could be deployed to a particular area to deter crime there and direct it toward other areas in which MP forces using covert tactics are working. To date, efforts to direct criminal activity to areas or targets where MP forces are set up to make apprehensions have only been tried on a sporadic basis. However, this appears to be a promising approach to crime control and warrants greater attention in the future. It can be viewed as the creative use of crime displacement.

B-113. Many of the tactics mentioned previously could not be effectively used by MP patrols that are responsible for handling routine calls for service. It is obvious that the effectiveness of decoy operations, physical stakeouts, and suspect surveillance would be destroyed if the MP officers had to interrupt these activities to handle calls for service. It is also generally unwise to have MP officers in civilian clothes respond to calls for service.

B-114. The importance of crime analyses in identifying crime problems and suggesting appropriate tactics deserves repetition. Specialized patrol units should be deployed to address clearly identified crime patterns, and they should choose their tactics on the basis of an analysis of these patterns and a comprehensive knowledge of the area of occurrence. The nature and characteristics of a particular crime pattern should be the driving force behind the selection of tactics. None of the tactics of specialized patrol can be effectively used unless crime patterns have been identified and analyzed.

## PUBLICITY CAMPAIGNS

B-115. Public-information campaigns are an essential part of every crime-prevention effort. The installation PAO can provide assistance in—

- Informing the public of the magnitude of the local crime problem.
- Disseminating information on crime circumstances.
- Generating the interest and enthusiasm necessary to initiate and sustain community crime-prevention programs.

B-116. The PAO should be a member of the installation's crime-prevention council and must be involved from the start in planning crime-prevention efforts. There are many information vehicles at the installation level available to carry the message to the public, including—

- Digital marquees.
- Radio.
- CCTV.
- Installation and unit newspapers.
- Posters and leaflets.
- Commanders calls and similar meetings.
- Town-hall meetings.

B-117. Not all media are equally effective in reaching a particular segment of the post population. To ensure that the media campaign is effective, identify the segment of the population requiring the information and the exact message to be communicated. The more specific the target audience, the more effective the media campaign will be.

B-118. Whatever media is used, it is important to provide coordinated input to the PAO well in advance of the desired publicity campaign date to allow for writing, rewriting, and publishing. For material to be published in a magazine, a minimum lead time of six months is normally required. For installation newspapers, a lead time of a week may be sufficient. Check with the PAO to determine the correct lead times for local publications.

B-119. When addressing civic groups or school audiences, films are often more effective than straight lectures because of the animation and other special techniques used to illustrate critical points. Several Army films are available as are a large number of commercial films.

## RESIDENTIAL-SECURITY SURVEYS

B-120. A security survey is an in-depth, on-site examination of a physical facility and its surrounding property. The survey is conducted to determine a residence's security status, to identify deficiencies or security risks, to define the protection needed, and to make recommendations to minimize criminal opportunity. Because of several common characteristics, one expert has likened the security survey to the traditional criminal investigation. This comparison hinges primarily on the facts that both techniques are systematic in nature; are aimed at identifying the method of a criminal act; and are, in effect, more an art than a science. It should be recognized, however, that the survey has two other advantages. First, it can be conducted before the commission of a crime; second, it can offer protection against, rather than just remedial action after, criminal victimization.

### THE RESIDENTIAL-SECURITY SURVEY AS A CRIME-PREVENTION TOOL

B-121. A residential survey is not a substitute for an aggressive neighborhood watch. It supplements these efforts and should be established after Neighborhood Watch and Operation ID programs are established.

B-122. The security survey is the primary tool used in crime prevention to recognize, appraise, and anticipate potential loss in residential areas. It is often defined as the backbone of a local crime-prevention program. In practice, it combines the security experience, training, and education of the local crime-prevention officer and focuses on a single element—the analysis of a residential facility.

B-123. The inherent value of surveys has been proven by nearly 300 civilian police departments across the country that have established crime-prevention bureaus or units. An even broader endorsement of the survey technique, however, was provided by the National Advisory Commission on Criminal Justice Standards and Goals, which stated that, "Every police agency should conduct, upon request, security inspections of businesses and residences and recommend measures to avoid being victimized by crime."

B-124. In short, the security survey is a tool that informs a homeowner of the particular areas in which his home is susceptible to criminal victimization along with steps that can be taken to reduce and minimize that potential. Further, the survey is a tangible document that reflects the efforts of the MP force not only to be responsive to community needs, but to get the community more directly involved in the criminal-justice process.

## THE CRIME-PREVENTION SURVEY—A PROCESS OF INVESTIGATION

B-125. An important factor in understanding the residential-security survey is that it must be considered as an ongoing process. While a particular survey will result in specific recommendations, each survey will provide a foundation for future action. In combination, these surveys will provide a database that can be used in the analysis of the community's crime problems and guide action toward the resolution or reduction of the problem on a community-wide basis.

B-126. As a starting point, five steps must be used in carrying out the actual survey, while four additional steps must be remembered afterward. These steps are—

- Analyzing the overall environment (neighborhood, block, and so forth).
- Assessing the general vulnerability of the premises.
- Defining the specific points of vulnerability.
- Recommending specific security procedures.
- Including specific remedial hardware recommendations.
- Urging the implementation of the recommendations.
- Conducting a follow-up to ensure that recommendations have been implemented.
- Keeping crime statistics to evaluate the survey's effect and the implementation of recommendations.
- Conducting a second survey of the premise's statistical analysis to determine the alteration in criminal activity in the areas surveyed.

B-127. This process is the continued involvement and participation of the police with the community. Once the survey is completed, the job is not finished. In fact, if this posture is assumed, it may later be learned that the recommendations were not implemented and that the work was done in vain. This can easily lead to a loss of community-wide public confidence in the programs. Similarly, even in those cases when recommendations are implemented, additional crime might be experienced. Attention to this fact, coupled with immediate follow-up, will be essential to avoid losing the confidence gained from the original survey. Through prompt actions, the proposal of additional tactics may become the final step needed to substantially reduce criminal opportunity. In short, the security-survey process is not a one-shot operation and it is not to be looked upon as a simple, straightforward task. It is a continuing, difficult, rigorous, yet effective approach to reducing crime that has not been systematically applied by police agencies in this country.

## THE ROLE OF POLICE IN IMPLEMENTING CRIME-REDUCTION PROGRAMS THROUGH SECURITY SURVEYS

B-128. At a minimum, there are two ways to encourage people to improve their personal security. First, you can organize, conduct, and participate in broad-based public-information and -education programs that make use of such media as radio, television, and the press. Second, you can organize, undertake, and follow up on a series of person-to-person security surveys. Clearly, both of these techniques have their advantages. The security survey has a unique quality that does not exist in the public-education program. It provides an added incentive on the part of citizens to implement recommendations because of the personal relationships and respect established during the actual survey by a crime-prevention officer.

B-129. For example, consider a homeowner listening to someone on television recommending improvements of security by installing better locks or alarm systems. You might ask yourself these questions: How secure am I? Are the locks that I have on my doors adequate? If they are inadequate, what kind of locks should I install? Do I really need an alarm system? What can the MP officers do for me in terms of making recommendations about security? In short, you would be aware of the possible need for improved security, but you would not know enough of the specifics to warrant real action. Because home security is a complicated matter requiring careful analysis and the installation of carefully tailored systems, implementation must be approached in a personal way. Certainly, public-education programs can assist in some aspects of crime prevention (such as prevention against schemes, auto theft, and personal protection), but they are not as effective in causing improved security as is personal contact.

B-130. Moreover, for the first time, police are placed in a position where they can actually provide a different kind of advice or service that can be offered in an environment where a crisis has not yet occurred. This is a truly unique opportunity for crime-prevention officers.

## CONDUCTING THE SURVEY

B-131. To develop a proper perspective of the types of crimes that a PSI will most frequently be trying to reduce, a review of the cases is necessary to get a broader feel for the actual conditions in your area. During this review process, pay specific attention to photographs in the files. Study crime scenes in an effort to identify the type of security device that was defeated. In particular, if a door was used as a point of entry, note whether it was of adequate construction; if the door frame was broken or separated; if hardware, such as strike plates and door trim was inadequate; and so forth. In addition, review photos to determine if lock cylinders were pulled or if door chains fastened to trim moldings were simply pulled away to permit easy entrance. If photographs of the crime scenes are not available, visit as many crime scenes as possible. While doing so, photograph security risks you can study later and use as examples in future presentations to community groups.

B-132. By becoming familiar with the MO of persons committing such crimes as burglary and larceny, you will be better-equipped to understand potential risk situations and to point them out to potential victims. Quite surprisingly, many of the cases investigated were invited by some obvious crime-risk hazard that was overlooked by a resident. Also, many additional crime risks existed at a particular crime scene that a burglar could have exploited. Such vulnerability might be an indication of other crime targets within the community to which you should pay particular attention in your survey work.

B-133. To be an effective security surveyor, it is necessary to develop an intimate knowledge of the crime factors in your community. You do not have to become a statistician; however, the more you know and understand about crime problems, the methods used in your community, and security devices or systems that were defeated, the better you will be equipped to analyze the potential crime when surveying a home.

B-134. In addition to the general types of crimes that occur in your community, it is necessary to develop an understanding of the details of particular types of offenses. For example, with regard to residential housebreaking, you should be familiar with the types of burglaries and approaches used in particular sections of the community.

B-135. On the surface, it might appear as if this would be a monumental task. However, in terms of housebreaking, you might pull the files for the last quarter and carry out the following steps:

- Tally the number of times entry was made through the front door, rear door, or through a window.
- Identify the approach used for entry (kicking the lock, throwing a shoulder through the door, jimmying the lock, and so forth).
- Determine how often a window was broken or removed or a mechanism was used to force the latch (when entry was made through a window).
- Attempt to determine whether security devices were used in residences (such as alarms, special lighting, or other systems).
- Identify the burglar's general escape route (down a back alley, through a schoolyard, and so forth).

B-136. In developing an understanding of the MO of crimes, statistics that illustrate exactly what is happening in your community will be valuable tools. Not only will you be able to use this information in explaining crime risks while you are surveying the site, but it will be invaluable in making public presentations.

B-137. Only when you have developed the ability to visualize the potential for criminal activity will you have become an effective crime-scene surveyor. It is important that when you arrive on a survey site, you are prepared to give a property owner sound advice on the type of security precautions he should consider.

B-138. In summary, to be a good crime-prevention surveyor, you have to be a good investigator. You must understand the criminal's method of operation and the limitations of standard security devices. In addition, you must be knowledgeable about the type of security hardware necessary to provide various degrees of protection. (Specific information on conducting surveys is contained in Chapter 11).

## JUVENILE CRIME PREVENTION

B-139. Law enforcement has long recognized the importance of establishing personal contact with youth. This need was first met with the "Officer Friendly" concept. The goal of this concept was to enhance the image of police among younger children and attempt to negate the unfavorable image other segments of society provided children. This concept has evolved into today's Drug Abuse Resistance and Education (DARE) program which has spread to most law-enforcement agencies.

B-140. Crime-prevention officers can have an impact on juvenile crime by establishing a positive interaction with juveniles. Many areas suffer a high crime rate of which over 50 percent is attributable to juveniles. Nationally, it is known that youth 17 years of age and under (who make up just 16 percent of the population) commit 42 percent of the crimes that cause injury or loss of property. As the punitive aspects of the juvenile justice system today are of questionable value, it is logical that crime-prevention officers should develop programs aimed at juvenile crime.

B-141. The basic problem in establishing any juvenile crime-prevention program is the destruction of preconceived ideas by both youths and law-enforcement personnel. This obstacle can only be overcome by the building of an honest rapport with individual youths. Just as a crime-prevention officer must sell crime-prevention procedures to the public, he must also sell himself and his program to the youth.

B-142. One method of establishing the needed rapport with youth reverts to a similar system used in the Officer Friendly concept. Classroom presentations in high schools and junior high schools can be used to destroy the preconceived ideas held by youth regarding police and law enforcement in general. Once an MP officer enters the classroom, he discovers that the majority of students obtain their ideas of police one of two ways—

- From television shows that portray every officer as a slow-witted moron (such as the old "Car 54" show).
- From their peers at school who relate every negative aspect of law enforcement.

B-143. However the students gain their information, it seldom resembles fact and must be changed to effectively establish a juvenile crime-prevention program.

B-144. In making classroom presentations for crime-prevention purposes, the crime-prevention officer must display honesty at all times. When confronted by uncomfortable questions, the MP officer should render the law-enforcement's point of view and make a truthful explanation. Obviously, some answers will not always be well received, but they will establish the credibility of his presentation. Further, when making these presentations, many areas dealing with specific areas of crime prevention will arise (such as rape prevention, property engravement, residential and apartment security, vehicle theft, and burglary prevention).

B-145. In dealing with schools, the crime-prevention officer can establish many positive contacts with individuals inside the school system that will prove to be of great value. Individuals such as administrators, counselors, and faculty members offer the crime-prevention officer assistance in many different ways. Perhaps the most valuable assistance is in the area of establishing a juvenile crime-prevention program that is based on peer influence.

B-146. A crime-prevention program aimed at peer influence is essential to a crime-prevention unit. Peer influence is directly related to the Determination Theory of Delinquency Causation. Simply stated, the theory holds that delinquency occurs as a result of external influences on youth. While attempting to control all external influences on youths is an impossible task, the control of peer influence is possible and perhaps the most important external influence to be controlled.

## PEER-INFLUENCE CRIME-PREVENTION PROGRAM

B-147. A peer-influence crime-prevention program can be established after the crime-prevention officer establishes himself in the junior/senior high schools as described earlier. The basic concept of the program is the negation of undesirable peer influence on predelinquent youths and the establishment of a positive peer influence.

B-148. The peer-influence program functions within the school district. A problem youth is detected at the junior high school level by his teacher, counselor, or the crime-prevention officer. When it is believed that the peer-influence program could serve a youth, his counselor arranges a conference with him. During the conference, the counselor attempts to determine the basic problem experienced by the youth and has him complete an application for admittance into the program.

B-149. When the application is complete, the counselor contacts the youth's parent or guardian and explains the program. If the parent agrees to allow participation, the application is sent to the parent for his signature.

B-150. When the completed application is returned to the counselor, it is forwarded to a senior high school counselor for review. If necessary, the senior high school counselor may contact the junior high school counselor personally to discuss the youth in more detail. The high school counselor then selects a suitable senior high school student who has been accepted for the peer-influence program and matches the student with the junior high school youth.

B-151. Once the match has been made, the senior high school student is given a briefing on the junior high school student by the crime-prevention officer. The senior youth is instructed to contact the junior youth (preferably at the junior youth's home) and introduce himself to the youth and to the youth's parent or guardian. After the initial contact has been made, the senior youth is instructed to plan activities with the junior youth and spend as much time as possible with him.

B-152. Each senior youth that volunteers for the program is required to complete an application. This application gives personal background information and references that are thoroughly checked by the crime-prevention officer. The youth's academic record is examined, but prime consideration is given to citizenship factors rather than grades. The crime-prevention officer personally interviews each applicant.

B-153. When a senior youth is selected for the program, he is given a basic outline to follow. He is warned not to preach to the junior youth but to lead him by allowing him to observe the senior youth's actions at different activities. The senior youth must devote the necessary amount of time to limit the junior youth's opportunity to continue the negative associations he had in the past.

B-154. While the peer-influence program is simple in nature, the results can be remarkable. During a three-year period in Irvin, Texas, approximately 600 students participated in the program. From this number, seven arrests were made from the group with one juvenile accounting for four of the seven arrests. As about 30 percent of the juveniles participating were legally judged delinquent, the relapse rate was officially established as 0.1 percent.

B-155. While juveniles judged as delinquent can participate in the program, the crime-prevention officer should concentrate his efforts on those juveniles identified as predelinquent by counselors or faculty members. Again, the personal contact of the crime-prevention officer and the school faculty is essential to the program's success. A personal rapport should be cultivated and maintained.

B-156. Operationally, the peer-influence program can be administered by the collective efforts of school personnel and MP officers who are committed to the program. Once the program is established, it will probably not require the services of an officer on a full-time basis (depending on the number of students involved). However, constant contact is necessary; the more time that can be devoted, the more success will be apt to follow.

## CRIME-PREVENTION PRESENTATIONS FOR ELEMENTARY AND JUNIOR HIGH SCHOOLS

B-157. As the traditional American family structure has experienced a significant decline over recent years, more and more emphasis is being placed on schools to provide guidance for children. As a crime-prevention officer who is frequently invited to speak to students, you can play an important role in assisting the school to provide this guidance. To stress the importance of the above statement, many experts in child behavior believe that the school is second only to the family in molding a child's socialization and behavior. Many people feel that the school is increasingly replacing much of the family socialization process. For this reason, schools are the most logical environment to begin a juvenile crime-prevention program.

B-158. Crime-prevention officers can have an impact on the most common types of juvenile crime by preparing specific lectures for particular grade levels in the elementary and junior high schools. Classes discussing shoplifting, vandalism, and bicycle theft are well received by students and have a wide-reaching effect on the community as a whole due to the children relaying the information to their parents.

B-159. A number of films are currently on the market that assist in making crime-prevention presentations in the classroom. In addition to specifically addressing juvenile crimes, the films directly relate to the peer-influence causation of crime.

## OVERVIEW OF SELECTED JUVENILE CRIME-PREVENTION PROGRAMS

B-160. In addition to the programs described above, there are a variety of other juvenile programs. You will notice that some of the programs are designed to offer activities for the juvenile while at the same time attempting to provide one or more of the following:

- Positive peer influence.
- Education.

- Police/juvenile cooperation.
- An understanding of law-enforcement and MP duties.

B-161. One program deals with the enforcement aspect, with the goal being truancy reduction. Also mentioned are programs designed for child safety and victimization reduction.

**DARE Program**

B-162. DARE is a drug-abuse educational program that gives children and parents the skills to recognize and resist the subtle and overt pressures that cause them to experiment with drugs, alcohol, and violence. This unique program uses uniformed law-enforcement officers to teach a formal curriculum to students ranging from kindergarten to twelfth grade. DARE gives special attention to fifth through ninth graders to prepare them for entry into middle/junior high and high school where they are most likely to encounter pressures to use drugs or alcohol.

**Police Explorer Programs**

B-163. Explorer programs are run in conjunction with the Boy Scouts of America and a sponsoring police agency. The programs offer a wide range of flexibility in that they may be structured along the lines of a police cadet program with all activities centered around law-enforcement training. This type of explorer program usually seeks juveniles who have an active interest in pursuing a law-enforcement career. Being an explorer in this type of program will offer the student the opportunity to view the duties of a police officer and to take part in minor police activities (such as traffic direction and observation riding in patrol cars). The selection and training of the students for this type of program strictly depend on the sponsoring agency.

B-164. Other police-sponsored explorer programs may be designed to provide the members with other nonpolice activities (such as hiking, camping, and canoeing). At the same time, this type of program will offer the student the chance to meet with police officers and learn more about police functions, though this is a secondary purpose. Selection for this program would not be as stringent as the above-mentioned program. A local council office of the Boy Scouts can provide additional information on this program.

**Alateen**

B-165. The Alateen program is run in cooperation with the Alcoholics Anonymous organization. This program's intended purpose is to provide assistance for a child's emotional needs if he or she is faced with family problems caused by alcoholism of one or both parents. Your local chapter of Alcoholics Anonymous can provide more information on this program.

**Truancy-Enforcement Program**

B-166. Various cities have successfully adopted truancy-enforcement programs that focus attention on identifying and apprehending youths that are on unexcused absence from school. The primary goal behind this program is to prevent or reduce the incidence of daytime residential burglary. The San Angelo, Texas, Truancy Enforcement Program selected a target area in an effort to see how effective the program would be in reducing daytime burglaries. The initial evaluation of the program showed positive results. The San Angelo program uses uniformed crime-prevention officers in marked police cars patrolling around the school and known hangouts. All street personnel in the patrol division are also encouraged to check school-age individuals. Those students under 17 years of age who are suspected of truancy are taken to the school office while those over 17 years of age are detained only long enough to obtain their name and other identifying information, which is then turned into the school office. The assistant principal in charge of attendance completes the investigation and takes the appropriate action. The action varies (depending on the circumstances) but can include counseling with the student, the teacher, and the parents or, in extreme cases, suspension for a short period.

B-167. The emphasis behind these programs is to inform youths of the facts of the law. It was shown that many of the problems with the youthful offenders would be eliminated if they were presented the facts of the law to which they are subject. The courses can be adapted to fit the needs and desires of the individual school system, and the extent of the areas covered will depend on the commitment and availability of the local MP officers. Various civilian school districts have incorporated these law-enforcement courses as part of their regular curriculum.

### Criminal Victimization Reduction

B-168. Programs designed toward this goal normally are structured to reduce the possibility of a child becoming a victim of a child molester. Most commonly, officers give a verbal presentation to groups of children using visual aids that help to act as a positive reinforcement to learning. Some effective methods have been the use of "talking traffic lights," "talking bicycles," and "talking motorcycles." These items are usually equipped with a tape-recorded message that can be controlled by the officer conducting the program. The talking traffic light uses a full-size traffic light that flashes red or green according to the type of person that is being described.

## VANDALISM

B-169. Today's vandals often attack their own territory. School vandalism—the illegal and deliberate destruction of school property—is committed by students themselves. They break so many windows that in large districts the funds spent annually on replacing broken windows could pay for a new school. Vandals destroy about $3 million worth of school-bus seats annually, and they commit enough arson to account for 40 percent of all vandalism costs.

B-170. School vandalism outranks all other assaults on private and public property. At the end of the 1973 school year, the average cost of damages from vandalism was estimated at $63,031 per school district. That figure could have paid the salaries of eight reading specialists or could have financed a school breakfast program for 133 children for one year. A typical school's chance of being vandalized in a month are greater than one in four, and the average cost of each act of vandalism is $81. Yet, these figures do not include the hidden costs of school vandalism—increased expenses for fencing, intrusion and fire detectors, special lighting, emergency communications equipment, and vandal-resistant windows. Every dollar spent in replacing a window or installing an alarm cannot be spent on education.

B-171. School vandalism can also have enormous social cost. The impact of an 89-cent can of spray paint used to cover a wall with racial epithets far exceeds the monetary cost of removing the paint. An abusive word scrawled across a hallway wall can destroy student morale, disrupt intergroup relations, undermine the authority of an administration, or even close the school. Incidents with high social costs damage the educational process as much as those with high monetary costs. Today's vandal is not a hardened, war-scarred veteran. Instead of grizzled whiskers, he sports peach fuzz. He is almost literally the boy next door. In fact, the typical vandal differs quite dramatically from the typical juvenile delinquent.

B-172. It is significant that vandals fall into a well-defined and relatively narrow age group. They are usually early adolescent males who are highly subject to group pressures and transitory impulses. It is not at all unusual for adolescents to act out whatever is controlling them at the moment—rage, boredom, pent-up energy, or the sheer joy of "wreckreation." While there are conditions that may predispose or provoke a youth toward vandalism, the problem seems to be almost human nature. Few among us have never written on a sidewalk or scrawled initials on a school desk. Vandalism cuts across all strata of society, all geographic regions, and all racial lines.

B-173. The causes of vandalism remain obscure. Though research addressing the "why" of vandalism is growing, it has yet to yield clear-cut answers. Among the motivating factors often cited are anger, frustration, hostility, bitterness, alienation, futility, inequality, restricted opportunity, emotional pain, failure, prejudice, revenge, and the need for attention. Although much of the research is convincing, the fact remains that many vandals do not appear to be among the most angry, frustrated, hostile, alienated, or needy youth. Only a small fraction of the youngsters who fall into that category actually commit acts of vandalism. So, while most experts agree that vandalism is not totally senseless, they do not claim to fully understand its causes. In fact, vandalism is often not understood by vandals themselves. Many vandals report that they do not know why they did it. Many others, according to case reports, offer the unsolicited observation that destruction is fun. Still others express satisfaction and exhilaration. Few consider themselves criminals. For the time being, we can conclude only that motives for vandalism are diverse. But the whys notwithstanding, the vandal profile suggests that our task is, in large part, to anticipate and redirect the impulses of young teenagers.

B-174. Schools are by no means the helpless victims of early adolescence. Many school factors, most of which are amenable to change, influence the amount of vandalism that schools experience. The following characteristics are typical of schools that suffer high property damage or loss:

- Vandalism is higher when there is poor communication between the faculty and the administration (such as when the principal fails to define policy or makes policy decisions unilaterally).
- Hostility and authoritarian attitudes on the part of teachers toward students often result in students "taking it out" on the school.

- Limited contact between teachers and students reduces student involvement with the school and increases the likelihood of vandalism.
- Schools characterized by intense competition for leadership positions suffer greater property damage and loss.
- The chances for vandalism increase when the students do not value their teachers' opinions of them.
- Schools at which students strive to get good grades experience more vandalism.
- Parents of students in high-damage schools express less favorable attitudes toward their schools than do other parents.
- The school is a convenient target for vandalism when it is close to students' homes.
- The chances for vandalism are increased when grades are used as a disciplinary tool.
- Damage is greater in larger schools where there is more property to destroy. This correlation between school size and vandalism prevails regardless of whether the school is located in an urban, suburban, or rural setting.
- Fewer offenses occur when rules are well understood by students and are consistently and firmly enforced by teachers and administrators.

**Vandalism Prevention**

B-175. If the special problems of early adolescence, often intensified by social or personal pressures, interact with school conditions to produce vandalism, then preventive measures must address the nature of both the child and the school. Furthermore, prevention must include both physical and human responses. At present, most vandalism-prevention or -reduction programs rely on physical security—bigger and better electronic alarm systems, patrol guards, dogs, tamperproof locks, and window grilles. These techniques help, but they address only 20 percent of the problem—those incidents involving breakage. These incidents usually occur when school is not in session and in the absence of witnesses. The techniques have little effect on the day-to-day trashing of the school or on the disruptive acts aimed at the school's routine (bomb threats, the setting of fires, and false fire alarms) that are committed during school hours. The most sophisticated physical and electronic barriers are not sufficient to keep vandals from what they consider an attractive target. In fact, it has been argued that alarms and armed guards, besides lowering student and staff morale, often themselves become a challenge, inviting rather than deterring vandals. Vandalism prevention requires not a narrow or piecemeal approach, but a varied and comprehensive effort that includes both physical and human components geared to the school's specific problems. Furthermore, an effective long-term program must involve the community, parents, neighbors, police, and civic groups as well as students, teachers, and school administrators.

B-176. Schools are an easy target for vandals. Most are public, secular, and often unoccupied. Most will remain public and secular; but they need not remain unoccupied, unprotected, or unobserved. The following are techniques that have made some schools less vulnerable to vandals. These are especially effective against problems occurring during nonschool hours.

B-177. **Occupy the School.** Employ a custodial force around the clock. In most schools, the entire custodial force works at one time, leaving the school at night. As an alternative, custodians can be assigned staggered shifts so that the school is always occupied. Twenty-four-hour custodians are particularly appropriate in schools suffering sporadic property damage that demand more than a roving patrol but less than permanent security guards.

B-178. Invite police to use the school buildings at night. Police can be issued keys to the schools in their patrol areas so that they can use school offices to write their reports. This places a police officer in the school when it might otherwise be unoccupied, and it places a police car in front of the school.

B-179. Bring the community into the school. The school is an excellent place for recreational programs; health clinics; adult-education classes; counseling centers; community gatherings; and Boy Scout, Girl Scout, and Parent-Teacher Organization (PTO) meetings. The presence of people in the school building not only reduces the opportunity for vandalism, but also stimulates community and student interest in the school.

B-180. **Watch the School.** Use school neighbors as eyes and ears. Ask nearby homeowners to watch the school and report suspicious activities. Emphasize careful observation and rapid reporting, but discourage direct involvement in any situation observed. Such programs work best if they are organized but based on informal involvement, if they are accompanied by overall involvement of parents and community with the school, and if they offer some sort of prestige to participants.

B-181. **Employ Roving Patrols.** A uniformed patrol used instead of or in conjunction with an alarm system can deter vandalism. The individuals who patrol should establish rapport with neighborhood youths and open communication with community leaders. They should also vary their patrol patterns.

B-182. Hire student patrols during the summer and on weekends. The school district or community can provide its youth with part-time or summer employment and, at the same time, curb vandalism by paying students to patrol the school grounds during weekends, holidays, and summer vacations. These students should be paid an adequate wage and considered an integral part of the school's security force.

B-183. **Control Access to the School by Using an Alarm System.** Alarms are the most expensive vandalism control measure a school can use. While they can detect vandals, they cannot apprehend them; they can merely signal the alarm-system monitor, which may be miles away. They can, however, be an efficient deterrent and should be considered as part of any comprehensive plan to control vandalism. If alarm systems are linked with a surveillance camera, the chances of apprehending intruders are greatly increased.

B-184. **Design the School With Vandalism Prevention in Mind.** The following designs for preventing vandalism can be implemented when building a school:

- Limit ground-to-roof access.
- Eliminate low, overhanging roofs.
- Avoid unnecessary exterior fixtures.
- Plant trees that cannot be climbed near buildings.
- Consider raising as much of the school building as possible from ground level.
- Build the school at some distance from residential areas. While it should be located near the homes of most of those it serves, it will suffer less property damage if there is a buffer zone between it and surrounding residential areas.
- Design the school with plenty of defensible space so that the normal flow of school traffic allows continuing, casual surveillance of the premises.
- Use vandal-resistant surfaces. Use harder surfaces in damage-prone areas. For walls, use epoxy paint or glazed tiles that are easily and inexpensively replaced or repaired; use small wall panels and keep replacement panels in stock; and place permanent signs, building names, and decorative hardware at a level that cannot be reached from the ground. Replacing damaged areas immediately shows a sense of pride in the appearance and helps to eliminate copycat acts of vandalism.
- Plan windows carefully. Avoid windows that are vulnerably placed. Use small panes of glass to simplify replacement; use thick, tempered glass, thick acrylic, or Plexiglas® for windows in heavily traveled or hangout areas. Avoid useless windows in student stores, administrative offices, and industrial-arts storage areas.
- Plan entries with multiple uses in mind. Install flexible internal gates to block off specific areas or corridors when necessary. Provide separate exterior entries for community and student use. Inside the building, create areas for informal gatherings near entrances and exits by installing soft-drink machines and benches.
- Locate or relocate the playground's access roads to provide better surveillance by roving patrols.
- Consider outdoor lighting. Opinions on this issue are divided. Many schools report a decline in vandalism after installing hardened exterior night lighting. Others report that elimination of all night lighting reduces vandalism, presumably because young adolescents are afraid of the dark. If lighting is used, it should be directed away from windows to keep vandals from seeing the process of destruction or its outcome.
- Channel graffiti. Graffiti artists will usually select light, smooth surfaces rather than dark, rough surfaces. Therefore, school officials can channel graffiti onto one or two walls designed to withstand such treatment. Students or maintenance staff can paint most walls at regular but not too frequent intervals. One wall can be officially designated a "legitimate" graffiti wall, though this approach removes some of the challenge inherent in informal graffiti.

B-185. **Make the Target Less Attractive.** The school is not only an easy target but also an attractive one. It represents enforced learning, discipline, and mandatory attendance to students who are, simply by virtue of their age, rejecting adult standards and striving to achieve independence from adult control. Students are additionally provoked if the school functions in an impersonal, undemocratic, repressive, or manipulative manner, further increasing the likelihood of vandalism.

B-186. **Revise the Curriculum.** The list of characteristics associated with vandals and vandalized schools indicates that property damage and loss are higher when competition for rewards is intense, the availability of rewards is limited, or the distribution of rewards is unfair. All of this suggests that school policy and atmosphere have a direct bearing on school vandalism. Below are changes in school governance that can help remove the features that make a school an attractive target for vandals. These procedures, especially in combination, have proven effective against all forms of vandalism, including those that most commonly occur while school is in session.

- **Alternative schools.** Though originally designed to perform educational functions, alternative schools have proven effective in reducing school violence and vandalism. They provide an option to students who are not benefiting from the regular academic classroom. These schools operate within or alongside the traditional school. They are characterized by low student-to-teacher ratios, intense individual counseling, and extensive use of the community as a learning resource. They offer an alternative to suspension, personalize the learning environment, and provide successful experiences to students who have performed poorly in the past.
- **Law-related education and police-school liaison programs.** In many communities, the police department has assigned school resource or liaison officers to local junior and senior high schools. These officers may on occasion assume policing duties, but their primary function is to counsel students and teach law-related courses. These activities acquaint students with the positive role that law plays in our society and personalize the relationship between the "cop on the beat" and "the kid on the corner."
- **Action learning.** This term refers to apprenticeship programs as well as training in practical aspects of adult life. The former allows students academic credit for community work (such as tutoring and assisting physicians, lawyers, or other professionals). The latter provides instruction in skills such as checkbook balancing, comparative shopping, and applying for a job. Both address the boredom and frustration that are linked to truancy, violence, and vandalism.
- **Reward distribution.** The school's reward structure is related to school crime. Although the school may offer attractive incentives, most students receive very little in the way of rewards. Many of those who lose out still care about the competition. They become frustrated, and they vent their anger on the apparent source of their problems, the school. It should be possible to spread the rewards around without compromising performance standards, perhaps by recognizing improvement as well as achievement or by acknowledging forms of achievement other than scholastic, athletic, and social.

B-187. **Change Administrative Policy and Organizational Structure.** According to the National Institute of Education's Safe-School Study, the principal's leadership is a critical factor in reducing or preventing school crime. Strong principals who are visible, available, and responsive to students and staff appear to be most successful in eliminating violence and vandalism. Those who are less successful are often described as unavailable and ineffective.

B-188. The Safe-School Study also found that the exercise of discipline through clear enunciation and even-handed enforcement of rules is associated with a low incidence of school crime. To increase the likelihood that students will find school a fulfilling experience, many districts are establishing minischools—schools small enough to allow the individual student to feel significant. Similarly, several large schools are currently functioning on a house basis—the school is divided programmatically into several smaller units with which students can more easily identify.

B-189. **Involve the Students.** Establishing a vandalism fund uses peer pressure to the school's advantage. The community or the school district puts aside a certain amount of money and announces that the funds will be used to cover the costs of vandalism. Any money left over reverts to the students to be used as they choose. This plan works because it educates students about the costs of vandalism, allows them to see the positive results of curbing property damage and, most important, gives them full responsibility for the problem.

B-190. Several school districts have established voluntary student security advisory councils that conduct workshops and small group discussions focusing on vandalism and violence prevention. These committees increase awareness of the school's problems, generate recommendations for action, and give students opportunities to participate in school decision making.

B-191. Older students can be helpful in influencing younger students. In several communities, junior and senior high school students visit fourth- and sixth-grade classrooms where they show a film about vandalism and then lead a discussion on the causes and consequences of vandalism.

B-192. School beautification projects involve students in the care of the school building and grounds in an attempt to increase their pride in the responsibility for the school. The more effective projects are selected by the students themselves and continue throughout the year. The projects focus on marginal students rather than school leaders.

B-193. **Involve the Parents.** An open-door policy allows parents to go directly to their child's classroom whenever they wish without securing a special visitor's permit from the office. This simple policy offers parents concrete evidence that they are indeed welcome at school.

B-194. In some schools, parents serve on antitruancy committees along with teachers and students. They visit youngsters and their families in an effort to resolve truancy problems. In other schools, parents serve as hall monitors, supervise extracurricular activities, and help in classrooms. Parents are also the school's best source of guest speakers and contacts for work-study or apprenticeship programs. Some school districts have initiated faculty men's clubs to acquaint fathers with the male teaching staff and to encourage them to assume more responsibility for their children's progress in school.

B-195. **Involve the Community.** In some communities, students and law-enforcement, school-district, and city personnel sponsor daylong forums on vandalism. The forums introduce citizens to the causes and costs of vandalism and give them an opportunity to voice their concerns and initiate preventive programs.

B-196. Police departments can initiate public-relations programs within schools and youth-service agencies. In addition, they can enlist the help of youth in preventing vandalism through police-sponsored groups, such as the Police-Youth Service Corps, which recruits adolescents from high-crime areas to work as public-safety aides. Similarly, law students from neighboring universities can be brought into the schools at minimal or no cost to discuss the legal implications of vandalism.

## Picking up the Pieces

B-197. The preventive measures listed on the preceding pages can function as part of a long-range, proactive response plan. However, they do not address the question of immediate response. What should the school do, right away, about 20 broken windows, a cherry bomb in the toilet, or recurring racial graffiti on the wall? The answer is to repair the damage immediately. Quick repair keeps perpetrators from admiring their handiwork, lessens the epidemic effect of vandalism, and minimizes any social impact the act may have.

B-198. Initiate formal record-keeping procedures, and ensure that they are followed. Schools faced with serious problems should begin recording all acts of vandalism. They should also consult law-enforcement personnel about when police should and should not be called. When a school begins to have problems, it should work with the juvenile justice system so that the two institutions can coordinate their efforts with regard to school-age offenders.

B-199. Careful record keeping allows a school to plot the incidence of vandalism to find out precisely where and when each type of offense is occurring. For example, using incident analyses, the National Institute for Education's Safe-School Study found that—

- Fire and bomb threats most often occur on Tuesdays.
- School-property offenses tend to occur with greater frequency toward the end of each semester, especially in November and December.
- Break-ins and school-property offenses occur most often on weekends and Mondays.

B-200. This type of information is invaluable in planning a vandalism reduction and prevention plan.

## Restitution

B-201. Restitution is a set of legal and administrative procedures through which the school receives payment from vandals for damages they cause. While it seems reasonable to require payment for damages, restitution does not appear to be worth the effort. In the first place, most vandals are not caught. In the Los Angeles School District (which has an aggressive restitution program), only 30 percent of the offenders are ever identified.

From this 30 percent, most restitution is paid before matters get to court. Going through lengthy legal processes to obtain the rest is simply not cost-effective. However, a parent faced with the possibility of a court case may make a greater effort to keep his or her child out of trouble.

### CONCLUSION

B-202. The goal of crime prevention is to reduce crime through public awareness and education. The skills that crime-prevention officers acquire can eventually benefit all segments of our communities. It is logical then for crime-prevention officers to attack one of the major crime problems in our society—juvenile delinquency. Efforts of crime-prevention units will certainly be lacking if the juvenile-crime problem is not given a high priority and if juvenile-crime programs are not established.

## FRAUD

B-203. Two types of fraud can be affected by crime-prevention efforts—fraud against the government and fraud against individuals in the Army. Since the countermeasures for each of these categories of fraud are different, they will be discussed separately.

### FRAUD AGAINST THE GOVERNMENT

B-204. Fraud of this type is a loss to the Army due to manipulation of systems from within the government with criminal intent. Typical examples of fraud are—

- The diversion or theft of government property by falsifying documents such as purchase orders, shipping documents, and so forth.
- False claims for temporary-duty (TDY) pay, for reimbursement for losses due to the movement of household goods, or for reimbursement for material reported as stolen.
- Overcharges or underproduction on contracts with the Army.
- Bribery.
- Kickbacks to secure purchase orders.
- Use of one's official position for personal gain.

B-205. This list is not all-inclusive. There are many more ways to defraud the government. Some of the factors that make it easy to perpetrate a fraud are—

- The lack of independent verification of records, transactions, and reports.
- The lack of adequate supervision.
- Unrealistic budgeting and acquisition requirements.
- Failure to correct deficiencies identified by existing systems.
- Concentration, at the operational level, of responsibility and authority for an entire process in one individual.

B-206. There are built-in measures to discourage fraud for most Army material-control systems. These programs are routinely examined by the inspector general (IG) along with other command inspectors. A crime-prevention officer's main function is to encourage the reporting of fraud by ensuring that the community understands what type of activities should be reported and by ensuring that the numbers are posted in work areas where personnel are in a position to detect fraud.

## FRAUD AGAINST INDIVIDUALS

B-207. The second type of fraud is con games or consumer fraud. This crime is perpetrated by professional con men, unscrupulous businesses, and amateurs who see a chance to make a fast buck. Most of the time, the operator offers a "something for nothing" deal. By the time the victim realizes that he has been duped, the con man is long gone. Frequently, the victim is so embarrassed that he has been duped that he does not report the crime. The only way that a crime-prevention officer can help prevent this crime is to publicize common con games and illegal business practices and to encourage victims to report them. When a case of fraud occurs, the specific con technique that was used should be advertised widely to prevent others from being fooled by the same ploy.

### Repair Fraud

B-208. Repair frauds are simple to execute but difficult to detect. Some unscrupulous repairmen do not fix the problem, but they charge you anyway. Some use inferior parts; others charge you for work that you did not expect or need. Some also do "insurance" work; they will repair one thing but ensure that something else will soon go wrong. The following are methods that you can use to protect yourself from repair fraud:

- Shop around. Ask friends, neighbors, or coworkers for references. When you find repairmen you trust, stick with them.
- Do not try to diagnose the problem yourself unless you are an expert. The mechanic may take your advice, even if it is wrong. If you know exactly what the problem is, do not tell the mechanics; wait and see if their recommendations agree with your diagnosis. That way you will know whether needless repairs are suggested.
- Try to get several detailed written estimates before any work is done. Compare job descriptions and materials to be used. Be sure to ask if there is a charge for an estimate.
- Ask for the old parts to ensure that replacements were really installed.
- Ensure that you get a guarantee on any work done.
- Make sure that the work was done before you pay. Take your car for a test drive. Plug in the refrigerator. Test the television.
- Ask your local consumer affairs office about the laws in your state and what specific protection it gives regarding professional services to be licensed or certified.

### Home-Improvement Fraud

B-209. Home repairs and improvements can be costly. Be cautious if somebody offers to do an expensive job for an unusually low price, if a firm offers to make a free inspection, or if the workers "just happened to be in the neighborhood." These are the favorite tricks of dishonest home-repair firms.

Some firms offer a price you just cannot resist. Once you sign the contract, you learn why—they never deliver the service. Others send door-to-door inspectors to do a free roof, termite, or furnace inspection. These free inspections will turn up plenty of expensive repairs. Some fly-by-night companies will offer to do the work on the spot. When they leave, you may be left with a large bill and a faulty repair job. To avoid home-improvement fraud, try the following:

- Get several estimates for every repair job. Compare prices and terms. Check to see if there is a charge for estimates.
- Ask your friends for recommendations, or ask the firm for references and check them.
- Check the ID of all inspectors.
- Call the local consumer affairs office or the Better Business Bureau to check the company's reputation before you authorize any work.
- Be suspicious of high-pressure sales tactics.
- Pay by check, never with cash. Arrange to make payments in installments—one-third at the beginning of the job, one-third when the work is nearly completed, and one-third after the job is done.

### Land Fraud

B-210. Real estate can be a great investment. The enterprising real estate salesperson knows how anxious you are to find just the right property, especially for an investment or a retirement home—a nice warm climate, not too crowded, or a new development. Some dishonest agents will promise you anything—a swimming pool, country club, or private lake—to get your name on the contract. Even if the sales agent promises you luxury, they may not guarantee the basics such as water, energy sources, and sewage disposal.

B-211. Most land developers offering 50 or more lots (of less than 5 acres each) for sale or lease through the mail or by interstate commerce are required by law to file a Statement of Record with the Housing and Urban Development Administration (HUD). This document tells you almost everything you need to know about your future home—legal title, facilities available in the area (such as schools and transportation), availability of utilities and water, plans for sewage disposal, and local regulations and development plans. All of this information must be given to you in a property report prepared by the developer. Always ask to see this report before you sign anything.

B-212. If the developer does not give you a copy of the property report for the lot you are considering, you can obtain it from HUD for a fee. Send requests to: Department of Housing and Urban Development, Office of Interstate Land Sales Registration, 451 Seventh Street South West, Washington, DC 20410.

### Door-to-Door Sales

B-213. Beware of the following door-to-door sales tactics:

- "Nothing like it in the stores!" This is a true statement. The vacuum cleaners in the stores are probably of better quality and come with a better warranty.

- "Won't find this price anywhere." This is also a true statement. The prices in the stores are probably lower.
- "Easy credit!" This is another true statement. They do not care what your credit rating looks like. Once you sign for the purchase, paying for it is your problem. Be wary of low monthly payments. Find out the total amount you will pay over the life of the loan, then subtract the actual cost of the item itself. The difference is what you will pay in interest. Your bank, your credit union, or a local legal-aid society can tell you if the interest rate is fair.

B-214. Be cautious of these words, and be firm if the salesperson pressures you to buy. If you do get trapped, you are protected by a Federal Trade Commission regulation. When you make a purchase in your home totaling $25 or more, the salesperson must give you a written contract and two Notice of Cancellation forms. You have three days to change your mind and use one of these forms to cancel your contract.

### Charity Fraud

B-215. Charity fraud does a lot of harm. The swindler takes advantage of a person's good will and takes his cash—money that was meant for people in need. You can ensure that any money you give gets into the right hands. Remember these pointers when somebody asks you for a donation:

- Ask for ID, either the organization's or the solicitor's. Find out what the charity's purpose is and how the funds are used. Ask if contributions are tax deductible. If you are not satisfied with the answers, do not give money.
- Give to charities that you know. Check out those you have never heard of or others whose names are similar to a well-known charity.
- Do not fall for high-pressure tactics. If solicitors will not take no for an answer, tell them no anyway, but do not give them your money.
- Be suspicious of charities that only accept cash. Always send a check made out to the organization, not to an individual.
- If a solicitor reaches you by telephone, offer to mail your donation. Shady solicitors usually want to collect quickly.

### Self-Improvement Fraud

B-216. Con artists know that everyone wants to look better, feel better, and be a better person. Selling worthless plans and cures is one of the easiest ways for them to make a "quick buck." The following ads can look tempting:

- "Miracle reducing plan."
- "Look like a model in only five days."
- "Learn to speak Spanish while you sleep."
- "You can have a new, dynamic personality."

B-217. What can you do? Be careful! Read the small print. Know what the product contains. You should check with your doctor before you embark on any diet or exercise program.

### Medical and Health Fraud

B-218. Most of us do not know much about medicine; that is why we go to doctors. It is also why we fall for miracle cures and other phony health products and services. Patent medicines, health spas, and mail-in lab tests should be warning signs for the potential consumer. A laboratory in Texas advertised nationally that it had perfected a fail-safe urine test for cancer. More than 15,000 tests were made at $10 each before authorities stopped this fraudulent company.

### Unsolicited Merchandise

B-219. Cagey con artists will send you a gift in the mail—a tie, a good luck charm, or a key chain. What do you do with it if you did not order it? If you are the kind of person they are looking for, you will feel guilty and pay for it, but you are not obligated to. If you—

- Have not opened the package, mark it "return to sender." The post office will send it back at no charge to you.
- Open the package and do not like what you find, throw it away.
- Open the package and like what you find, keep it, free of charge. This is a rare instance where the rule of "finders, keepers" applies unconditionally.

B-220. Whatever you do, do not pay for it. Look at your gift as an honest-to-goodness way of getting something for nothing. Do not get conned if the giver follows up with a phone call or a visit—by law the gift is yours to keep.

### Mail Fraud

B-221. The following are examples of mail fraud:

- **The contest winner.** "You've won! This beautiful brand-name sewing machine is yours for a song. To claim your prize, come to our store and select one of these attractive cabinets for your new machine. Bring this letter with you and go home with a new sewing machine for next to nothing." Treat an offer like this carefully. Shop around before you claim your prize. Chances are, the cost of the cabinet will be more than the machine and cabinet are worth.
- **The missing heir.** You have just received a very official looking document. The sender is looking for the rightful heirs to the estate of someone with your last name. It could be you. To find out, just send $10 for more information. There may be thousands of people with your last name, and letters like these are often mailed nationwide. Even if there really was claimed estate, it is highly unlikely that you would be an heir. Save your money; why help a swindler get rich?

B-222. These are just two examples of mail fraud. Many of the other frauds described in this section can be handled through the mail. When they are, the US Postal Service can launch a full-scale investigation. If you think you have been cheated in a mail-fraud scheme—

- Save all letters, including envelopes.

- See if your neighbors or business associates received the same material.
- Contact your local postmaster who can direct you to your regional Postal Inspector's Office.

### SUMMARY

B-223. The following are recommendations to avoid being duped:

- Do not believe something-for-nothing offers. You get what you pay for.
- Be suspicious of high-pressure sales efforts.
- Take your time. Think about the deal before you part with your money.
- Get all agreements in writing. Insist that agreements are made in "plain English," not "legalese."
- Read all contracts and agreements before signing. Have a lawyer examine all major contracts.
- Compare services, prices, and credit offers before agreeing to a deal. Ask friends what their experiences have been with the firm or service in question.
- Check the firm's reputation with your local consumer affairs office or the Better Business Bureau.

## INTERNAL THEFT

B-224. Employee theft is a major problem in most large organizations. It is estimated that one-third of all business bankruptcies are a result of theft. Of course, in the Army there is no danger of the organization going out of business; however, waste and internal theft can divert significant amounts of critical resources from mission-essential activities. Unlike many businesses, the Army has recognized that internal losses can be a problem, and most areas have adequate controls mandated by regulation. As crime-prevention officers, we should monitor reports of survey, physical-security inspection reports, and results of crime-prevention inspections to ensure that control measures are being followed. When it is apparent that the correct measures are not being followed, the problem must be identified for the senior commander's action.

B-225. Research has shown that an organization's atmosphere is just as important as management controls and physical security in preventing employee theft. To control this problem, the following items are essential:

- Employees, both civilians and soldiers, must believe they are part of a professional organization that expects superior performance from all of its members. When second-rate work is accepted, the lax attitude carries over into property-control procedures and losses increase. Standards must be high, but fair.
- Leadership must set the example. If leaders, supervisors, and managers take advantage of their positions to use government material or services for their own benefit (even in minor ways), their employees will also feel justified in diverting Army resources to their own use.

- The organization must show a genuine concern for the problems of its personnel. If an employee feels he has been treated unfairly, it is easy for him to justify stealing from the organization. He believes that he is only taking what he would be getting if the organization was fair.
- The organization must take appropriate disciplinary or administrative actions in cases of theft. There can be no acceptable level of internal theft. If employees believe that an activity considers a certain level of loss to be acceptable, then material theft will grow rapidly. Everyone will consider the material he takes to be within the 2 percent that the activity expects to lose.
- Policies on internal thefts must be enforced. Frequently, a thief is a long-term employee whose honesty has always been above question. In these cases, the temptation is very strong to let the offender go with a "slap on the hand." However, this is a clear signal to other employees that diversion of Army resources for personal use is acceptable to management. If an employee chooses to steal and is caught, he should be prosecuted.
- Publicity campaigns using posters and other media should be used to disseminate command policies on internal theft. Crime hot lines are also useful in increasing reports of employee thefts.

## PILFERAGE

B-226. The protection of property, including the prevention of pilferage of government supplies and equipment, is one of the primary functions of the MP and civil service security forces. This function may include protecting supplies and equipment while in storage areas, during the issue process, or while they are in transit.

B-227. Pilferage is probably the most common and annoying hazard with which security personnel are concerned. It can become such a financial menace and detriment to operations that a large portion of the security force's efforts may have to be devoted to its control. Pilferage, particularly petty pilferage, is frequently difficult to detect, hard to prove, and dangerous to ignore.

B-228. Military property loss throughout the world would increase millions of dollars each year if subjected to uncontrolled pilferage. However, the risks incurred cannot be measured in terms of dollars alone. Loss of critical supplies for tactical units could result in loss of life or a danger to national defense. In some areas, losses could assume such proportions as to jeopardize an installation's mission. All installations and facilities can anticipate loss from pilferage. Actual losses will depend on such variable factors as the type and amount of materials, equipment, and supplies produced, processed, and stored at the facility; the number of persons employed; social and economic conditions in surrounding communities; command attitude; and physical-security measures used. Because these factors differ greatly in various types of installations and in different geographical locations, each must be considered separately.

B-229. To determine the severity of this hazard at any given installation or facility, it is necessary to determine the amount of loss that may be occurring.

Unfortunately, this is not always an easy task. Accounting methods may not be designed to pinpoint thefts; consequently, such losses remain undisclosed or they are lumped together with other shrinkages, thus effectively camouflaging them.

B-230. Inventory losses may be inaccurately labeled as pilferage for the following reasons:

- Failure to detect shortages in incoming shipments.
- Improper stock usage.
- Poor stock accounting.
- Poor warehousing.
- Improper handling and recording of defective and damaged stock.
- Inaccurate inventories.

B-231. In some cases, inventory losses may be impossible to detect because of the nature and quantities of materials involved. Stock inventory records may not be locally maintained, or there may be no method for spot checks or running inventories to discover the shortages.

## PROFILE OF PILFERERS

B-232. Physical-security personnel must be able to recognize and counteract two types of pilferers—casual and systematic. A casual pilferer is one who steals primarily because he is unable to resist the temptation of an unexpected opportunity and has little fear of detection. There is usually little or no planning or premeditation involved in casual pilferage, and the pilferer normally acts alone. He may take items for which he has no immediate need or foreseeable use, or he may take small quantities of supplies for the use of family or friends or for use around his home. The degree of risk involved in casual pilferage is normally slight unless a very large number of persons are involved.

B-233. Casual pilferage occurs when the individual feels the need or desire for a certain article, and the opportunity to take it is provided by poor security measures. Though it involves unsystematic theft of small articles, casual pilferage is nevertheless very serious. It may have a great cumulative effect if permitted to become widespread, especially if the stolen items have a high cash or potential value.

B-234. There is always the possibility that casual pilferers, encouraged by successful theft, may turn to systematic pilferage. Casual pilferers are normally employees of the installation and usually are the most difficult to detect and apprehend.

B-235. A systematic pilferer is one who steals according to preconceived plans. He steals any and all types of supplies to sell for cash or to barter for other valuable or desirable commodities. He may work with another person or with a well-organized group of people, some of whom may be members of a cleaning team or even be in an advantageous position to locate or administratively control desired items or remove them from storage areas or transit facilities.

B-236. The act of pilferage may be a one-time occurrence, or such acts may extend over a period of months or even years. A large quantity of supplies with great value may be lost to groups of persons engaged in elaborately planned and carefully executed systematic-pilferage activities. Systematic pilferers may or may not be employees of the installation; if they are not, they frequently conspire with employees.

## TARGETS FOR PILFERAGE

B-237. Both casual and systematic pilferers have certain problems to overcome to accomplish pilferage objectives. These problems include the following:

- A pilferer's first requirement is to locate the item or items to be stolen. For the casual pilferer, this may be accomplished through individual search or even accidental discovery. In systematic pilferage, more extensive means are generally used. These may consist of surveilling by members of the group or checking shopping and storage areas or documents by those who have access to them.
- The second requirement is to determine the manner in which he can gain access to and possession of the desired item. This may involve something as simple as breaking open a box, or it may be as complex as surveying security factors (such as physical safeguards or security procedures) for weaknesses. It may also involve attempting to bribe security forces, altering or forging shipping documents or passes, or creating disturbances to divert the attention of security personnel while the actual theft is taking place.
- The third requirement is to remove the stolen items to a place where the thief may benefit from his act. Articles of clothing may be worn to accomplish this. Small items may be concealed in many possible places on the thief's body or in vehicles. Through falsification of documents, whole truckloads of supplies may be removed from their proper place without immediate discovery.
- Finally, to derive any benefit from his act, the pilferer must use the item himself or dispose of it in some way. The casual pilferage of supplies is intended primarily to satisfy the need or desires of the thief. The systematic pilferer usually attempts to sell the material through "fences," pawnbrokers, or black-market operations.

B-238. Detection of use or disposal can help prevent similar pilferage through investigation and discovery of the means used to accomplish the original theft. Similarly, each of the problems faced by would-be pilferers offers an opportunity for constructive preventive measures. Careful study of the possible opportunities for the pilferer to solve his problem is essential in security work.

B-239. The primary concern of a systematic pilferer in selecting a target is its monetary value. Since he steals for personal profit, the systematic pilferer looks for items from which he can realize the greatest financial gain. This means he must also have or be able to find a ready market for items he may be able to steal. He pilfers small items of relatively high value (such as drugs, valuable metals, or electronic items). However, we cannot discount the possibility that a systematic pilferer may, if the profit is substantial, select a target of great size and weight. As a rule, bulk-storage areas contain most of the material that may be selected by systematic pilferers.

B-240. The casual pilferer is likely to take any item easily accessible to him. Since he normally will remove the item from the installation by concealing it on his person or in his POV, size is also an important consideration. Monetary value and available markets are not of any great concern to the casual pilferer, because he usually does not have any idea of selling the property he steals. Storage areas containing loose items are more likely to tempt casual pilferers than bulk-storage areas.

## METHODS OF PILFERAGE

B-241. There are many ways that pilfered items may be removed from installations. Because the motives and targets likely to be selected by systematic and casual pilferers are very different, the methods of operation for each are very different.

B-242. As stated above, the casual pilferer steals whatever is available to him and generally removes it from the installation by concealing it on his person or in his automobile. The methods of the systematic pilferer are much more varied and complex. The means he may use are limited only by his ingenuity.

B-243. Shipping and receiving operations are extremely vulnerable to systematic pilferage. It is here that installation personnel and truck drivers have direct contact with each other and readily available means of conveyance. This offers a tempting opportunity for collusion.

B-244. One individual must not have control of all shipping and receiving transactions. Obviously, this procedure invites manipulation of government bills of lading and inaccurate storage and movement procedures through failure of one activity to compare bills and invoices with another activity. The opportunities for monetary kickbacks increase without a sound system of checks and balances.

B-245. Railway employees assigned to switching duties on the installation can operate in a similar manner. However, this operation is more difficult because a railway car normally cannot be directed to a location where stolen property can be easily and safely removed.

B-246. Tanker trucks used for shipping petroleum products may be altered to permit pilferage of the product. Trash- and salvage-disposal activities offer excellent opportunities to the systematic pilferer to gain access to valuable material. Property may be hidden in waste materials to be recovered by a accomplice who removes trash from the installation.

B-247. Other methods that may be used by systematic pilfers to remove property from military installations include—

- Throwing items over fences to be retrieved later by themselves or by accomplices.
- Packaging property and sending it to outside addresses through mail channels.
- Conspiring with security personnel.

- Wearing loose-fitting clothing to conceal small items.
- Removing items by using vehicles belonging to outside contractors and vendors.

## CONTROL MEASURES FOR CASUAL PILFERAGE

B-248. Specific measures for preventing pilferage must be based on careful analyses of the conditions at each installation. The most practical and effective method for controlling casual pilferage is to establish psychological deterrents. This may be accomplished in a number of ways, such as—

- Searching individuals and vehicles leaving the installation at unannounced times and places.
- Conducting spot searches occasionally to detect attempts of theft.
- Making employees aware that they may be apprehended if they attempt to illegally remove property.

B-249. Care must be taken to ensure that personnel are not demoralized nor their legal rights violated by oppressive physical control or unethical security practices. An aggressive security-education program is an effective means of convincing employees that they have much more to lose than to gain by engaging in acts of theft. Case histories may be cited where employees were discharged or prosecuted for pilferage. Care must be taken in discussing these cases to preclude the identification of individuals because of possible civil suits for defamation of character. Also, it is generally poor policy to publicize derogatory information pertaining to specific individuals. It is important for all employees to realize that pilferage is morally wrong no matter how insignificant the value of the item taken.

B-250. It is particularly important for supervisory personnel to set a proper example and maintain a desirable moral climate for all employees. All employees must understand that they have a responsibility to report any loss to the proper authorities. Adequate inventory and control measures should be instituted to account for all material, supplies, and equipment. Poor accountability, if it is commonly known, provides one of the greatest sources of temptations to the casual pilferer.

B-251. Identifying all tools and equipment by some mark or code (where feasible) is necessary so that government property can be identified. Installation tools and equipment have counterparts on the civilian economy and cannot otherwise be identified as government property. Another control method requires individuals to sign for tools and equipment. The use of the signature control method reduces the temptation to pocket the item.

B-252. In establishing any deterrent to casual pilferage, physical-security officers must not lose sight of the fact that most employees are honest and disapprove of theft. Mutual respect between security personnel and other installation employees must be maintained if the facility is to be protected from other more dangerous forms of human hazards. Any security measure that infringes on the human rights or dignity of others will jeopardize rather than enhance the overall protection of the installation.

## CONTROL MEASURES FOR SYSTEMATIC PILFERAGE

B-253. Unlike the casual pilferer, the systematic thief is not discouraged by psychological controls. Nothing short of active physical-security measures are effective in eliminating loss from this source. Some of these measures include—

- Establishing security surveillance at all exits from the installation.
- Establishing an effective package- and material-control system.
- Locating parking areas for POVs outside of the activity's perimeter fencing.
- Eliminating potential thieves during the hiring procedure by careful screening and observation.
- Investigating all losses quickly and efficiently.
- Establishing an effective key-control system.
- Establishing adequate security patrols to check buildings, grounds, perimeters, and likely locations for clandestine storage of property removed from its proper location.
- Installing mechanical and electrical intrusion-detection devices where applicable and practical.
- Coordinating with supply personnel to establish customer ID to authenticate supply-release documents at warehouses and exit gates.
- Establishing appropriate perimeter fencing, lighting, and parking facilities and effective pedestrian, railway, and vehicle gate-security control.

## HOW TO STOP EMPLOYEE THEFT

B-254. No matter what it is called—internal theft, peculation, embezzlement, pilferage, inventory shrinkage, stealing, or defalcation—thefts committed by employees are behind at least 60 percent of crime-related losses. So many employees are stealing so much that employee theft is the most critical crime problem facing businesses today.

B-255. Although employee theft results in part from factors beyond control, the extent of employee theft in any business is a reflection of its management—the more mismanagement, the more theft. An effective stop-employee-theft policy must include at least the following:

- Pre-employment screening.
- Analysis of opportunities for theft.
- Analysis of how employees steal.
- Management-employee communication.
- Prosecution of employees caught stealing.

B-256. Each employer must reduce losses as much as possible. A police state need not be created. Large monetary expenditures need not be made.

B-257. The best way to stop employee theft is to simply not hire those employees inclined to steal. The best way is also impossible. The employer must set up a screening process that will weed out obvious security risks. Many experts believe that personnel screening is the most vital safeguard against internal theft. The following are some basic guidelines for the employer:

- Have the applicant fill out a written application. Ensure that the written application does not discriminate and that it conforms to any applicable laws.
- Solicit references, but keep in mind that those contacted will give favorable opinions. Ask primary references for secondary references. In contacting the latter, make it clear that the applicant did not refer them.
- Interview. During an interview, assess the applicant's maturity and values. Observe his gestures.
- Use psychological deterrents. Inform the applicant that your business routinely runs a background security check or that fingerprints will be taken. The hope is that the dishonest applicant will not be back.
- Obtain credit-bureau reports, but only after following guidelines set forth in the Fair Credit Reporting Act.

### Opportunities, Methods, and Control

B-258. Cases of employee theft have been documented in almost every conceivable phase of business operations, from theft of petty cash to theft of railroad cars. An infinite variety of methods have been used. Some of the areas that are most vulnerable are—

- Shipping and receiving.
- Inventory.
- Accounting and record keeping.
- Cash, check, and credit transactions.
- Accounts payable.
- Payroll.
- Facility storage units.

B-259. Some of the methods used include—

- Pilfering (one item at a time).
- Stealing from cash registers or altering cash-register records.
- Issuing false refunds.
- Using the back door or trash containers.
- Taking advantage as a supervisor.
- Avoiding package controls.
- Embezzling.
- Forging checks.
- Stealing credit cards.
- Manipulating computers and stealing computer time.
- Conspiring with night cleaning crews.
- Duplicating keys or using a master key that is not properly controlled.
- Conspiring with outsiders (such as inflating insurance claims).

B-260. Too many opportunities exist for employees to exploit. Reduce these opportunities and the losses will be reduced. Reduce opportunities by using the following controls:

- Perform random spot checks on all phases of business in addition to regular, comprehensive audits.
- Check the payroll. Make sure that you are not paying a fictitious or dead employee.
- Take physical inventory seriously.
- Know what you own. Be able to identify it.
- Do not allow one employee to perform all functions. Separate receiving, purchasing, and accounts payable. Separate accountants from cash.
- Control payment authorizations.
- Keep blank checks secured. Do not presign or use uncoded, unnumbered checks.
- Reconcile cancelled checks with original invoices or vouchers.
- Secure exits. Restrict employees to one exit, preventing exit from the rear of buildings.
- Establish strict package control.
- Inspect cash-register receipts daily. Inspect the tape and ensure that the employee is identified on the slip. Deposit money daily.
- Issue ID badges to decrease employee presence in unauthorized areas.
- Simplify red tape; make it harder for the employee to disguise theft.
- Locate employee parking away from the business area.
- Establish a usage schedule of supplies to isolate irregularities.

## Management-Employee Communication

B-261. Leadership must be firm, yet reasonable. Most employees pattern their values after their leaders, so a good example must be set. If you expect employees to remain honest, do not take home office supplies or goods.

B-262. Train new employees, advising them of the company's values and the standards by which they will be expected to perform. Explain all security procedures, stressing their importance. Emphasize that any deviations will be thoroughly investigated.

B-263. Establish grievance procedures; give your employees an outlet for disagreement and be receptive to all grievances submitted. Ensure that employees are aware of grievance procedures and that no reprisals are taken.

B-264. Regularly evaluate employee performance and encourage employees to evaluate management. Unrealistic performance standards can lead either to desperation and anger (resulting in dishonesty) or to get-even attitudes. Regularly review salaries, wages, and benefits—do not force employees to steal from you.

B-265. Delegate responsibility. Unless decision making exists among lower and middle levels, there is a tendency for development of an "it's-us-against-them" attitude. Delegate accountability as well; no decision is valid if it is lost in a "pass-the-buck" routine.

## SECTION IV — ARMY PROPERTY AT THE LOCAL LEVEL

B-266. Proper accountability by commanders and subordinate personnel cannot be overemphasized. To ensure accountability of property, commanders must establish, implement, and supervise an installation's, activity's, or organization's security program.

B-267. Weaknesses in security procedures at the installation, activity, or organizational levels involving military property create vulnerability supported by criminal activity. Criminal activity includes—

- Theft.
- Fraud.
- Property diversion.
- Property manipulation.

B-268. Commanders and subordinate personnel must conduct a risk analysis; identify military property and; in the interest of monetary value and mission accomplishment, design mandatory security measures for specific property.

B-269. Security doctrine (as outlined in this manual) should be used to the maximum extent in securing Army property vulnerable to theft, destruction, or manipulation. Certain categories of property (such as in Table B-2, page B-62) must be assessed for security vulnerability and protective treatment. Security protective measures addressing this military property should be documented in the installation's physical-security plan. If the security measures recommended in Table B-2 are implemented using established doctrine, they should eliminate or reduce the property's vulnerability. This will reduce the incidents of theft, pilferage, and manipulation at the installation.

## MOTOR VEHICLES

B-270. Security of tactical vehicles should be based on a uniform and cost-effective approach. For example, to ensure that a tactical vehicle without a locking device is properly secured, install a clamp, chain, and locking device as illustrated in Figure B-7, page B-63. To install the security device properly while maintaining safety, refer to Technical Bulletin (TB) 9-2300-422-20. Army motor-vehicle security should also incorporate the use of the following:

- Key/lock security and accountability.
- Protective lighting.
- Fencing.
- Walking patrols, as appropriate.
- Frequent observation and visits by mobile patrols or unit personnel (such as the charge of quarters [CQ], the staff duty officer [SDO], and the staff duty noncommissioned officer [SDNCO]).

**Table B-2. Recommended Security Measures**

| Property | Inventoried by— | | | | | Inspected by— | | | | | | | |
|---|---|---|---|---|---|---|---|---|---|---|---|---|---|
| | Security Plan | Hand Receipt/Property Book | Secured by IDS | PS Plan | Regulation | Unit Leaders | PS Officer | SDO | Maint Officer | Supply Officer | Dining-Facility Officer | CQ | SDNCO |
| Arms/ammunition | x | x | x | x | x | x | x | x | x | x | | | x |
| Small arms | x | x | x | x | x | x | x | x | x | x | | | x |
| Explosives | x | x | x | x | x | x | x | x | x | x | | | x |
| Communications/electronic equip | x | x | x | x | x | x | x | x | x | x | | | x |
| Hand tools, tool sets/kits, and shop equip | x | x | x | x | x | x | x | | x | x | | | x |
| Subsistence items | x | x | | x | x | x | | | | x | x | | x |
| Controlled substances, precious metals, and tax-free items | x | x | x | x | x | x | x | x | | x | | | x |
| Accounts | x | x | x | x | x | x | | | | | | | |
| POL products | x | x | | x | x | x | x | x | x | x | | x | |
| Repair parts | x | x | x | x | x | x | x | | x | | | | |
| Aircraft | x | x | * | x | x | x | x | x | x | | | | |
| Vehicles | x | x | | x | x | x | x | x | x | x | | x | x |
| Towed weapon systems/components | x | x | x | x | x | x | x | x | x | | | | |
| Carriage-mounted weapon systems | x | x | x | x | x | x | x | x | x | | | | x |
| Construction material | x | | * | x | | x | x | | | x | | | x |
| Special-issue clothing (CTA) | x | x | x | x | x | x | x | x | | | x | | |
| Individual clothing and equipment | x | x | x | x | x | x | | x | | x | | | |
| Organizational equip/components | x | x | * | x | x | x | x | x | x | x | | x | x |
| Compasses, binoculars, and flashlights | x | x | | x | x | x | x | | | x | | | x |
| Medical-unique items | x | x | x | x | x | x | x | x | | x | | | x |
| Housekeeping supplies and equipment | x | x | x | x | x | x | | | | x | | x | |
| Housing furniture | x | x | | x | x | x | | | | x | | x | |
| Mess equipment | x | x | x | x | x | x | x | | | x | x | | x |
| Office machines | x | x | x | x | x | x | x | | | x | | | x |
| Expendable/consumable supplies | x | | x | x | ** | x | | x | | x | x | | x |

*** Depends on facility availability and cost effectiveness**
**** Depends on local policy**

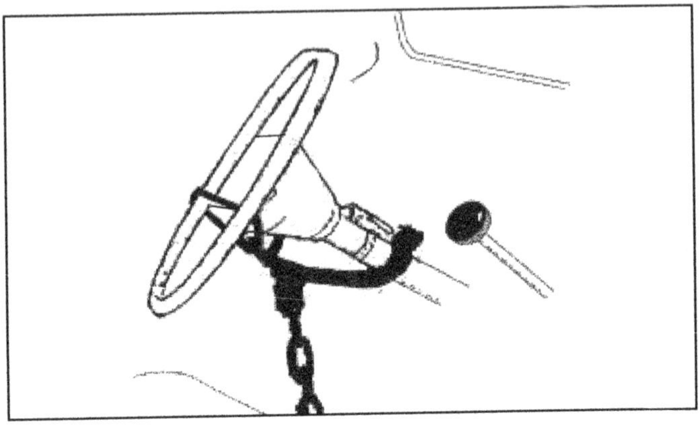

Figure B-7. Typical Clamp and Chain Installation

## CONSUMER OUTLETS

B-271. The lack of initiative at the management level within operational consumer outlets does little to prevent or reduce pilferage. Such shortcomings are identified as—

- Failure to present a professional image.
  - Lack of continuing interest, motivation, and direction.
  - No alertness to internal control of pilferage.
- Failure to institute and implement methods of operational effectiveness and efficiency. These methods include—
  - A clearly defined delegation of responsibility.
  - Insistence on stringent accountability
  - Orientation and training programs for subordinate supervisors and current and new employees.
- Failure to emphasize and enforce established criteria for continual employment.
  - Rules of conduct.
  - Standards of job performance. (Officially request appropriate action for employees guilty of criminal acts or infractions conducive to criminal acts.)
  - Inattentive job attitudes of subordinate supervisors.
  - Inadequate personal checks of established accounting and inventory procedures. **NOTE: Checks on both a regular and unannounced basis tend to control access to official stock records and to ensure careful and organized storage or stocking of merchandise.**
  - Infrequent observation of an employee's job performance.

- Failure to report misconduct, criminal or otherwise, to superiors or responsible law-enforcement personnel in the activity.
- Failure to implement recommendations made during physical-security inspections or crime-prevention surveys.

## PILFERAGE

B-272. Pilferage may be accomplished by individual employees, by a team of employees, or by employees and patrons in collusion. These actions can be greatly reduced by tightening supervision and security in the following areas:

- **Merchandise display or dispensing areas.** The following measures can reduce merchandise pilferage in display areas:
  - Detecting unauthorized price reductions.
  - Preventing or making it difficult to alter price tags.
  - Checking procedures for declaring merchandise old, shopworn, damaged, or salvage.
  - Providing more unpackaged items for personal consumption.
  - Discouraging the careless waste of foods and other perishable items.
- **Cash registers.** The following are methods of pilferage from cash registers:
  - Stealing directly from an unattended register.
  - Rerunning register tapes at lower figures (this is preventable if the reset key is maintained by the supervisor).
  - Clearing the register at a lower total figure than actual receipts for the operational period.
  - Reporting overrings and refunds falsely.
- **Theft of merchandise.** The following methods enable merchandise to be pilfered:
  - Underringing.
  - Reusing cash-register tapes. This occurs when employees fail to provide patrons with tapes or when patrons allow employees to retain tapes. The tapes allow employees to package merchandise and remove it from the premises.
  - Removing items from bags or containers carried out by employees.

## SHOPLIFTING

B-273. Shoplifting is usually confined to sales areas and is committed by casual and systematic pilferers. Items that are most frequently pilfered—

- Are relatively small in size.
- Have a high degree of consumer desirability.
- Are easily carried in pocketbooks or secreted on the person.

B-274. Amateur, adult shoplifters share the following characteristics:

- Theft is from a sudden temptation (impulse theft). There is success in the initial theft, which leads to more temptation, stronger impulses, and more thefts.
- They rarely have a genuine need for the item.

- They usually have enough money to pay for the item.
- They display symptoms of nervousness and uneasiness.

B-275. Juvenile shoplifters have the following traits:

- They act on a dare or "to belong."
- They may be coached or directed by an adult.

B-276. Professional shoplifters share these characteristics. They—

- May be talkative and are usually polite and deliberate.
- Look for opportunities continually.
- Do not take many chances.
- Are very capable of spotting security personnel.
- Steal for resale.
- Have "fences" (usually).
- Steal "to order" (often). They may have a list describing the items to be pilfered.
- Use innovative techniques.

B-277. Genuine cases of kleptomania are rare. Kleptomaniacs share the following characteristics. They—

- Take items without regard to their value or use.
- Steal compulsively and often openly.
- Are nervous and shy.

B-278. Narcotic addicts as shoplifters have the following characteristics. Only MP/security personnel should attempt apprehension. These types of shoplifters—

- Are desperate for money and they fear imprisonment.
- Take big chances.
- Take the merchandise and exit the premises quickly.
- Steal when they are at their lowest physical and/or psychological ebb.
- Are dangerous if you try to apprehend them (sometimes violent).

B-279. Alcoholics and vagrants as shoplifters share the following traits. They—

- Usually steal because of need.
- Are often under the influence of liquor at the time of theft.
- Take the merchandise quickly, then exit the premises.
- Are less likely to steal regularly at a single location.

B-280. The most pilferage occurs when employee coverage is low or when employees are untrained, inexperienced, or indifferent to the issue. The ineffective use of floor space aids shoplifters by creating congestion in the patron's traffic flow. Allowing an emphasis on small rooms or partitioned areas causes congestion that clusters, isolates, or partially hides displays. Shoplifting involves the use of one or more of the following means to obtain items:

- Palming or placing an open hand on a small article, squeezing the muscles of the hand over the article to grasp it, and lifting the still open and apparently empty hand.
- Using fitting rooms to put on tight or close-fitting garments under clothing worn into the store.
- Trying on hats, gloves, sweaters, and jackets, then exiting the store.
- Stepping around counters and removing items from unlocked showcases.
- Handling several items at once and replacing all except the item pilfered.
- Using accomplices to create a diversion of employee attention when secreting items on the person. Such items include:
  - Clothing.
  - Pocketbooks or handbags.
  - Umbrellas.
  - Various items placed in packages or paper sacks containing merchandise paid for at other departments.

## ARSON

B-281. On an installation there is little incentive for a professional arsonist to operate since the government owns the buildings and insurance fraud in collusion with the property owner is not possible. However, arson can still be a problem. In the civilian community, most deliberate fires are not set for profit. They are set to "get even" with the property owner or just for the excitement of watching something burn. Military installations are susceptible to this type of crime. On many Army posts, there is a large number of empty, wooden structures that are ideal targets for revenge seekers or vandals.

B-282. Arson is an easy crime to perpetrate, and it is relatively difficult to collect the information needed to convict an arsonist. However, this does not mean we are helpless in combating this crime. Several major cities, including Seattle, Denver, Houston, and Philadelphia, have developed successful programs to reduce the number of arsons. Some of the successful, proactive measures that have been developed are—

- Securing or disposing of materials that could be used to start fires. Ensuring that the regulations on the storage of gasoline, paint, and solvents are enforced and that paint lockers are locked.
- Enforcing command policies on the police of the post. Removing piles of trash, scrap lumber, and other material that burns easily.
- Securing empty buildings (especially empty wooden structures) and posting them as off-limits areas.
- Patrolling areas susceptible to arson.
- Encouraging participation in neighborhood watches, taxi patrols, and other community programs that increase surveillance.
- Encouraging the reporting of suspicious activities through the establishment of crime hot lines.

- Establishing a close working relationship between fire fighters and law-enforcement personnel to ensure that fires of suspicious origin are reported and thoroughly investigated.
- Offering rewards for information leading to the apprehension of the arsonist (if there is an outbreak of arson).

## SECTION V — COMMUNITY CRIME-PREVENTION PROGRAMS

B-283. There are a number of crime-prevention programs set up in communities to help deter and detect crime. These programs are supported and often headed by members of the community in conjunction with local law-enforcement agencies. The programs are as diverse in nature as they are in number, yet they complement one another.

### NEIGHBORHOOD WATCH PROGRAM

B-284. A neighborhood watch program is an organized network of citizens interacting with other neighbors and the police in preventing and detecting crime in their neighborhood. Law-enforcement efforts to reduce crime cannot be accomplished effectively without the support and cooperation of all citizens. A strong community involvement with neighbors helping themselves and other neighbors in becoming more alert to activities in the neighborhood, protecting their property, and reporting suspicious activities is essential to an effective crime-prevention program.

B-285. The Army Neighborhood Watch Program is designed to encourage Army service members and their families to actively participate in protecting their own property and the property of their neighbors, joining community crime-prevention programs, and reporting suspicious activities to MP officers. The program is designed to develop the following:

- The awareness of community crime trends and prevention efforts.
- The knowledge of quarters' security procedures.
- A cooperative system of surveillance over each neighbor's property.
- Accurate observation and reporting of suspicious activities.
- The establishment of a reliable, two-way information link between the community and MP forces.

B-286. Most neighbors know the routines of the other families that live near them. They know what cars are normally parked in the neighborhood and when families are on vacation or out of the area. Neighbors are in a very good position to recognize burglars and other intruders. Also, residents are in a good position to recognize safety hazards and crime-conducive conditions near their homes.

B-287. To capitalize on these advantages, neighborhood watch programs organize blocks in family-housing areas or floors in troop billets to improve police and community interaction. They also disseminate information on crime problems and countermeasures.

## BLOCK CLUBS

B-288. Block clubs are the basic components of an installation-wide neighborhood watch program. The geographical size of a block club may vary widely depending on the population's density and the nature of the terrain. The key factor is that the terrain organized into a single block club should promote a feeling of unity and mutual assistance. In a troop billet area, block clubs could be organized along company lines, by individual barracks, or by floor. In a family-housing area, a block club could cover one high-rise apartment building, one block, or one or more streets that are so situated that the residents identify themselves as a subcommunity within the housing area.

### Organization

B-289. Individual residents, community-service organizations, or MP units can initiate block clubs. Existing organizations (such as the PTO) have frequently established contacts within the community and have sponsored the organization of block clubs. Regardless of the approach used to organize the installation, every family or resident in the area should be contacted and encouraged to attend a block club (see Figure B-8).

B-290. A block captain and a deputy block captain should be elected at the initial block meeting. These individuals serve as "spark plugs" to sustain interest in their geographic area. They also represent their block club at district meetings. The effectiveness of the presentation at the initial block-club meeting is critical. A PMO representative should explain the crime problems on the installation and clearly outline the functions of a block club and how it can affect the crime problem. The main objective should be to generate enthusiasm and build a foundation upon which an effective neighborhood watch can be built. The use of camcorders or other audiovisual programs can generate interest. Association with programs in the civilian community may also help to generate interest.

B-291. The functions of a block club should be clearly identified. Block clubs—

- Serve as the "eyes and ears" of the police.
- Encourage the implementation of individual countermeasures, such as marking personal property.
- Disseminate crime-prevention information.
- Improve police and community relations

### Block-Club Districts

B-292. A block-club district should be organized as an intermediate link between individual block clubs and the installation's crime-prevention council. These districts should cover at least one housing area and should be composed of the block captains and deputies from each block club in the district's area. District leaders should be elected from the block captains and should serve as members of the installation's crime-prevention council. The primary functions of districts are to transmit information from the crime-prevention council to the block clubs and to develop incentive awards to recognize effective participation by the clubs.

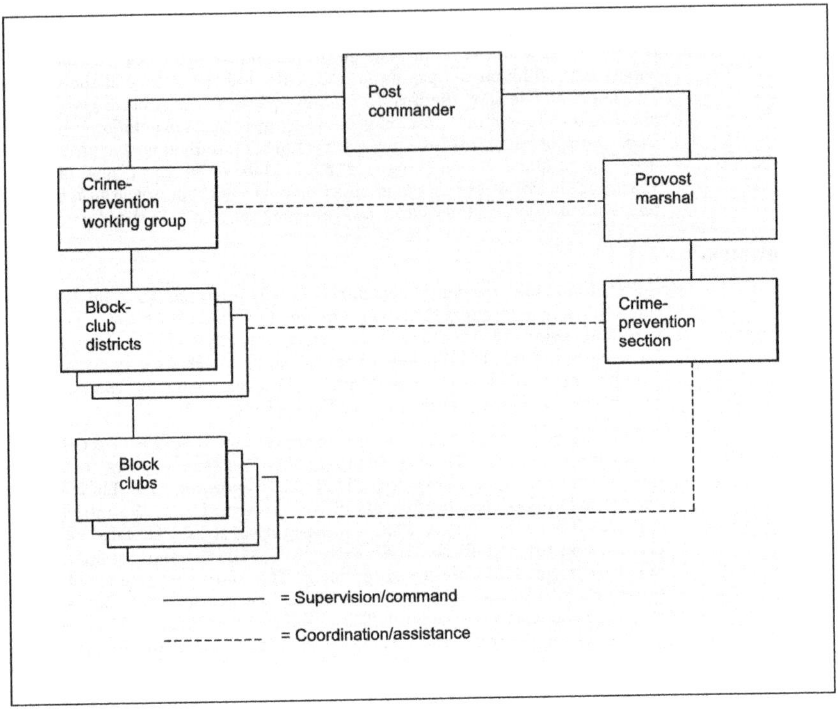


**Figure B-8. Block-Club Organization**

## Maintaining Interest

B-293. Maintaining interest in neighborhood watch programs is a major problem. Typically, a particular incident generates the level of interest required to initially organize a system of block clubs and districts; but as the particular problem is overcome, interest wanes and the clubs gradually dissipate. A review of programs that have maintained their effectiveness for extended periods indicates that successful programs have the following characteristics:

- They are a formal organization with elected block captains and district leaders.
- They address a wide range of problems.
- They feature an incentive award program to recognize individuals and clubs that participate effectively.

- Their members attend periodic workshops to train leaders on various aspects of crime prevention.
- They have clear-cut, attainable objectives. The members need to see accomplishments.
- They are actively supported by the local police.

### Block-Club Operations

B-294. Block clubs may become umbrella organizations that conduct the entire range of community crime-prevention programs, including foot and mobile patrols. However, each of these programs will be discussed in a separate section. In their most basic form, block clubs disseminate information on residential security and report suspicious activity in their neighborhood. Members of a neighborhood watch program should exchange names, addresses, and telephone numbers to enhance communication among neighbors. They must learn how to record and report suspicious activities. Some activities that block-club members should be alert for and, when observed, report to the MP officers are—

- A stranger entering a neighbor's house when it is unoccupied.
- Someone screaming.
- An offer of merchandise at a ridiculously low price (it could be stolen).
- Persons entering or leaving a place of business after duty hours.
- The sound of breaking glass or an explosion.
- A person going door to door and then going into a back or side yard, or a person trying a door to see if it is locked.
- A person loitering around a school, a park, a secluded area, or in the neighborhood.
- A person carrying property at an unusual hour or in an unusual place.
- A person exhibiting unusual mental or physical symptoms (he may be in need of medical help).
- A vehicle being loaded with valuables when parked by a closed business or untended residence.
- A business transaction conducted from a vehicle.

## ARMY NEIGHBORHOOD WATCH MEETINGS

B-295. A PMO representative should attend the initial neighborhood block meeting to explain the relationships of crime prevention and the neighborhood watch programs. The meeting should be publicized with handout invitations announcing the time, location, and purpose of the meeting. If possible, the meeting should be held in the neighborhood (such as the training room in the troop billets, a local school, or a residence in family quarters). In addition to explaining the program concepts, the initial meeting could include the following:

- Installation and community criminal statistics concerning the nature and volume of housebreakings, larcenies, and other crimes.
- An exchange of names and telephone numbers (if applicable) of attendees. This information should also be placed on a neighborhood block sheet that is a geographical diagram of the block showing the

location of each room, apartment, and building address number in the block. The names and phone numbers of participants would be added to the address number of each residence drawn on the diagram. This block sheet should be distributed to block members at a subsequent meeting.

- Determination of the second meeting date and location.

B-296. Other agenda items and activities for subsequent meetings could include the following discussions:

- Residential security procedures and the conduct of quarters security inspections.
- Operation ID—marking and recording of personal property.
- The observation and report of suspicious activities.
- Other crime-prevention measures and procedures being implemented at the installation (such as rape prevention and citizen escort patrols).
- Fire prevention, personal safety, and other related activities.

# OPERATION ID

B-297. Operation ID, which entails marking of property to make it identifiable and traceable to its owner if lost or stolen, was initiated by the Monterey Park, California, Police Department in 1963 and has been adopted by more than 80 percent of the police departments in the US. Operation ID is a low-cost, highly effective crime-prevention program. However, its success is contingent upon the willingness of individuals and communities to actively participate in marking and identifying their personal property.

B-298. Operation ID is designed to encourage Army service members and their families to mark their personal property with a standard Army-wide, owner-applied number. This numbering system permits the positive ID of the property and determines the location of the owner in case of theft or loss.

B-299. The principal advantages of Operation ID are theft deterrence and recovery of personal property. Marked stolen property is more difficult to dispose of, and illegal possession can result in the prosecution of a thief. Recovered lost or stolen property can only be returned if there is some means of identifying and locating the rightful owner.

## METHODS OF IDENTIFYING PROPERTY

B-300. Various methods of establishing positive ID and ownership of property in case of loss are available for individuals. Each method has advantages and limitations, and a combination of these methods would be required to ensure the ID of all high-value personal property.

- Inscribing the owner's applied number with an etching or engraving tool would allow the recovering agency to visually identify the number inscribed on the property for notification and subsequent return of the property. Electrostatic markers are available for use at no cost to the individual. However, some personnel are reluctant to use this method since it can mar the property. The inscription should be made in a location that can be readily seen by the recovering agency but which would not deface the property's appearance or reduce its value. Some high-value personal items such as coins, jewelry, and silver cannot be inscribed with an owner's applied number. Another method of identifying these items would be required.

- Using invisible fluorescent ink, powder, or paste to mark the property can make it easier for the agency recovering the property to use an ultraviolet light and identify the owner. This marking will not mar the property. However, the fluorescent markings and ultraviolet light are an additional cost, and many agencies do not have ultraviolet lights to inspect recovered property.
- Using a laser photographic process to identify diamonds. Every diamond emits a unique reflection when penetrated by a low-level laser light. A laser photographic process has been developed to record a diamond's pattern of light reflection on film. Several jewelers throughout the country have the laser photographic equipment available. To register a diamond, two photographs are taken. One is provided to the owner and the other to a central registry. If a diamond is lost or stolen, the recovering agency can take the diamond to a jeweler that maintains the laser photographic equipment for print. This photograph would then be forwarded to the registry for owner ID.
- Photographing the personal item. Individuals can photograph personal high-value items that cannot be engraved. Although the agency recovering the lost or stolen items could not identify the owner, use of photographs could assist in verifying ownership if it is known that the item has been recovered. In addition, a photograph would assist in submitting claims against the government or private insurance companies, as appropriate.

## RECORDING PERSONAL PROPERTY

B-301. Individuals should record identifying data (such as brand name, model, serial number, and value of the personal items), even if they use other methods of identifying property. This information would assist in determining what items may be lost, stolen, or damaged through fire, explosion, or other hazards. This information can also be used in claims against the government or private insurance companies, as appropriate.

## IDENTIFICATION NUMBERING

B-302. There are various types of owner-applied numbering systems used to mark personal high-value items. Criteria that should be considered in determining which owner-applied numbering system to incorporate include—

- Uniqueness, where no two people have the same identifier.
- Permanence, so that the owner-applied number will not change.
- Ubiquity, so that an identifying number is available to any individual who desires one.
- Availability, where the identifying number can be easily obtained and remembered.
- Indispensable, so that there are incentives requiring an individual to have the number.

- Privacy, so that the number is not a means of infringing upon an individual's right to privacy.
- Uniformity, so that the owner-applied number would be readily recognized by law-enforcement agencies who handle or come into contact with the recovered property.
- Traceability, so that the property owner can be identified and located.

B-303. The most commonly used owner-applied numbering systems include—

- Driver's license number with the issuing state abbreviation prefix.
- Social security number.
- Personal numbers assigned to individuals by a local law-enforcement agency.
- Personal numbers with the marking agency's National Crime Information Center originating-agency ID number.
- A private numbering system maintained by a commercial organization.

## STANDARD ARMY NUMBERING SYSTEM

B-304. To ensure that criteria for property ID numbering are met, the standard Army-wide, owner-applied numbering system is designated as the service member's social security number with a "USA" prefix. Upon recovery of lost or stolen property by other military or civilian law-enforcement agencies, the USA prefix, owner-applied number would alert the recovering agency that the property belongs to a member of the Army. The recovering agency can then contact the nearest Army installation's PMO/security office concerning the property and the owner-applied number inscribed. The service number can then be identified and located through the Army worldwide locator system. Since this locator system lists only social security numbers and locations of active Army service members, family members should also use the service member's social security number with the USA prefix when marking their personal items.

## IDENTIFYING AND LOCATING OWNERS OF RECOVERED PROPERTY

B-305. The installation's PMO/security officer should be the initiator of tracer actions to identify and locate the owner of recovered property marked by the standard Army Operation ID numbering system. Continuous liaison should be maintained with local civilian law-enforcement agencies and other military installations to ensure that they are cognizant of the standard Army owner-applied numbering system and will contact the PMO/security office upon recovery of private property marked with this system. The PMO/security officer should readily accept custody of the private property if the recovering agency is willing to release the property and it is not required as evidence for criminal prosecution.

B-306. Upon notification of recovered property and the inscribed Army standard owner-applied number, the PMO/security officer should contact the servicing military personnel office (MILPO) for assistance in determining the name and location of the property owner by using the Army's worldwide locator microfiche.

B-307. If the owner-applied social security number is not listed on the Army's worldwide locator microfiche and the servicing MILPO is unable to provide the requested information, the PMO/security office should contact the following offices:

- For enlisted members: USA Enlisted Records and Evaluation Center, ATTN: PCRE-RF-L, Fort Benjamin Harrison, Indiana 46249.
- For officers: PERSCOM, ATTN: TAPC-MSR-S, 200 Stovall Street, Alexandria, Virginia 22332-0444.
- For warrant officers: PERSCOM, ATTN: TAPC-OPW, 200 Stovall Street, Alexandria, Virginia 22332-0444.

B-308. In case the owner-applied number cannot be identified by either the servicing MILPO or the USA Enlisted Records and Evaluation Center, the Army Reserve Personnel Command (AR-PERSCOM) should be contacted for assistance. The AR-PERSCOM retains files on separated, retired, and Reserve component Army members. Written requests should be forwarded to: Commander, AR-PERSCOM, ATTN: ARTC-PS, 1 Reserve Way, St. Louis, Missouri 63132.

B-309. When the name and location of the service member associated with the owner-applied number inscribed on the property is verified, the PMO/ security office should notify the service member in writing that the property has been recovered. The notification should ascertain if the recovered property belongs to the service member, if the service member ever reported the property as lost or stolen, and if a claim was submitted to the SJA claims service for loss or theft of the property. Notification should also state where the property is located and a point of contact that the owner can deal with for the return of the property. If the property is to be retained as evidence for legal proceedings, the owner should be informed that the property will be returned upon completion of the proceeding. A copy of this letter should be provided to the US Army Claims Service, Fort Meade, Maryland 20755. The Army Claims Service will advise the PMO/security office if a claim was or was not submitted.

B-310. If the owner cannot be located, recovered property in the custody of the PM/security offices should be disposed of according to AR 190-22 and DOD 4160.21-M.

### USE OF THE NATIONAL CRIME INFORMATION CENTER

B-311. Stolen articles may be entered into the National Crime Information Center (NCIC) file if a theft report has been made, if the item is valued at $500 or more, and if it has a unique manufacturer's assigned serial number and/or owner-applied number. Entering stolen personal property items meeting the above criteria into the NCIC or other police information systems as outlined in AR 190-27 is encouraged.

## NEIGHBORHOOD WALKS

B-312. While most people are unwilling to participate more actively than as observers in a neighborhood watch, there are some individuals who want to become more actively involved in securing their neighborhoods. For this segment of the population, the organization of "neighborhood walks" provides a welcome opportunity to make a more active contribution. The basic idea is simple; residents patrol on foot through their own neighborhoods to observe and report crime. In practice, it is a little more complicated; however, neighborhood walks can have a dramatic impact on the crime rate, so the effort expended is worthwhile. The points that must be considered are—

- **Patrol composition.** Both adults and teenage children can volunteer to participate in neighborhood walks. However, there should be at least one adult in each party. There should be at least two people in each patrol group. Groups of four to six individuals are desirable since, by their numbers alone, they discourage attacks on the walkers. Larger groups are also more fun, and this is important when volunteers are providing the manpower.
- **Times/patrol duration.** As with other crime-prevention programs, maintaining a high level of interest can be a problem. A successful walk program in Philadelphia schedules groups for one 2-hour patrol per month. More frequent tours caused high drop-out rates among participants. Neighborhood walks should be conducted only during those times when the crime rate is the highest. Normally, there are not enough volunteers to conduct walks at times other than peak crime periods.
- **Functions.** Members of neighborhood-walk groups must understand that they are to observe only and not actively intervene in criminal acts. Participants and the government are legally liable for their actions during walks. When a crime or a suspicious activity is spotted, the neighborhood walkers should report it to the police. In Philadelphia, the walkers are equipped with horns. When a crime is spotted, they activate their horn and go to the nearest house to call the police. When residents hear the walker's horn, those in the immediate area turn on all of their lights and sound their own horns. The noise and increased lighting invariably causes the criminal to flee.
- **Neighborhood escorts.** In addition to observing and reporting criminal activity, neighborhood patrols can escort children and older persons between community-service facilities and residences. They can also request that owners secure property when they find it unsecured; for example, when there are unsecured bicycles parked on a front lawn.

## VIGILANTISM

B-313. While active community participation is essential, vigilantism must be discouraged at all costs. Both formal law codes and US common law offer few protections for private citizens who take the law into their own hands. Block captains and installation crime-prevention officers must be alert for indications that neighborhood patrols are doing more than observing and take swift remedial action when required. Of course, all nonpolice participants must be prohibited from carrying weapons of any type while engaged in crime-prevention programs. Experience in neighborhoods having much higher violent crime rates than found on Army installations has demonstrated that passive devices like horns or whistles were adequate to discourage attacks.

These devices, plus the assignment of four to six individuals to a neighborhood patrol, provide sufficient protection.

## MOBILE PATROLS

B-314. Some communities with high street-crime rates have been successful in organizing private citizens into mobile-patrol programs. Like neighborhood foot patrols, these mobile patrols serve as the "eyes and ears" of the police but do not actively intervene when they spot a crime in progress.

B-315. In a typical program, block captains or police crime-prevention officers assign specific patrol areas to each private mobile patrol. In addition, each patrol receives training on the patrol's functions, communications procedures, and emergency actions. Normally, patrols are instructed to blow their auto horns steadily when they observe a crime in progress. This is usually sufficient to drive off the criminal.

B-316. Most private citizen patrols use cellular telephones or citizen's band (CB) radios as direct communications links with the supporting police department. Installation CB radio clubs are often willing to sponsor anticrime patrols under police supervision. Commercial taxi companies that operate on military installations are also excellent candidates to organize into patrols. The cab drivers normally cover most of the high-crime-rate areas on the installation. Because of the frequency with which they cover them, they are familiar with the routine conditions in each area and are quick to spot suspicious activity.

B-317. The government cannot provide gasoline for POVs used for anticrime patrols; however, it may often provide magnetic signs to affix to the vehicle for identifying it as part of the police-sponsored, neighborhood-patrol program. As in the case of foot patrols, the installation crime-prevention officer must be alert for signs of vigilantism and must take positive action to discourage it if it appears.

## PROJECT LOCK

B-318. Nearly one million automobiles are stolen in the US every year. The total value of cars stolen is around the billion dollar mark, making auto theft the nation's costliest crime involving property. Of even greater importance is the social impact of auto theft. For a growing number of young people each year, stealing cars represents the first step toward a life of crime.

B-319. Police agencies have been diligent in the apprehension of auto thieves and the recovery of stolen vehicles, but the auto-theft problem seems to be more amenable to improvement through prevention rather than punishment. Barring strict security, auto theft is one of the easier crimes to commit. All vehicles left unattended are vulnerable, and widespread prevention by police surveillance is a physical impossibility.

B-320. The problem has grown to serious proportions despite determined law enforcement because motorists continue to be negligent or unaware of their responsibility. As long as people invite theft by leaving their cars unlocked or leaving the key in the ignition, auto thefts will continue to climb. Almost half of all stolen cars each year had been left with keys in the ignition; nine out of ten of the stolen vehicles had been left unlocked.

B-321. If a significant reduction is to be made, the motorists themselves must make it. Widespread adoption of accepted and effective prevention practices by motorists presents the most logical and immediate improvement to this growing problem.

B-322. In 1963, the Boston police department conducted a broad information campaign with the assistance of the National Automobile Theft Bureau and the Insurance Information Institute. Since then, more than 525 "Lock Your Car" campaigns have been held in about 400 communities in 49 states.

B-323. In the months following these campaigns in such cities as Denver, Chicago, Atlanta, and San Francisco, significant reductions in the number of auto thefts (ranging from 9 to 54 percent) have been recorded. Undoubtedly, the "Lock Your Car" campaigns contributed to reducing auto-theft statistics.

B-324. Project Lock is designed to permit sponsoring groups to conduct one-day or one-week "Lock Your Car" campaigns. Its purpose is twofold; it—

- Alerts the public on the importance of locking cars and removing keys as a deterrent to auto theft.
- Contributes to the welfare of youths by preventing the commission of a first crime.

B-325. The following materials can be ordered to support Project Lock:

- Windshield flyers.
- Identifying insignia to be worn by inspectors.
- Tally cards for noting cars left with keys in ignitions or unlocked doors.

B-326. Groups considering sponsorship of this campaign should consider installation areas known to have car-theft problems. The PMO should assign patrolmen to accompany the teams on inspection day. Usually, the teams' routes can be arranged to fit the regular patrols of the MP officers. If uniformed police will not be available, the campaign should not be held.

## COMMUNITY ORIENTATION

B-327. If the commander desires, a community orientation meeting might be held a month or so in advance of the campaign. This would be more desirable for a week-long rather than a day-long campaign. If such a meeting is planned, the PMO should issue invitations to representatives of the installation crime-prevention council, service clubs, women's clubs, PTOs, high schools, and churches.

B-328. The meeting should be opened with an introduction of the installation commander. After appropriate comments, the commander would read a proclamation setting the date for Lock-Your-Car Week. If possible, a representative of the National Automobile Theft Bureau should be asked to address the meeting. An alternative would be to have the PM review national and local trends in auto thefts stressing the importance of the forthcoming campaign. In conclusion, a representative of the sponsoring group might review the schedule of activities for the campaign.

## PROGRAM ACTIVITIES

B-329. If the campaign is conducted in one day, a kick-off breakfast might be substituted for the orientation meeting. The program and the attendance might be similar if a weeklong campaign is planned and the orientation meeting were held. The kick-off breakfast might have a simple program with attendance limited to the sponsoring group and police representatives. An alternative to the breakfast would be a luncheon at which results to the moment are reported. Whether the campaign will last a week or a day, the general activities will be similar.

B-330. With the assistance of the MP officers, the installation should be zoned according to the established MP patrol areas where possible. The size and number of the zones will depend on the number of police and volunteer personnel that will be available. However, the zones should cover most of the installation's service areas, troop billet areas, and family-housing areas.

B-331. The inspection teams are each composed of a uniformed patrolman and three to five sponsoring group members, cover their assigned zones and place flyers under the windshield wipers of all cars found to have keys in the ignition or to be unlocked. Under no circumstances should the flyers be placed inside the cars, even through open windows.

B-332. It is suggested that flyers be ordered early enough to allow a local printer to inscribe an overleaf statement such as "This public service is provided as a courtesy of the MP force." The tally cards should be used to record the number of cars inspected, the number that were unlocked, and the number with keys in the ignition.

## SCHOOL PROGRAM

B-333. The crime-prevention office might sponsor a poster contest for art students. It should be announced at least a month before the campaign to allow the entries to be placed in public in advance of the date. The PMO should present awards in an office ceremony. The ceremony should take place at noon on the campaign day or midweek if the campaign is longer.

## PUBLIC ADDRESSES

B-334. Close to or during the campaign date, addresses by crime-prevention professionals should be scheduled for programs of service clubs, women's clubs, PTOs, and other civic groups. Well in advance of the campaign, the PAO should be visited by the PM and the other sponsors to develop comprehensive internal-information programs in support of the installation's Project Lock. Plans for the public-affairs program should include advance publicity for the campaign, coverage of events during the campaign, and wrap-up coverage following its completion. During the inspection days, findings on how many cars were unlocked and how many had keys in the ignition should be reported regularly to a headquarters (preferably to the MP unit).

## CONCLUSION

B-335. Project Lock has been outlined to provide basic suggestions for a "Lock Your Car" day or week. No procedure can be designed to fit all needs or circumstances, and variations often will be desirable. However, if the suggestions and the materials contained in this section are used, Project Lock will not be difficult to organize and conduct. It provides better publicity opportunities than most public-service projects. It presents an opportunity for active participation by a number of crime-prevention groups. It has been field-tested and found to be an outstanding success.

# SECTION VI — EVALUATION

B-336. Evaluations of installation crime-prevention programs (see Figure B-9, page B-80) do not rely on control groups, a tight control of variables, or elaborate statistical analyses to produce worthwhile results. Most often, the resources to conduct a formal evaluation that will stand up to rigorous academic scrutiny are not available.

## CRIME-PREVENTION PROGRAMS

B-337. Several inherent difficulties in data collection on crime-prevention programs make it difficult to determine with 100 percent accuracy that a particular reduction in the crime rate was a result of a particular crime-prevention measure (unless a very elaborate analysis and control system are used). These difficulties include the control of variables, the displacement effect, and unreported crime.

### CONTROL OF VARIABLES

B-338. In its simplest form, the type of evaluation most commonly used in academic or scientific settings seeks to determine the relationship between two variables. By varying the independent variable (for example, the dosage of a drug), the effect on the dependent variable (for example, a pulse rate) is determined while all other variables (such as food intake) are held constant.

B-339. Although more complex in form, the same model can be used to evaluate complex programs. Experimental and control groups can be selected. The "treatment" can be administered by researchers or those taught by researchers. The results can be analyzed for their statistical significance. However, since crime-prevention programs deal with human subjects, certain complications arise. The degree of success may have nothing to do with the efficacy of the program, but only with the way it was introduced or with the personal predilections of the groups involved. There is no standard population; human beings are not standardized as mice are for laboratory purposes. A program found successful at one installation may be a failure in another.

B-340. These considerations also apply in the evaluation of crime-control programs. This evaluation is further complicated by another problem—the people whose behavior is to be modified (the offenders) cannot be treated directly or separated into experimental and control groups; they will not stand up and be counted. Although public-health programs often encounter this problem, they often deal with physical cause-and-effect links between treatment and improvement. The same is not true for crime-control programs.

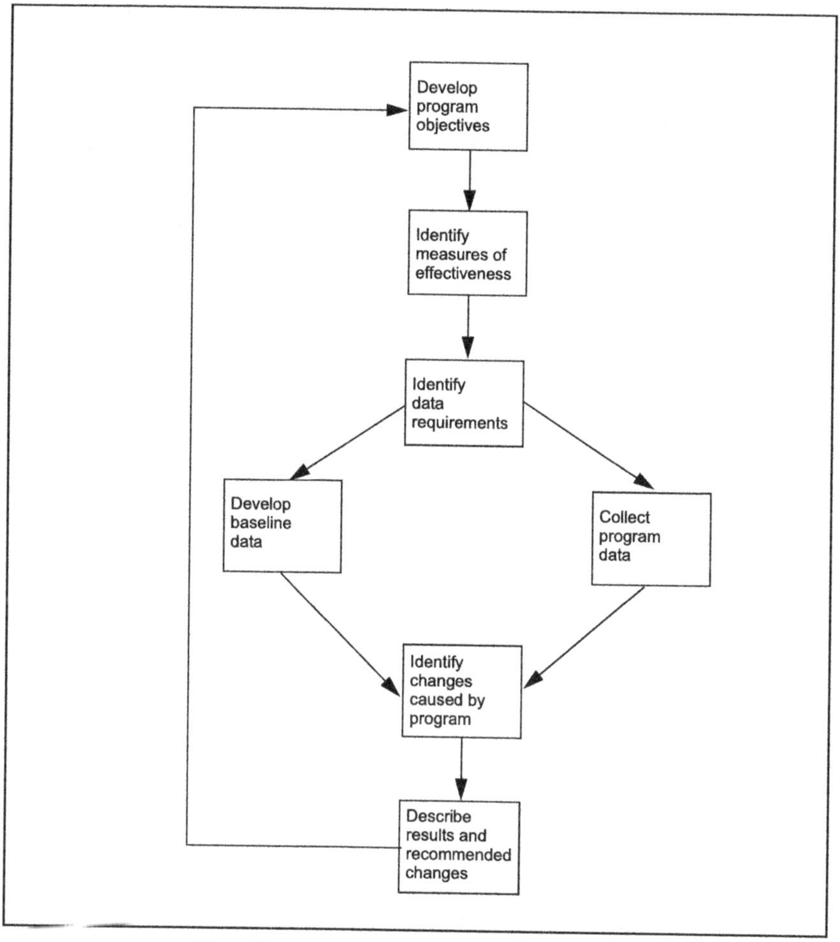


**Figure B-9. Crime-Prevention Program Evaluation**

The effectiveness of these programs is normally determined by looking at statistics of reported crime and arrests, which are more indirect indicators.

B-341. In a crime-control program, it may be impossible to classify variables as dependent and independent; they may all affect and be affected by each other. Furthermore, because of the difficulty in determining why people behave the way they do, a number of intervening and antecedent variables may go unnoticed. Police programs designed to reduce crime may have their most direct effect on the behavior of the general public toward the police, which in turn affects the crime rate.

B-342. Evaluations are not necessarily restricted to the analysis of objective crime data; they can also include subjective considerations and perceptions. These subjective evaluations can be of significant benefit in augmenting the statistical analyses of the program's results. They are especially helpful in assessing why and how a program worked and whether a statistical outcome is actually evidence that the program was successful. Interviews of participating agency personnel and residents of the program's area are usually used to supply this information. They can give the evaluator new insight into the actual program operation.

## DISPLACEMENT EFFECT

B-343. In many cases where crime reductions have been measured and attributed to programs, it is unclear whether there has been an actual reduction in crime or whether the crime has been displaced. The amount of displacement depends to an extent on the offender's characteristics. An opportunistic offender can be pictured as having a relatively elastic demand—if the risk is too high, he will forgo the crime. An addict offender is typically pictured as having a relatively inelastic demand for the product because of his inelastic demand for drugs—despite the risks, he needs the product.

B-344. Deterrents may have little effect on perpetrators of expressive crimes. These are crimes in which the perpetrator is emotionally involved and is expressing these emotions. Most assaults and homicides fit this category. On the other hand, deterrents may have a strong effect on instrumental crimes, those that are seen by the offender only as a means to an end (usually money). If alternative avenues to the same end are made more attractive by comparison, the offender may well be deterred. Deterrence may produce a diversion to legal alternatives to crime; it also may cause displacement to illegal alternatives.

### Displacement to Other Crimes

B-345. There is no immutable law that says a burglar cannot hold up a liquor store and a robber cannot burglarize a warehouse. If a specific crime or a set of crimes is the target of a crime-control program, offenders may decide to avoid the target crimes and ply their trade in other ways. Some offenders will be deterred from all crime if their crime specialty is the object of a crime-control program, but the extent of this deterrence should not be overestimated. The statutory categories of crime should not be confused with categories that serve to classify offenders.

B-346. In some cases, the result of displacing offenders to other crimes is beneficial. If the targeted crimes are more serious than the ones to which offenders are diverted, the net effect on the program may be the reduced danger to society. Of course, the converse may also be true; closing off the vulnerable and more easily protected targets of crime may cause an offender to commit more serious crimes with a net increase in the danger to society.

B-347. In some instances, the individual effect may be substantial but the overall effect may be negligible. Protecting a small fraction of premises against burglary will reduce the number of crimes committed against them, but the burglary rate against unprotected premises may go up.

### Displacement to Other Tactics and Targets

B-348. Offenders can change their manner of committing a crime when a new program is established to counter their activity. One example of this took place in 1969 in a section of the Bronx that was having a rapid increase in outdoor crime. The crimes took place primarily in the evening hours when people were returning from work. The program instituted by the police consisted of intensive sweeps of randomly selected city blocks, coupled with plainclothes police officers patrolling the streets. It succeeded in reducing the number of offenses committed during the evening hours, but at the expense of increasing the number taking place in the late afternoon when patrolmen were taking their lunch hours or were occupied with school crossings or shift changes.

### Displacement to Other Areas

B-349. The most frequently discussed type of crime displacement is from one area to another. For instance, it has been cynically suggested that the goal of the New York subway police is to chase crime into the streets where it belongs. More seriously, some recent police-helicopter-program evaluations have been questioned because they did not consider possible area displacements.

B-350. One type of boundary of interest to crime displacement is the jurisdictional boundary between the installation and surrounding cities. It has been conjectured that the crime reduction experienced in some central cities has been at the expense of the surrounding suburbs that have experienced increased crime rates.

B-351. An initial study of crime displacement was performed for the Washington, DC, area. It concluded that, although the decrease in Washington's crime rate was concurrent with an increase in the suburban crime rate, there is no evidence that the reduction in reported crime in Washington, DC, has resulted in a corresponding crime increase in the nearby suburbs.

B-352. The area-displacement effect can be measured with some degree of reliability. Three zones can be defined for the purposes of the measurement—the area containing the crime-control program (zone 1), a border around the area (zone 2), and the area chosen as the control area (zone 3). The width of the border may depend on the type of program implemented. If the program involves police helicopters, a quarter-mile-wide border may be necessary; for a patrol car, one or two blocks may suffice.

B-353. Crime rates before program initiation should be determined for all three zones. If zone 2 records a greater increase in crime than zone 3 while zone 1's crime rate decreases, then the increase in zone 2 can be attributed to two factors—

- The general increase in crime rate verified by any increase in zone 3.
- The increase caused by a displacement of crime from zone 1.

B-354. A displacement of this crime does not mean that the program is ineffective. It may suggest that the program should be expanded for all three zones.

## UNREPORTED CRIME

B-355. Crime statistics are based on crimes reported to the police. It is well known that many crimes go unreported. Victimization studies can determine the extent of unreported crime and its change from year to year by area of the country or the reasons for failure to report them. These victimization studies are best suited to determining long-term effects. They are not well suited to most crime-control evaluations in which short-term changes must be assessed.

B-356. The amount of unreported crime is important but not for planning crime-control programs affecting police activities. The extent of unreported crimes is of little significance unless a program affects it. If a program encourages reported crime, the reported-crime rate may increase despite the program's effectiveness.

B-357. Ironically, a lowered reported-crime rate may be the direct result of an increase in the actual crime rate. Taking reports from victims of crimes occupies a substantial amount of a patrolman's time. Many of these crimes are minor and have no potential for solution. In an effort to increase the police department's time on patrol, some police chiefs have stopped the practice of sending a patrolman to get reports from the victim of a minor crime. This requires the victim to travel to the police station to report the crime. If the crime is minor or is seen by the victim to be unsolvable or the theft is not covered by insurance, the victim may decide not to inconvenience himself by going to the police station to report the crime. Therefore, the number of crimes reported to the police may drop. This may lead to a larger number of unreported crimes and prevent a complete picture to the crime-prevention counsel. Conversely, an actual decrease in crime due to the increased effectiveness of the police may produce an increase in the reported crime rate.

# CRIME RATES

B-358. It has been pointed out that the crime rates, as presently calculated, do not reflect the true situation. For example, the rape rate should be calculated by dividing the annual number of rape cases by the number of women (since they are the population at risk). You would expect that the rate of commercial burglaries would be less in a residential area than a commercial area. When calculated on the basis of "per thousand people," this would be true; however, these rates should be obtained by dividing the number of cases by the number of commercial establishments (the population at risk) in each area.

B-359. The victim or the target is only one aspect of the crime. The offender can also be calculated into the rate. For example, the potential offenders in stranger-to-stranger crimes are usually considered to be males between 16 and 25 years of age. Therefore, one would expect fewer of these crimes in a city full of pensioners and retirees than in a city of the same population but with a higher proportion of young men. This fact is of minor importance in evaluating crime-control programs, since the age distribution of people in a city or a section of a city does not normally change greatly over the evaluation period. However, the former factor (the population at risk of becoming victimized) can be misleading if it is not taken into account. If possible, crime rates in experimental and control areas should be compared to the population that risks becoming victims of the target crimes.

## MEASURES OF EFFECTIVENESS

B-360. The goals of the program determine the criteria that are used to measure its effectiveness. These goals and criteria should not be seen as confining; the evaluator should be amenable to broadening the criteria, especially if the program to be evaluated is a new one. For example, the program might be beneficial in some unforeseen way, wholly outside the original criteria. Conversely, the program may be an overall failure but a success according to the evaluation. It may be that the specified measures were the wrong ones to use for the program or should not have been used alone.

B-361. Programs aimed at controlling crime should not be evaluated solely for their effect on crime. Most programs cannot, by their very nature, focus on one specific objective alone. They normally are multifaceted in their effect and should be evaluated with respect to all of their facets. Similarly, the measures of effectiveness discussed in this section may not be adequate for every crime-control program, but they comprise some of the more useful measures that can be used.

B-362. This section concentrates on the two evaluation types—internal and external. Internal and external refer to whether the evaluation is conducted on the program's inner workings and logic or the external effect of the program (which depends on the program type). An internal evaluation of a crime-control program involving the use of new police patrol techniques would include the analysis of police response time and how it was effective in controlling crime or why it was successful in one area and not in another. The external evaluation would focus only on the effectiveness of the program in reducing crime rates or solving crimes, not on how or why or the conditions under which the results were achieved.

B-363. Evaluating how well a program achieved its goals is not the only purpose of an evaluation; how and why the results were achieved are of equal importance. External measures relate to the former evaluation; internal measures are concerned with the latter. The following examples will further serve to highlight the differences between these measures:

- Many crime-control programs depend on good community relations in order to achieve their goals. In these cases, a public-affairs campaign is often instituted concurrent with the crime-control program. The success of the public-affairs campaign should not be interpreted as program success. It may be a necessary part of the program, but it does not substitute for the results of the program in controlling crime. Testimonials from people involved in the program should also be considered only as a supplement to the evaluation based on external measures.

- A study undertaken for the President's Crime Commission showed that for certain types of incidents, the probability of arrest increased as the response time decreased. As a result of this finding, many police departments purchased new equipment or tried novel techniques to reduce response time without first determining whether their workloads included enough of the incidents for which quick response is useful. If this measure (response time) is to be used, it should be recognized as an internal measure and not substituted for the external evaluation.

## INTERNAL MEASURES

B-364. Each program will have its own internal measures of effectiveness based on the logical elements of which it is constituted. This section covers only the internal measures of effectiveness that are common to most crime-control program evaluations. The measures covered include the crime rate, the clearance rate, the arrest rate, the crime-seriousness index, and the fear of crime.

### CRIME RATE

B-365. The crime rate (the number of a specified type of crime committed per resident in a specified time period) is normally considered to be a measure of deterrence. If the crime rate decreases, it is presumed that potential offenders have modified their behavior to some extent and have committed fewer crimes. This is based on the assumption that the program has made the target crimes unattractive by increasing the actual or perceived risk of apprehension, by reducing the expected return from the crime, or by making alternative forms of behavior more attractive than the target group of offenses.

B-366. These deterrent effects use different means for their accomplishment. Most crime-control programs are police-oriented and concentrate on the risk-related aspects of deterrence. Victim-oriented programs focus on reducing the expected return. Many social and recreational programs deal with making alternatives more attractive. Regardless of the orientation of the programs, their deterrent effects are determined by measuring reported crime rates.

B-367. Reported crime rates can be changed by a number of factors, some of which are misleading. The public may feel that the police are becoming less effective in dealing with crimes and, therefore, report them less often. Conversely, if the public perceives that the police are becoming more effective, they may begin to report crimes that previously would have gone unreported. Another apparent crime-rate reduction may be due to the police not recording crimes that have been reported to them. Displacement effects that can produce misleading crime-rate reductions were discussed earlier.

B-368. There may also be an actual reduction in crime due to a program's deterrent effect. In some cases, the reduction in crime can be attributed to psychological deterrence. That is, the police department may have instituted some change (such as painting all police cars canary yellow) in preexisting patterns of operation that may cause a change in the behavior patterns of potential offenders. This type of deterrence is rarely long-lived.

B-369. On the other hand, there may have been a change instituted by the police that had the desired effect of increasing the actual risk of apprehension and, therefore, reducing the number of target offenses. An example of this is the police-operated burglar alarm of commercial establishments. In the experimental program, the number of alarms were increased almost tenfold compared to the (nonalarmed) control establishments. There was 1 capture in 36 control-group burglaries (2.8 percent), while there were 12 captures in 46 experimental-group burglaries (26 percent). Crime displacements to other crimes, tactics, targets, and areas reduced the actual effectiveness of the program, but this example shows that a significant change can be made in the actual risk of apprehension. Preliminary results indicate that the rate of increase of commercial burglaries has been decreased from about 15 percent per year to about 0 percent (at the expense of a greater increase in residential burglaries).

B-370. It is difficult but useful to distinguish between actual deterrence (due to an actual increase in risk) and deterrence that is purely psychological in nature (due to a perceived increase in risk). If it is suspected that part of the deterrent effect may be transient, a long-term study would be of benefit. In this way, the half life of the psychological deterrence can be gauged, which can give some indication of the extent to which resources should be committed to the program.

B-371. Some forms of psychological deterrence are almost entirely counterproductive. They may appear effective to those who would not commit a crime and ineffective to those who are "in the business" and study the presumed deterrent more closely. For example, a tear-gas pen may give a person a sense of security that is entirely without foundation. It may be dangerous to him if he actually attempts to use it when faced with an assailant.

B-372. One investigator has pointed out that for given criminal situations, nondelinquents perceive a higher risk of apprehension than do delinquents; in all probability, the delinquents have a more realistic assessment of the situation. A purely psychological deterrent may have the unfortunate effect of making only a cosmetic improvement. This gives the general population the impression that there has been a change for the better, while in reality the situation may not have changed or may have changed for the worse because of the division of resources to a nonexistent solution.

B-373. The crime rate can be used as a measure of effectiveness. However, the evaluator should delve into the determination of the crime rate to see if any change in the rate reflects a change in reporting procedures or the deterrent effect (with tangible evidence).

## CLEARANCE RATE

B-374. The clearance rate is normally considered to be a measure of the ability of police to solve crimes. A cleared crime is one in which the police have identified the offender and have sufficient evidence to arrest him. The clearance rate is the percentage of total crimes that were cleared.

B-375. This measure of effectiveness should be used with care. A decreasing clearance rate may not mean that a police department is becoming less effective, and an increasing clearance rate may not mean that it is becoming more effective. This is due to a number of factors, primarily the public's conception of the role of the police with respect to crime and the present method of collecting crime data.

B-376. Often overlooked in discussion about crime is the role of the public in assisting the police. Police rely on community support to legitimize their authority as well as to help them carry out their work. If a segment of the community becomes alienated from the police (for whatever reason) and offers them little assistance in pursuing offenders, crime rates in these areas may rise. However, it is not only alienation of community groups that reduces the ability of the police to deal with crime; the profit motive is also to blame. Many store owners that have been robbed refuse to give their clerks time off (with pay) to help the police in their investigation. They absorb the loss of a robbery easily (it rarely comes close to the amount lost from shoplifting, employee theft, and damaged goods) and are unwilling to increase it by helping the police. They may feel that the chances of apprehending the offender are too slim, or they may be afraid of retribution if the offender discovers their assistance. They may also be afraid that their insurance will be cancelled.

B-377. If a police department begins a drive to increase its clearance rate, the increase may be forthcoming without any real change in police effectiveness. A survey of three police departments found that arrests for felonies were not made by the police in about 43 percent of the cases, in which there was probable cause, while the police were accompanied by witnesses. Making arrests in such instances would inflate the clearance rate quite easily. However, it should be noted that the police officer has a great deal of discretion in the exercise of his power of arrest. He may feel that the arrest charges will not hold up. One measure of the arrest quality is the percentage of arrests that leads to prosecutions.

B-378. In summary, clearance rate can be a measure for determining the effectiveness of crime-control programs. Its use can be increased by careful selection and specification of the crime categories that are studied, by determining the manner in which the crimes were cleared, and by determining if there has been a change in where the police draw the line in the exercise of their discretion.

## ARREST RATE

B-379. Another measure of effectiveness that is often used as a determinant of crime-control effectiveness is the arrest rate, calculated either per police officer or per resident for a specified time period. Most of the considerations concerning the clearance rate (discussed above) also apply to the arrest rate. However, the arrest rate is distinguished from the clearance rate by an additional factor—it is not related to the total number of offenses. For example, the number of arrests for drug violations has risen considerably over the past few years. However, this increase is indicative of the extent of the problem, not of the effectiveness of the solution. It has been described how drug arrests may be traded off against arrests for other offenses and vice versa, especially when informal arrest quotas are established. Therefore, the use of the arrest rate by itself does not appear to be appropriate as a measure of the effectiveness for most crime-control programs.

## CRIME-SERIOUSNESS INDEX

B-380. Among the many criticisms of crime statistics is the contention that, even if the data were reliable and complete, we would still have only a count of the number of incidents without an indication of their relative seriousness. The crime-seriousness index was proposed to include some of the major disutilities of crime typically committed by juveniles. Crimes are weighted according to the degree and nature of injury to the victims—whether they were intimidated and the nature of the intimidation or whether premises were forcibly entered, and the kind and value of property stolen. The weights were determined by requesting a sample of people to estimate the relative seriousness of various crimes.

B-381. All of the factors used to determine the weight are (or should be) included in offense reports. It would not be difficult to calculate an incident-seriousness score based on these reports, either for a specific evaluation or as a matter of course. Use of the seriousness index has also been proposed to measure the relative performance of law-enforcement agencies.

B-382. The crime-seriousness index is not the ultimate weighting scheme. The seriousness appears to be calculated more from the offender's viewpoint and the event than from the victim or society's viewpoint. For example, most people would consider the murder of a robbery victim by his assailant to be more serious than the murder of one spouse by the other. With regard to property loss, there is a difference between loss suffered by an individual who is insured and one who is not covered.

B-383. The loss relative to the individual's income is also an important factor; the theft of a $100 television from a low-income family has a much greater impact than the loss of $10,000 of jewels from a wealthy family. Perhaps a better index of the relative value of property loss to the victim would be the value of the loss in relation to the amount of the individual's discretionary income (that is, income not used for the basic necessities of life). Of course, such information is not available on police crime reports.

## FEAR OF CRIME

B-384. It has been pointed out that the perceived risk of crime is greater than the actual risk of crime, and that this perceived risk does not seem to be correlated with the actual crime rate. Unless the public feels safer in proportion to its increased actual safety, the full potential of improvements will not have been reached. Therefore, the goal of a crime-control program can be broadened to include not only improved public safety (deterrence), effectiveness (clearance rate), and reduced crime impact (seriousness); but also improved, more accurate, public perceptions of safety as well.

B-385. Measurements of perceived safety can be both direct and indirect. Public-opinion surveys with regard to perceptions about crime and safety have been made frequently. It is also possible to gauge the effect of this fear using indirect measures by observing what people do rather than what they say. The number of patrons of movie theaters and restaurants at night (or other observations of this type of activity) could be used to gauge the fear of crime.

B-386. A reliable measure of the public's perception of public safety has not been developed. Additional research is being done and needs to be done before this type of measure of effectiveness can be used with confidence.

## Appendix C

# Intelligence, Counterintelligence, and Threat Analysis

Intelligence and counterintelligence make up the first line of defense in an antiterrorism program. A well-planned, systematic, all-source intelligence and counterintelligence program is essential. The role of intelligence and counterintelligence in antiterrorism is to identify the threat. Additionally, counterintelligence provides a warning of potential terrorist attacks and provides information for counterterrorism operations. This appendix provides the elements of the intelligence cycle that have particular importance in a viable antiterrorism program. Effective intelligence and counterintelligence support requires effort, planning and direction, collection and analysis, production, investigation, and dissemination. The entire process provides decision makers with information and timely warnings upon which to recommend antiterrorism actions.

## INFORMATION SOURCES

C-1. The primary sources of intelligence and counterintelligence for the antiterrorism program are open-source information, criminal information, government intelligence and counterintelligence, and local information.

- **Open-source information.** This information is publicly available and can be collected, retained, and stored without special authorization. The news media is an excellent open source of information on terrorism. The media reports many major terrorist incidents and often includes in-depth reports on individuals, groups, or various government counterstrategies. Government sources include congressional hearings; publications by the Defense Intelligence Agency (DIA), the FBI, the Central Intelligence Agency (CIA), the Department of State (DOS), and the national criminal justice reference services. Additionally, there are private data services that offer timely information on terrorist activities worldwide. Terrorist groups and their affiliates may also have manuals, pamphlets, and newsletters that reveal their objectives, tactics, and possible targets.
- **Criminal information.** Both military and civil law-enforcement agencies collect criminal information. Because terrorist acts are criminal acts, criminal information is a major source for terrorist intelligence. Commanders must work through established law-enforcement liaison channels because the collection, retention, and dissemination of criminal information are regulated. Local military criminal investigative offices of the CID; the Naval Investigative Service Command (NISCOM); the Air Force Office of Special Investigations (AFOSI); and Headquarters, US Marine Corps,

Criminal Investigations Division, maintain current information that will assist in determining the local terrorist threat.

- **Government intelligence and counterintelligence.** The Community Counterterrorism Board (CCB) is responsible for coordinating with the national intelligence agencies concerning combating international terrorism. These agencies include the CIA (the lead agency), the DIA, the National Security Agency (NSA), the DOS, the Department of Justice (DOJ), the FBI, the Department of Energy (DOE), the Federal Aviation Administration (FAA), the Department of Transportation (DOT) (including the USCG), and the DOD. Service intelligence and counterintelligence production organizations include the US Army Counterintelligence Analysis Center; the Navy Antiterrorism Analysis Center (NAVATAC); Headquarters, US Marine Corps, Counterintelligence; and the US AFOSI Operations Center. These organizations compile comprehensive intelligence and counterintelligence for distribution on a need-to-know basis throughout the services. In combatant commands, the J2 is responsible for the intelligence fusion center. The Counterintelligence Support Officer (CISO) provides interface between the combatant command, the component commands, and the joint staff.
- **Local information.** Other valuable sources of information are the soldiers, civil servants, family members, and individuals with regional knowledge (such as college faculty or members of cultural organizations). Local crime or neighborhood watch programs can be valuable sources of information and can serve as a means to keep individuals informed in dispersed and remote areas. Intelligence exchanges with local government agencies through cooperative arrangements can also augment regional information.

## RESPONSIBILITIES OF US GOVERNMENT LEAD AGENCIES

C-2. The FBI is responsible for collecting and processing domestic terrorist information. Overseas, terrorist intelligence is principally a CIA responsibility; but the DOS, the DIA, and the HN are also active players. The MI activities are conducted according to Presidential executive orders, federal law, status of forces agreements (SOFAs), memorandums of understanding (MOUs), and applicable service regulations. Responsibilities of intelligence activities include the following:

- The combatant commander (through the commander's J2 and the CISO in coordination with the DIA, the CIA, the embassy staff, the country team, and applicable HN authorities) obtains intelligence and counterintelligence specific to the AO. The commander issues intelligence and counterintelligence reports, advisories, and assessments to the units within his command or those operating within his command's AO. This network is the backbone for communicating intelligence and counterintelligence information and advisories and for warning of terrorist threats throughout the region.
- The Secretaries of the military departments were asked (in DOD Directive 2000.12) to ensure that a capability exists to receive and evaluate data from a service perspective and that the capability exists

to disseminate all relevant data on terrorist activities, trends, and indicators of imminent attack. To accomplish this task, each Secretary appoints an MI agency (INSCOM, NISCOM, or AFOSI) to conduct intelligence and counterintelligence activities directed against terrorists and to detect, neutralize, or deter terrorist acts. To accomplish this mission, the military department intelligence agency establishes counterintelligence offices on an area basis to collect and disseminate information to combatant commanders. Each military department's intelligence agency—

- Coordinates with appropriate US and HN agencies.
- Provides overall direction and coordination of the service counterintelligence effort.
- Operates a 24-hour operations center to receive and disseminate worldwide terrorist threat information to and from the combatant command's J2, applicable service staff elements, subordinate commands, and national agencies.
- Provides service commanders with information on terrorist threats concerning their personnel, facilities, and operations.
- Investigates terrorist incidents for intelligence, counterintelligence, and force-protection aspects (with the FBI or HN authorities).
- Provides terrorist threat information in threat briefings.
- Conducts liaison with representatives from federal, state, and local agencies (as well as HN agencies) to exchange information on terrorists.
- Provides international terrorism summaries and other threat information to supported commanders. On request, provides current intelligence and counterintelligence data on terrorist groups and disseminates time-sensitive and specific threat warnings to appropriate commands.

- Service criminal investigative services (such as the CID, the NISCOM, and the AFOSI) collect and evaluate criminal information and disseminate terrorist-related information to supported installation and activity commanders as well as to the service lead agency. As appropriate, criminal investigative elements also conduct liaison with local military or security police and civilian law-enforcement agencies.
- Intelligence staff elements of commanders at all echelons—
  - Report promptly all actual or suspected terrorist incidents, activities, and early warnings of terrorist attacks to supported and supporting activities, to the local counterintelligence office, and through the chain of command to the service lead agency.
  - Initiate and maintain liaison with the security police or PMO; local military criminal investigative offices; local counterintelligence offices; security offices; HN agencies and; as required, other organizations, elements, and individuals.
  - Develop and present terrorism threat-awareness briefings to all personnel within their commands (in cooperation with the local counterintelligence offices).
- Law-enforcement staff elements will—

  - Report all actual or suspected terrorist incidents or activities to their immediate commander, supported activities, and service lead agency through established reporting channels.
  - Initiate and maintain liaison with local counterintelligence offices and military criminal investigative offices.
  - Maintain liaison with federal, HN, and local law-enforcement agencies or other civil and military antiterrorism agencies as appropriate.
- Installation, base, unit, and port security officers—
  - Report all actual or suspected terrorist incidents or activities to their immediate commander, supporting military law-enforcement office, other supported activities, local counterintelligence office, and local military criminal investigation office.
  - Conduct regular liaison visits with the supporting military law-enforcement office, counterintelligence office, and local criminal investigation office.
  - Coordinate with the supporting military law-enforcement office and counterintelligence office on their preparation and continual updating of the threat assessments.
  - Assist in providing terrorism threat-awareness training and briefings to all personnel and family members as required by local situations.

## INFORMATION REQUIREMENTS

C-3. To focus the threat analysis, intelligence and counterintelligence officers develop information requirements to identify targets using the following terrorist considerations—

- Organization, size, and composition of group.
- Motivation.
- Long- and short-range goals.
- Religious, political, and ethnic affiliations.
- International and national support (moral, physical, and financial).
- Recruiting methods, locations, and targets (students).
- Identity of group leaders, opportunists, and idealists.
- Group intelligence capabilities and connections with other terrorist groups.
- Sources of supply and support.
- Important dates, such as religious holidays.
- Planning ability.
- Internal discipline.
- Preferred tactics and operations.
- Willingness to kill.
- Willingness for self-sacrifice.
- Group skills, demonstrated or perceived (for example, sniping, demolitions, masquerade, industrial sabotage, airplane or boat

operations, tunneling, underwater, electronic surveillance, poisons, or contaminants).

- Equipment and weapons (on hand and required).
- Transportation (on hand and required).
- Medical-support availability.
- Means and methods of $C^2$.
- Means and method of communication.

## THREAT ANALYSIS AND ASSESSMENT

C-4. The preparation of the terrorist threat analysis is a continual process of compiling and examining all available information to identify terrorist targeting of US interests. A vulnerability analysis is a continual process of compiling and examining information on a facility's security posture. The threat analysis is then paired with the facility's vulnerability analysis to create the threat and vulnerability assessment. Threat analysis is an essential step in identifying the probability of a terrorist attack. To enhance the capability to collect and analyze information from many sources, the DIA maintains a terrorism database. The combatant command's J2 and CISO (in coordination with the DIA) focus this database information and regional information toward the intelligence and counterintelligence needs specific to the security of the command. However, this terrorism database is limited to foreign terrorist groups because of limitations on US intelligence-collection operations. A country's threat assessments, information and biographies about terrorist organizations, and incidents in the database can be disseminated to commands. Commands at all echelons then augment or refine the DIA's threat analysis to focus on their area of interest. This process, operative across the full range of military operations, promotes coordination between all levels of the intelligence, counterintelligence, and law-enforcement communities; broadens acquisition channels; and enhances timely distribution of information to the supported commander.

C-5. Several factors complicate intelligence and counterintelligence collection and operations. The small size of terrorist groups, coupled with their mobility and cellular organization, make it difficult to identify the members. Unlike other criminals, terrorist cadres often receive training in counterintelligence and security measures from foreign intelligence agencies or other terrorists. Additionally, the traditional orientation of police organizations is toward individual criminals, while MI organizations focus on conventional forces. Terrorist activity, therefore, requires some degree of reorientation for police and MI and counterintelligence collection and operations.

C-6. An intelligence system's ability to provide critical and timely information to the user depends not only on efficient collection and processing, but also on the ability to organize, store, and retrieve this information rapidly. This capability, coupled with early warning, careful observation, and assessment of threat activity, enhances the probability of accurately predicting the types and timing of terrorist attacks.

C-7. Commanders must carefully exercise judgment in estimating both the existing terrorist threat and the need for changes in antiterrorism measures. Key questions are—

- What has changed (mission, political climate, installation and unit personnel or equipment, terrorist capabilities)?
- What affect will the changes have on the security posture?

C-8. Extraordinary security measures, unless part of a deliberate deception during critical or high-threat situations, draw attention and detract from mission accomplishment. Sound physical security, personnel who are aware, accurate threat and vulnerability assessments, and well-rehearsed response plans reduce the probability of a successful terrorist venture. The goal is to make an attack too difficult or the level of risk unacceptable to the terrorist.

## DETERMINATION OF THE THREAT LEVEL

C-9. This threat-analysis methodology is used by the DIA, the joint staff, and the unified and specified commands for selecting the level of threat for an installation. It is applicable in an overseas setting, but fails to address issues unique to the sustaining base within CONUS. In CONUS there is a lack of intelligence and counterintelligence information from the CIA, the DIA, or the military services for the CONUS-based threat. That information must be extracted from law-enforcement channels at the local, regional, and national levels. In that context, the factors mentioned in this appendix are not as clear as they are within the intelligence process in place overseas. A modified version of this appendix should be considered for assessing threat levels in CONUS.

C-10. Threat levels within CONUS historically have been either low or negligible. This trend will most likely continue at least through the next decade. Domestic groups not covered by DOD intelligence reporting pose the greatest threat to the CONUS-based military. Due to the lack of reporting and information on these groups, the domestic terrorist groups are not currently factored into the current terrorist threat program. Therefore, the law-enforcement community must become a key player in establishing the threat levels in the context of the recommended model. A possible methodology would be the establishment of an additional threat level between low and medium—one that allows local commanders more flexibility in implementing additional security measures. Table C-1 shows the procedure for determining the threat level.

**Table C-1. Threat Levels**

| Threat Level | |
|---|---|
| Critical | Factors 1, 2, and 5 are present. Factors 3 or 4 may be present. |
| High | Factors 1, 2, 3, and 4 are present. |
| Medium | Factors 1, 2, and 4 are present. |
| Low | Factors 1 and 2 are present.<br>Factor 4 may be present. |
| Negligible | Factors 1 and/or 2 may be present. |
| **Explanation of Factors** | |
| Factor 1: Existence. A terrorist group is present, assessed to be present, or able to gain access to a given locale. | |
| Factor 2: Capability. The acquired, assessed, or demonstrated level of capability to conduct terrorist attacks. | |
| Factor 3: Intentions. Recent demonstrated anti-US terrorist activity or stated and/or assessed intent to conduct such activity. | |
| Factor 4: History. Demonstrated terrorist activity over time. | |
| Factor 5: Targeting. Current credible information on activity indicative of preparations for specific terrorist operations and/or specific intelligence that shows an attack is imminent. | |

## Appendix D

# Crisis-Management Plan

The following pages highlight areas of concern in crisis-management planning. This plan is not meant to be all-inclusive or rigidly followed. Figure D-1 is a sample format only. It does not reflect a format developed and approved for use with OPLANs or contingency plans (CONPLANs) prepared by the CINCs to fulfill tasks assigned in the Joint Strategic Capabilities Plan (JSCP) or as otherwise directed by the Chairman of the Joint Chiefs of Staff. Figure D-2, page D-4, is a sample of the Crisis-Management-Plan Checklist, which is Annex A or Appendix H to the crisis-management plan. This checklist will help ensure that the plan is sound.

Copy No. ________
Issuing Headquarters
Place of Issue
Date of Issue

**Crisis-Management Plan**

Ref: Maps, charts, and other relevant documents.

Time Zone: X

Task Organization: List units organized to conduct antiterrorism operations. Include attachments, supporting roles, and the delegation of operational control as necessary.

1. Situation. Identify essential information to understand ongoing events.

a. Terrorist force. Identify the terrorist's composition, disposition, methods of operation, and estimated strengths and capabilities that could influence the crisis-management operation. Refer to an appropriate annex.

b. Response force. Explain the response force's abilities and responsibilities. The response force's abilities can influence the crisis-management mission.

c. Attachments and detachments. Address here or refer to an annex.

d. Assumptions. Provide assumptions used as a basis for this plan (for example, the strength of the response force to be supported and the support available from other agencies).

**Figure D-1. Sample Crisis-Management Plan**

(1) Tactical-situation possibilities. Obtained from the commander's planning guidance.

(2) Personnel situation. Provided by the personnel officer.

(3) Logistics situation. Provided by the logistics officer.

(4) Legal-situation possibilities. Provided by the SJA.

2. Mission. Identify the antiterrorism mission (for example, detect, deter, contain, and neutralize terrorist threats and actions aimed at the disruption of the installation).

3. Execution.

a. Concept of operations. State the commander's tactical plan. The purpose is to inform. It may address how the commander will conduct combat-terrorism operations. It provides enough detail to ensure proper action by subordinates in the absence of specific instructions. If the required details are extensive, address them in an annex. If an operation involves two or more distinct phases, designate each phase and use subparagraphs (for example, Phase I and Phase II).

b. Tasks. Identify specific tasks for each command element charged with executing a crisis-management mission. When giving multiple instructions, itemize and indicate the priority or sequence.

c. Coordinating instructions. Include coordination and control measures applicable to two or more command elements.

4. Service Support. Provide a statement of service-support instructions and arrangements supporting the crisis-management operation. Use the following subparagraphs as required:

a. General. Outline the general plan for service support.

b. Materiel and services. Address supply, transportation, labor, and services required.

c. Medical evacuation and hospitalization. Provide the plan for evacuating and hospitalizing sick, wounded, or injured personnel. Address evacuation responsibilities and the air-evacuation policy.

d. Personnel. Provide required information and instructions to supporting unit personnel.

(1) Maintenance of unit strength.

(a) Strength reports. Provide instructions for submitting status reports. Include requirements for routine and special reports.

(b) Replacements. Address validating existing personnel requisitions, instructions for submitting requisitions, and instructions for processing and removing replacements.

(2) Personnel management. Address military and civilian personnel and civilian detainee management procedures.

(3) Development and maintenance of morale.

(a) Morale and personnel services. Provide postal and financial services, religious activities, personal hygiene, and special services activity information.

(b) Graves registration. Include evacuation procedures and handling personal effects.

**Figure D-1. Sample Crisis-Management Plan (continued)**

(4) Maintenance of discipline, law, and order. Obtain this guidance from the PMO/security officer.

(5) Miscellaneous. Include personnel administrative matters not specifically assigned to another coordinating staff section or included in preceding subparagraphs.

e. Miscellaneous. Provide special instructions or special reports not covered in preceding paragraphs.

5. Command and Signal. Provide instructions for the command and operation of communications-electronics equipment. Communications-electronics instructions may refer to an annex but should list the index and issue number of the command, control, and communications ($C^3$) operation instructions in effect. If not already issued, give instructions for the control, coordination, and establishment of priorities in the use of electromagnetic emissions. Command instructions include subordinate and higher unit CP locations and designated alternate CPs.

6. Acknowledgement Instructions.

/s/

Commander

Annexes as applicable

Distribution:

**Figure D-1. Sample Crisis-Management Plan (continued)**

## Annex A

## Crisis -Management Plan Checklist

| Yes | No | |
|---|---|---|
| | | 1. Intelligence and/or Counterintelligence. |
| ____ | ____ | Does the plan allow for the threat-analysis process (collection, analysis, production, and dissemination) to help identify the local threat? |
| ____ | ____ | Does the plan consider restrictions placed on the collection and storage of information? |
| ____ | ____ | Does the plan indicate an awareness of sources of information for the threat-analysis process (MI, counterintelligence, federal agencies, and state and local authorities)? |
| ____ | ____ | Does the plan allow for liaison and coordination of information (such as establishing a committee)? |
| | | 2. Threat Assessment. |
| ____ | ____ | Does the plan identify the local threat (immediate or long term)? |
| ____ | ____ | Does the plan identify other threats (such as national and international groups that have targeted or might target US installations)? |
| ____ | ____ | Does the installation incorporate factors for assessing the threat? Does it address — |
| ____ | ____ | Geography of the area concerned? |
| ____ | ____ | Law-enforcement resources? |
| ____ | ____ | Population cultural resources? |
| ____ | ____ | Communication capabilities? |
| ____ | ____ | Does the plan establish a priority of identified weaknesses and vulnerabilities? |
| | | Is the threat assessment periodically updated? |
| | | 3. Security Countermeasures. |
| ____ | ____ | Does the plan have specified THREATCONs and recommended actions? |
| ____ | ____ | Do security countermeasures include a combination of physical operations and sound-blanketing security measures? |
| ____ | ____ | Do the THREATCONs correspond to DOD 0-2000.12-H, Appendix BB? |

**Figure D-2. Sample Crisis-Management-Plan Checklist**

| Yes | No | |
|---|---|---|
| | | 4. OPSEC. |
| ___ | ___ | Have procedures been established that prevent terrorists from readily obtaining information about plans and operations (for example, not publishing the commanding general's itinerary and safeguarding classified material)? |
| ___ | ___ | Does the plan allow for in-depth coordination with the installation's OPSEC program? |
| ___ | ___ | Has an OPSEC annex been included in the CONPLAN? |
| | | 5. Personnel Security. |
| ___ | ___ | Has the threat analysis identified individuals vulnerable to terrorist attacks? |
| ___ | ___ | Has a training program been established to educate both military and civilian personnel in the proper techniques of personnel protection and security commensurate with the local threat and the type of position held? |
| | | 6. Physical Security. |
| ___ | ___ | Are special-threat plans and physical-security plans mutually supportive? |
| ___ | ___ | Do security measures establish obstacles to terrorist activity (such as guards, HN forces, lighting, and fencing)? |
| ___ | ___ | Does the special-threat plan include the threats identified in the threat statements of higher headquarters? |
| ___ | ___ | Does the physical-security officer assist in the threat analysis and corrective action? |
| ___ | ___ | Does the installation have and maintain detection systems and an appropriate assessment capability? |
| | | 7. Security Structure. |
| ___ | ___ | Does the plan indicate that the FBI has primary domestic investigative and operational responsibility in the US and US territories? |
| ___ | ___ | Has coordination with the SJA been established? |
| ___ | ___ | Does the plan allow for close cooperation between principal agents of the military, civilian, and HN communities and federal agencies? |
| ___ | ___ | Does the plan clearly indicate parameters for the use of force, including briefing any elements augmenting MP assets? |
| ___ | ___ | Is there a mutual understanding between all local agencies (military, local, FBI resident or senior agent-in-charge, HN forces, and local law enforcement) that might be involved in a terrorist incident on the installation regarding authority, jurisdiction, and possible interaction? |

**Figure D-2. Sample Crisis-Management-Plan Checklist (continued)**

| Yes | No | |
|---|---|---|
| ____ | ____ | Has the SJA considered the ramifications of closing the post (such as possible civilian union problems)? |
| ____ | ____ | Does the plan identify the DOS as having primary investigative and operational responsibilities overseas? |
| | | 8. Operations-Center Training. |
| ____ | ____ | Has the operational command and coordination center been established and exercised? |
| ____ | ____ | Is the operations center based on the needs of the installation while recognizing manpower limitations, resource availability, equipment, and command? |
| ____ | ____ | Does the plan include a location for the operations center? |
| ____ | ____ | Does the plan designate alternate locations for the operations center? |
| ____ | ____ | Does the plan allow for the use of visual aids (chalkboards, maps with overlays, and bulletin boards) to provide status reports and countermeasures? |
| ____ | ____ | Does the plan create and designate a location for a media center? |
| ____ | ____ | Have the operations and media centers been activated together within the last quarter? |
| ____ | ____ | Does the operations center have SOPs covering communications and reports to higher headquarters? |
| ____ | ____ | Does the operations center offer protection from a terrorist attack? |
| | | 9. Reaction-Force Training. |
| ____ | ____ | Has the force been trained and exercised under realistic conditions? |
| ____ | ____ | Has corrective action been applied to shortcomings and deficiencies? |
| ____ | ____ | Has the reaction force been formed and mission-specified trained (for example, building entry and search techniques, vehicle assault operations, countersniper techniques, and equipment)? |
| ____ | ____ | Has the reaction force been tested quarterly (alert procedures, response time, and overall preparedness)? |
| ____ | ____ | Has responsibility been fixed for the negotiation team? Has the negotiation team been trained and exercised under realistic conditions? |
| ____ | ____ | Does the negotiation team have the proper equipment? |
| | | 10. General Observations. |
| ____ | ____ | Was the plan developed as a coordinated staff effort? |
| ____ | ____ | Does the plan outline reporting requirements (logs, journals, and after-action reports)? |

**Figure D-2. Sample Crisis-Management-Plan Checklist (continued)**

| Yes | No | |
|---|---|---|
| ___ | ___ | Does the plan address the media's presence? |
| ___ | ___ | Does the plan include communication procedures and communication nets? |
| ___ | ___ | Does the plan consider the possible need for interpreters? |
| ___ | ___ | Does the plan consider the need for a list of personnel with various backgrounds to provide cultural profiles on foreign subjects and victims as well as to assist with any negotiation efforts? |
| ___ | ___ | Does the plan provide for and identify units that will augment MP assets? |
| ___ | ___ | Does the plan delineate specific taskings for each member of the operations center? |
| ___ | ___ | Does the plan provide for a response force for each phase of antiterrorism activity (initial response, negotiation, and assault)? |
| ___ | ___ | Does the plan designate service-support communications? |
| ___ | ___ | Does the plan make provisions for the notification of an accident-and-incident control officer? |
| ___ | ___ | Does the plan provide for EOD support? |
| ___ | ___ | Does the plan take into consideration the movement from various locations, including commercial airports, of civilian and military advisory personnel with military transportation assets? |
| ___ | ___ | Does the plan allow for the purchase or use of civilian vehicles, supplies, and food (if needed)? (This includes items used to satisfy a hostage demand.) Does the plan make provisions for paying civilian employees overtime if they are involved in a special-threat situation? |
| ___ | ___ | Does the plan take into consideration the messing, billeting, and transportation of civilian personnel? |
| ___ | ___ | Do appropriate personnel have the necessary language training? |
| ___ | ___ | Is WMD support available? |

**Figure D-2. Sample Crisis-Management-Plan Checklist (continued)**

# Appendix E

# Office Security Measures

The office environment should afford executives the greatest degree of physical security. Executives usually work in facilities where attackers must pass by guards, security checkpoints, office workers, aides, or secretaries before reaching them. Unfortunately, the high media value of attacking executives in security strongholds where they are clearly associated with government activity increases the value of such attacks to terrorists. Hence, there may be a need to add security measures to offset the escalating capability of attack on more secure office areas by terrorist groups.

## PHYSICAL-SECURITY SURVEY

E-1. A thorough physical-security survey of an office facility should be conducted. Offices of defense components attached to US embassies abroad should have these surveys performed by the DOS. Other DOD facilities should have surveys performed by the cognizant physical-security and facilities-engineering staffs. The best way to approach a physical-security site survey is to think like an intruder. Consider how approaches to the installation or facility could be made, how access to the building that houses executive offices could be gained, and how attacks on offices or other frequently used facilities could be mounted.

## SECURITY-ENGINEERING ASSESSMENT

E-2. The next step in evaluating the need for supplemental physical-security measures is a thorough and detailed assessment of the weapons and tactics that terrorists might use to attack the structure in which DOD executives work. Security engineers and architects need technical threat data or assessments containing the following information—

- The mode of attack, such as—
  - Standoff weapons (man-portable AT/antiaircraft weapons, sniper rifles, rock grenades, and mortars).
  - Close combat weapons (submachine guns, pistols, knives, and garrotes).
  - Contact weapons (bombs, incendiary devices, and mines).
- Perimeter penetration aids (such as power tools, hand tools, or explosives), if used.
- The time of attack.
- The attacking force's size.
- The anticipated degree of outside support or autonomy.

E-3. Engineering design requirements are developed from the security engineering assessments. The data is used to—

- Assess the ability of building components to resist the effects of the threat.
- Identify appropriate security window-glazing materials and window treatments to determine what is required to achieve the desired penetration resistance times for anticipated threats.
- Calculate the total amount of delay time. This time is achieved by using camouflage, deception, barriers, and security devices to permit response forces to reach the scene of a terrorist attack in time to thwart the attack, capture or eliminate the terrorists, and rescue executives and their staffs or dependents.

## TECHNICAL ASSESSMENT OF RESPONSES

E-4. After establishing a basic-design threat, engineers need data on the anticipated performance of response forces to be arrayed against the design threat and the expected or desired behavior of the protected executive. Some specific information needed includes—

- The response force's size, capability, supporting weapons, response time, and estimated effectiveness against the range of attacks.
- The desired options for the executive's protection—evacuate on warning, on detection, only if attacked, or only if forced to capitulate or do not evacuate.

E-5. Security planners need to know how long the structure that houses executives can withstand an attack before help arrives. Matching threat capabilities and anticipated operations by response forces establishes significant physical-security-system performance parameters. These can be quantified and used to develop detailed plans, drawings, and physical-security equipment-acquisition plans.

## PHYSICAL-SECURITY ENHANCEMENT MEASURES

E-6. Several physical-security measures intended to provide additional protection for executives can be considered based on the requirements defined through the detailed analyses outlined above. The primary purpose of such measures should be to increase the time required by persons outside an installation to reach the executives housed at an installation. A secondary purpose of such measures should be to reduce or eliminate hazards to executives that might result from violence in the vicinity. Examples of physical-security measures to consider are—

- Increase the threat-detection time by installing sensors on perimeters and barriers. This includes—
  - Combining surveillance systems including seismic, acoustic, and IR sensors at or beyond the outer perimeter.
  - Supplementing surveillance systems with CCTV/imaging IR systems tied into the alert response-force staging area.

    - Extending restricted areas or exclusion zones and relocating access-control points from the executive office area to a point closer to the installation's boundary.
    - Enlarging and extending intrusion-detection sensors from within the installation to its perimeter, allowing the IDS to collect additional data necessary and sufficiently classify and identify an intrusion before the response force arrives.
    - Enhancing both the number and the phenomenology of surveillance and detection systems within the executive office area as well as approaches leading to and from it in conjunction with measures listed below.
- Increase the threat's delay time between the perimeter and the executive office building. This includes—
    - Installing vehicle barriers and realigning roadways to eliminate straight, level stretches of road in excess of 50 meters in length.
    - Increasing concentric rings of fences, Jersey barricades, planters, bollards, and vehicle/personnel barriers.
    - Enhancing access-control areas supplemented by fire doors/security doors kept in a closed condition between the entrance to the building that houses executive offices and the executive office area.
- Confuse, camouflage, and deceive observers by hiding an executive's location. Accomplish this by—
    - Relocating executives to buildings not usually associated with office activities (barracks, motor pools, research and development [R&D] facilities, and so forth).
    - Constructing office areas in the barracks, motor pool, R&D facilities, and so forth.
    - Adding executive styles, decorative lighting, and window treatments to several different areas of office buildings to minimize the differences in external appearances between executive and nonexecutive offices.
- Increase the delay time between the entrance to the building that houses executives and the executive office area. Execute this by—
    - Adding fire doors, access-control points, dead-end corridors, and midcorridor physical barriers to complicate access to the executive office areas.
    - Adding security devices that, when activated, disrupt the intruder's ability to retain his thought processes (for example, flashing strobe lights, fog generators, noise generators, sirens, and fire-extinguishing systems).
- Increase the delay time by making access more difficult within the executive office structure. This may be accomplished by—
    - Substituting high-security doors and door frames for standard doors in areas leading to or from executive offices.
    - Installing high-security grating, wire mesh, or other materials to bar access to the executive office area through utility tunnels or conduits.

  - Strengthening walls, floors, and ceilings by substituting steel-plate, concrete-filled, steel-reinforced cinder blocks or other ballistic-resistant materials for plaster/lath or wallboard room dividers, thereby protecting against explosive devices that are used as tools to breach a barrier.
- Increase the protection for building occupants against weapons and explosives effects. This includes—
  - Substituting blast- or bullet-resistant panels for glass windows or adding a fragment-retention film at least 4 millimeters thick to the interior of glass windows.
  - Adding exterior screens/plates to cover window areas and protect against gunfire and grenade/bomb fragments.
  - Installing blast curtains, metal blinds, metal shutters, or other window treatments in executive offices to protect interior space from glass shards and other small projectiles.
  - Strengthening walls to resist weapons and explosives effects by adding steel plates, reinforced concrete, or other retrofitting measures.
  - Adding steel plates or other ballistic materials in crawl spaces above dropped ceilings or extending walls separating the executive office area from other portions of an office building from floor to floor, thereby preventing unobserved and undetected access to the space between dropped ceilings.
- Increase the hold time to contain penetrators by—
  - Adding positive-action controls to a facility's doors and gates so that gates default to a closed and locked condition unless manually released.
  - Adding positive-action controls to access-control areas so that persons inside an access-control area can neither advance nor withdraw without affirmative action by a security officer posted outside the access-control area.

E-7. These measures are used to facilitate the apprehension of terrorists. There may be some instances when defeating terrorist attempts to gain access to the executive enhances the security of the executive and the response force. This is accomplished by channeling the terrorists out of the facility and installation along one route, leaving alternative routes available to evacuate executives and other key personnel.

E-8. Install emergency executive-support facilities (including a safe haven and an emergency evacuation facility) by—

- Installing helicopter landing aids on a structure's roof or on an adjacent field far removed from parking areas.
- Installing a safe haven or other reinforced security structure adjacent to a helicopter landing facility to provide a secure waiting place for executives until a rescue helicopter with additional supporting air and ground units can extract the executives.

# Appendix F

# Physical-Security Plan

It is essential and in the best interest of security that each installation, unit, or activity maintains and uses a detailed physical-security plan. The plan should include at least special and general guard orders, access and material control, protective barriers/lighting systems, locks, and IDSs. All physical-security plans have the potential of being classified documents and must be treated accordingly. Figure F-1 depicts a sample physical-security plan.

Map Reference

Copy No. ________
Issuing Headquarters
Place of Issue
Date of Issue

**Physical-Security Plan**

1. **Purpose.** State the plan's purpose.

2. **Area Security.** Define the areas, buildings, and other structures considered critical and establish priorities for their protection.

3. **Control Measures.** Define and establish restrictions on access and movement into critical areas.

   a. Categorize restrictions as to personnel, materials, and vehicles:

      (1) Personnel access:

         (a) Establishment of controls pertinent to each area or structure.

         - Authority for access.
         - Criteria for access.
             - Unit personnel.
             - Visitors.
             - Maintenance personnel.
             - Contractor personnel.
             - National guard.
             - Emergency response teams (police, fire, ambulance, and so forth).

**Figure F-1. Sample Physical-Security Plan**

(b) Identification and control.

- Description of the system to be used in each area. If a badge system is used, a complete description covering all aspects should be used in disseminating requirements for ID and control of personnel conducting business on the installation.
- Application of the system:
  - Unit personnel.
  - Visitors to restricted areas.
  - Visitors to administrative areas.
  - Vendors, tradesmen, and so forth.
  - Contractor personnel.
  - Maintenance or support personnel.
  - Fail-safe procedures during power outages.

(2) Material control.

(a) Incoming.

- Requirements for admission of material and supplies.
- Search and inspection of material for possible sabotage hazards.
- Special controls on delivery of supplies or personal shipments in restricted areas.

(b) Outgoing.

- Documentation required.
- Controls, as outlined in paragraph 3a(2a).
- Classified shipment not involving nuclear/chemical material.

(c) Nuclear/chemical material.

- Controls on movement of warheads/chemicals on the installation.
- Controls on shipments or movement of training warheads/chemicals.
- Controls on pickup or delivery of warheads/chemicals outside the installation.

(3) Vehicle control.

(a) Policy on search of military and privately owned vehicles.

(b) Parking regulations.

**Figure F-1. Sample Physical-Security Plan (continued)**

(c) Controls for entrance into restricted and administrative areas:

- Military vehicles.
- POVs.
- Emergency vehicles.
- Vehicle registration.

b. Indicate the manner in which the following security aids will be implemented on the installation:

(1) Protective barriers:

(a) Definition.

(b) Clear zones.

- Criteria.
- Maintenance.

(c) Signs.

- Types.
- Posting.

(d) Gates.

- Hours of operation.
- Security requirements.
- Lock security.
- Barrier plan.

(2) Protective lighting system:

(a) Use and control.

(b) Inspection

(c) Action taken in case of commercial power failure.

(d) Action taken in case of failure of alternate power source.

(3) Emergency lighting system:

(a) Stationary.

(b) Portable.

**Figure F-1. Sample Physical-Security Plan (continued)**

(4) IDSs:

(a) Security classification.

(b) Inspection.

(c) Use and monitoring.

(d) Action taken in case of alarm conditions.

(e) Maintenance.

(f) Alarm logs or registers.

(g) Tamper-proof provisions.

(h) Monitor-panel locations.

(5) Communications:

(a) Locations.

(b) Use.

(c) Tests.

(d) Authentication.

(6) Security forces: General instructions that would apply to all security-force personnel (fixed and mobile). Detailed instructions such as special orders and SOP information should be attached as annexes. Security-force facets include—

(a) Composition and organization.

(b) Tour of duty.

(c) Essential posts and routes.

(d) Weapons and equipment.

(e) Training.

(f) Use of MWD teams.

(g) Method of challenging with signs and countersigns.

(h) Alert forces:

- Composition.
- Mission.
- Weapons and equipment.
- Location.
- Deployment concept.

**Figure F-1. Sample Physical-Security Plan (continued)**

(7) Contingency plans: Required actions in response to various emergency situations. Detailed plans for situations (counterterrorism, bomb threats, hostage negotiations, disaster, fire, and so forth) should be attached as annexes.

(a) Individual actions.

(b) Alert-force actions.

(c) Security-force actions.

(8) Use of air surveillance.

(9) Coordinating instructions. Matters that require coordination with other military and civil agencies such as—

(a) Adjacent installations or units.

(b) State and local agencies.

(c) Similar host-country agencies.

(d) Federal agencies.

The coordination/interaction allows for an exchange of intelligence information on security measures being used, contingency plans, and any other information to enhance local security.

On an installation, the host activity shall assume responsibility for coordinating physical-security efforts of all tenants, regardless of the components represented, as outlined in the support agreements and the host-activity security plan. Applicable provisions shall be included in, or be an appendix to, the support agreement. A formal agreement will contain definite assignment of physical-security responsibility for the items stored. The agreement should address—

- Maximum qualities to be stored.
- Physical safeguards to be used.
- Frequency of, and responsibility for, physical inventories or reconciliations.
- Reporting of losses for investigation.
- Lock and key control.
- The unit that has overall responsibility.

Procedures for authorization and ID of individuals to receipt for and physically take custody of Army property. The purpose of such coordination is protection in depth. Authority, jurisdiction, and responsibility must be set forth in a manner that ensures protection and avoids duplication of effort.

/s/ Commander

**Figure F-1. Sample Physical-Security Plan (continued)**

## ANNEXES

F-1. Annexes to the plan should include, but are not limited to, the following. More information can be found in AR 190-13.

- **Annex A.** The installation threat statement (intelligence). This annex should contain the Terrorism Counteraction Plan (refer to AR 190-13).
- **Annex B.** A bomb-threat plan. As a minimum, the bomb-threat plan should provide guidance for—
  - Control of the operation.
  - Evacuation.
  - Search.
  - Finding the bomb or suspected bomb.
  - Disposal.
  - Detonation and damage control.
  - Control of publicity.
  - After-action report.
- **Annex C.** An installation closure plan.
- **Annex D.** A natural-disaster plan. This plan will be coordinated with natural-disaster plans of local jurisdictions. At a minimum, the natural-disaster plan should provide guidance for—
  - Control of the operation.
  - Evacuation.
  - Communication.
  - Control of publicity.
  - After-action report.
- **Annex E.** A civil-disturbance plan. It is the commander's responsibility to formulate a civil-disturbance plan based on local threats. (For example, commanders of chemical facilities should anticipate the need to develop crowd-control procedures to handle antichemical demonstrations.)
- **Annex F.** A resource plan to meet the minimum-essential physical-security needs for the installation or activity.
- **Annex G.** A communication plan. This plan is required to establish communications with other federal agencies and local law-enforcement agencies to share information about possible threats. The communications plan should address all communication needs for annexes B through F above.
- **Annex H.** A list of designated restricted areas.
- **Annex I.** A list of installation MEVAs.
- **Annex J.** A contingency plan. In most instances, it will be necessary to increase security for AA&E and other sensitive property, assets, and facilities during periods of natural disasters, natural emergencies, or increased threat from terrorists or criminal elements. Therefore, CONPLANs should include provisions for increasing the physical-security measures and procedures based on the local commander's assessment of the situation. Such contingencies may include hostage negotiations, protective services, and special-reaction teams. These

provisions should be designed for early detection of an attempted intrusion, theft, or interruption of normal security conditions.

- **Annex K.** Work-stoppage plan. This is a requirement for conducting a physical-security survey.

## TACTICAL-ENVIRONMENT CONSIDERATIONS

F-2. In a tactical environment, the development of a physical-security plan is based on METT-TC (using the OPORD format and the higher headquarters' order). The order may be specific about the tasks the unit will perform. Time available may be limited and the scheme of maneuver may be dictated, but the leader must still evaluate the mission in terms of METT-TC to determine how MP elements can best carry out the commander's order.

F-3. Consider each of the following factors and compare courses of action to form a base for the physical-security plan. When the plan is firm, issue it as an order.

- Concepts for reconnaissance, coordination with adjacent and/or supporting units, and troop movement.
- Physical-security installation configurations and facilities. Areas to consider may include drop zones, landing zones, ranges, and training areas.

### MISSION

F-4. The mission is usually the emplacement of defensive security rings to protect the populace against insurgents. The number of defensive security rings depends on the particular site and situation. The following questions must be evaluated:

- What is the mission?
- What specific and implied tasks are there to accomplish the mission?
- What is the commander's intent?

### ENEMY

F-5. The commander identifies insurgent units operating in the area and tries to determine the type and size of the unit; the enemy's tactics, weapons, equipment, and probable collaborators; and the inhabitants' attitudes toward the insurgents. The following questions must be evaluated:

- What is known about the enemy?
- Where is the enemy and how strong is he?
- What weapons does the enemy have?
- What is the enemy doing?
- What can the enemy do in response to MP actions?
- How can we exploit the enemy's weaknesses?

### TERRAIN AND WEATHER

F-6. The commander can use observation and fields of fire, cover and concealment, obstacles, key terrain, and avenues of approach (OCOKA) to plan for the physical-security defensive sites. The following questions must be evaluated:

- How will the terrain and weather affect the operation?
- How fast can movement be accomplished, and how much space does the terrain and unit formations take up?
- Will the weather affect the terrain or personnel?
- Has the weather already affected the terrain?

## TROOPS

F-7. The commander must consider available equipment, the reaction time, reaction forces, communication assets, organization of troops, and medical support (if available). The following questions must be evaluated:

- What are the present conditions of vehicles and personnel?
- What is the status of ammunition and supplies?
- Who is best able to do a specific task?
- How much sleep have the soldiers had in the past 24 hours?
- What other assets are available to support the mission?
- How many teams/squads are available?
- What supplies and equipment are needed?
- What fire support is available and how can it be obtained?

## TIME AVAILABLE

F-8. This factor is critical since the inhabitants must be ready to respond to an insurgent attack with little or no warning. The following questions must be evaluated:

- How much time is available to conduct planning?
- How long will it take to reach the objective?
- How long will it take to prepare the position?
- How much time do subordinates need?
- How long will it take the enemy to reposition forces?

## CIVILIAN CONSIDERATIONS

F-9. The commander also must consider nonbelligerent third parties (such as dislocated civilians, personnel of international businesses and relief organizations, and the media). Every commander must prepare a site overlay that shows, as a minimum, the following:

- The attitude of the HN toward US forces.
- The population density near the objective.
- The condition of the local civilians.
- The possible effect of refugees and dislocated civilians on the mission.

# Appendix G

# Personal-Protection Measures

Terrorists frequently emulate military organizations as they develop, plan, train, and carry out terrorist attacks against DOD assets. Terrorists have a critical need for information regarding the whereabouts, habits, working environments, home environments, and other potential points of leverage against their targets. The three intelligence-collection methods used by terrorists against potential targets are human intelligence (HUMINT), photographic intelligence (PHOTINT), and signal intelligence (SIGINT).

## PERSONAL PROTECTION

G-1. The measures that follow are useful in providing personal protection for US government employees and DOD civilian contractors in CONUS or OCONUS facilities.

### OVERCOME ROUTINES

G-2. The reduced probability of success in kidnapping or killing a target makes the target far less desirable. Perform the following measures to prevent daily routines from being observed:

- Vary your route to and from work and your arrival and departure times.
- Vary your exercise schedule, using different routes and distances. It is best not to exercise alone.
- Do not divulge family or personal information to strangers.
- Enter and exit buildings through different doors, if possible.
- Avoid other routines.

### MAINTAIN A LOW PROFILE

G-3. Americans are easy to identify in an overseas area. Perform the following measures to reduce easy ID:

- Dress and behave in public in a manner consistent with local customs. Items that are distinctively American should not be worn or displayed outside American compounds.
- Reduce visibility in the local community.
- Avoid flashing large sums of money, expensive jewelry, or luxury items.
- Avoid public disputes or confrontations, and report any trouble to the proper authorities.
- Ensure that personal information (home address, phone number, or family information) is not divulged.

## PREPARE FOR UNEXPECTED EVENTS

G-4. All DOD personnel, contractors, and their family members should implement the following general measures:

- Get into the habit of checking in with friends and family.
- Know how to use the local phone system.
- Know the locations of civilian police, military police, government agencies, and the US embassy.
- Know certain key phrases in the local language.
- Set up simple signal systems that can alert family members or associates that danger is present.
- Carry ID showing your blood type and any special medical conditions.
- Keep personal affairs in good order.
- Avoid carrying sensitive or potentially embarrassing items.

# WORKING ENVIRONMENT

G-5. The working environment is not immune from attempted acts by criminals or terrorists. DOD installations in CONUS and OCONUS usually provide a level of basic security comparable or superior to the basic level of security provided in the surrounding community. The following are general practices that can help reduce the likelihood of a terrorist attack:

- Establish and support an effective security program.
- Discourage the use of office facilities to store objects of significant intrinsic value unless it is mission essential.
- Train personnel to be alert for suspicious activities, persons, or objects.
- Arrange office interiors so that strange or foreign objects left in the room will be recognized immediately.
- Provide for security systems on exterior doors and windows.
- Ensure that access-control procedures are rigorously observed at all times for access to—
  - The installation.
  - Buildings within an installation.
  - Restricted or exclusion areas within buildings.
- Use an ID badge system containing a photograph.
- Identify offices by room number, color, or object name and not by rank, title, or the name of the incumbent.
- Avoid using nameplates on offices and parking places.

## OFFICE PROCEDURES

G-6. In an office, the following steps can be taken to make intelligence collection and targeting more difficult for terrorists:

- Telephone and mail procedures:
  - When answering the telephone, avoid using ranks or titles.
  - When taking telephone messages, do not reveal the whereabouts or activities of the person being sought.

  - When leaving telephone messages, place them in unmarked folders; do not leave them exposed for observers to identify caller names and phone numbers, persons called, and messages left.
  - When opening mail, use a checklist to help identify letter bombs or packaged IEDs.
- Visitor-control procedures:
  - Place strict limitations on access to the executive office area.
  - Lock doors (from the inside) from the visitor-access area to executive offices or other restricted areas of a facility.
  - Ensure that receptionists clear all visitors before they enter inner offices.
  - Permit workmen or visitors access to restricted areas or exclusion areas under escort and only with proper ID. Confirm the work to be done before admitting workmen to restricted areas of the facility.
  - Limit publicity in public waiting areas to information that does not identify personnel by name, position, or office location.
  - Avoid posting unit rosters, manning boards, or photo boards where visitors or local contractors can view them.
  - Restrict the use of message boards, sign-in/-out boards, and other visual communications to general statements of availability.
- General working procedures:
  - Avoid carrying attaché cases, briefcases, or other courier bags unless necessary.
  - Avoid carrying items with markings that identify the owner by rank or title, even within the office environment.
  - Avoid working alone late at night and on days when the remainder of the staff is absent.
  - Ensure that office doors are locked when the office is vacant for any lengthy period, at night, and on weekends. If late-night work is necessary, work in conference rooms or internal offices where outside observation is not possible.
  - Ensure that the security office retains the office keys.
  - Ensure that papers, correspondence, communications materials, and other documents are not left unattended overnight.
  - Ensure that maintenance activity and janitorial services in key offices, production offices, or maintenance facilities are performed under the supervision of security personnel.
  - Prohibit the removal of property, material, or information stored on any media from the facility without proper written authorization.
  - Consider prohibiting the importation of property, material, or information stored on any media into the facility unless such items have been properly inspected.
  - Lock offices not in use to prohibit unauthorized access of stored materialthatcouldbeusedtohideIEDsorintelligence-collectiondevices.
  - Minimize the use of vehicles or vehicle markings that make it possible to readily identify the vehicle and its occupants as US-government or DOD-contractor personnel.

- Ensure that all personnel have access to some sort of duress alarm to annunciate and warn of a terrorist attack.
- Ensure that secretaries and guard posts are equipped with covert duress alarms that can be used to alert backup forces.
- Avoid placing office furnishings directly in front of exterior windows.

## SPECIAL PROCEDURES FOR EXECUTIVE ASSISTANTS

G-7. The following suggestions are intended to be a guide for secretaries and executive assistants who may find themselves performing personnel-security duties as collateral duty. Executive assistants and security personnel should regularly train and exercise procedures used in case they must evacuate mission-critical personnel to safe havens.

- Request the installation of physical barriers (such as electromagnetically operated doors) to separate offices of senior executives from other offices.
- Request the installation of a silent trouble-alarm button with a signal terminating in the security department.
- Admit visitors into the executive area when they are positively screened in advance or are personally recognized.
- Do not inform unknown callers of an executive's whereabouts, home address, or telephone number.
- Store a fire extinguisher, a first aid kit, and an oxygen bottle in the office area.
- Remain calm and listen carefully when receiving a threatening call.
- Do not accept packages from strangers until satisfied with the individual's identity and the nature of the parcel.
- Keep travel itineraries for all personnel confidential.
- Distribute daily schedules for senior officers and civilian officials on a limited basis.

# HOME ENVIRONMENT

G-8. The following discussion is intended to assist personnel in formulating plans to obtain housing outside US government compounds or DOD facilities. Personnel assigned to government housing may also find the antiterrorism and security tips presented below helpful in reducing the threat of violence and loss of property.

G-9. For general residential-security routines, discuss with family members the importance of—

- Varying routines in their daily activities.
- Blending in with the local environment.
- Avoiding unnecessary publicity and photographs that identify individual family members.
- Being alert to individuals, parked or abandoned vehicles, unusual utility work, or gatherings of people inconsistent with the residential environment.

## SECURITY PRACTICES AT HOME

G-10. The following measures are specifically recommended for residential implementation. These measures are an extension of office antiterrorism-security practices.

- Do not use nameplates or uniquely American symbols on the exterior of residences occupied by DOD personnel overseas.
- Do not use nameplates on parking places, and avoid parking private or government vehicles in the same location day after day.
- Ensure that all family members answer the telephone politely but that they provide no information as to the name of the occupants until the caller's identity has been established.
- Treat all telephone conversations as though anyone who wanted to listen in was doing so.
- Examine carefully all mail delivered to the residence.

## SOCIAL AND RECREATIONAL ACTIVITIES

G-11. DOD personnel are encouraged to participate in many social and recreational activities. The following precautions are recommended:

- Respond to formal social invitations in person (where possible) or by direct telephone contact.
- Be attentive to the security environment of social gatherings.
- Avoid the development of patterns with respect to time of arrival or departure at social events.
- Avoid prolonged presence at social functions where there is a high concentration of persons thought to be terrorist targets.
- Refrain from excessive use of alcohol at social functions; remain clearheaded and unimpaired.
- Vary routes to and from social events held at a central facility.
- Minimize appearances in uniform or formal attire.
- Decline invitations to appear in publicity photos.
- Participate in recreational activities within the American compound or at a DOD installation whenever possible.

**NOTE: Refer to DOD 0-2000.12-H, Graphic Training Aid (GTA) 19-4-3, and Joint Services (JS) Guide 5260 for further guidance and explanation regarding protective measures.**

## Appendix H

# Bombs

Terrorists have frequently used homemade devices or IEDs to carry out their attacks against DOD personnel, facilities, and assets. The IEDs are ideal terrorist weapons. They are relatively inexpensive to make, and the components of many IEDs are common items that can be obtained from many sources and are difficult to trace. The IEDs can be large or small and be designed so that they are transported to the attack site in components for last-minute assembly. Such design concepts make detection more difficult and provide an additional increment of personal safety to the terrorists.

## GENERAL

H-1. The use of IEDs can enhance the violence that gives terrorist groups their ability to intimidate or coerce a target population. The detonation itself creates a highly visual, newsworthy scene, even hours after the detonation occurs. Bombs can detonate anywhere, without apparent reason and without warning. The use of bombs in a terror campaign emphasizes the authorities' inability to safeguard the public and maintain law and order. Bombs are ideal weapons because they can be designed to give terrorists opportunities to escape from the scene of their crimes.

## CONCEALING BOMBS

H-2. Given the question, "Where have terrorists placed bombs in the past, and where should we look for them?" results in no easy answer. Table H-1, page H-2, lists a few obvious locations that should be examined. Terrorists who use bombs as their weapons of choice can be very creative in designing and placing their weapons.

H-3. Bombs can be found anywhere people can place them. Without becoming paranoid and seeing a bomb under every rock and behind every tree, the practical answer to the above questions is: "Where they can be easily placed without the bomber being caught."

## DAMAGE AND CASUALTY MECHANISMS

H-4. The IEDs and other explosive devices inflict casualties in a variety of ways, including the following:

- Blast over pressure (a crushing action on vital components of the body; eardrums are the most vulnerable).
- Falling structural material.
- Flying debris (especially glass).
- Asphyxiation (lack of oxygen).

**Table H-1. Potential IED Hiding Places**

| Outside Areas | |
|---|---|
| • Trash cans<br>• Dumpsters<br>• Mailboxes<br>• Bushes | • Street drainage systems<br>• Storage areas<br>• Parked cars |
| **Inside Buildings** | |
| • Mail parcels or letters<br>• Inside desks/storage containers<br>• Ceilings with removable panels<br>• Areas hidden by drapes or curtains<br>• Recent repaired/patched segments of walls, floors, or ceilings | • Restrooms<br>• Trash receptacles<br>• Utility closets<br>• Boiler rooms<br>• Under stairwells |
| **In Plain Sight** | |

- Sudden body translation against rigid barriers or objects (being picked up and thrown by a pressure wave).
- Bomb fragments.
- Burns from incendiary devices or fires resulting from blast damage.
- Inhalation of toxic fumes resulting from fires.

H-5. It is impossible to calculate a single minimum safe distance from an IED or other explosive device. The safe distance varies with each device and its placement. As a rule, the farther away from a bomb, the safer the intended or collateral targets are. Blast effects, fragmentation injuries, and injuries resulting from flying debris diminish greatly as the distance between a bomb and possible targets increase. The amount of material in the device, the type of explosive material, the manner in which the device is constructed, and the location or the container in which it is placed all have a bearing on the specific destructive potential for each IED.

H-6. The following are four general rules to follow to avoid injury from an IED:

- Move as far from a suspicious object as possible without being in further danger from other hazards such as traffic or secondary sources of explosion (such as POL storage).
- Stay out of the object's LOS, thereby reducing the hazard of injury because of direct fragmentation.
- Keep away from glass windows or other materials that could become flying debris.
- Remain alert for additional or secondary explosive devices in the immediate area, especially if the existence of a bomb-threat evacuation assembly area has been highly publicized.

H-7. Some terrorists have used two especially devious tactics in the past to intensify the magnitude of casualties inflicted by bombing attacks. In some instances, they have detonated a small device to lure media attention and curiosity seekers to the site; a larger, more deadly device has detonated some time after the first device, thereby inflicting a large number of casualties.

H-8. Other terrorists have used a real or simulated device to force the evacuation of a facility only to detonate a much more substantial device in identified bomb-threat evacuation assembly areas. These attacks are especially harmful because the evacuation assembly areas often concentrate government or commercial office workers more densely than they are when dispersed throughout their usual workplaces.

## TELEPHONIC THREATS

H-9. When receiving a telephonic threat, treat the call seriously. Often, an anonymous telephone call is made regarding a bomb or an IED. See Figure H-1, page H-4, for information to record/obtain when receiving these calls

H-10. When an anonymous warning or threat is received, initiate the bomb-threat data card and notify the PMO, security police, security forces, or other law-enforcement/security offices immediately. Local SOPs will determine subsequent actions. Immediate action may include a search without evacuation, the movement of personnel within the establishment, a partial evacuation, or a total evacuation. The following criteria helps determine what immediate action to take:

- Factors favoring a search before the movement of personnel:—
  - There is a high incidence of hoax telephone threats.
  - Effective security arrangements have been established.
  - Information in the warning is imprecise or incorrect.
  - The caller sounded intoxicated, amused, or very young.
  - The prevailing threat of terrorist activity is low.
- Factors favoring movement of personnel before searching:
  - The area (post or base) is comparatively open.
  - Information in the warning is precise as to the matters of location, a description of the device, the timing, and the motive for the attack.
  - A prevailing threat of terrorist activity is high.

## EVACUATION DRILLS

H-11. Evacuation and search drills should be performed periodically under the supervision of the installation's or unit's senior officer. The drills should be held in cooperation with local police if possible. Personnel in adjacent buildings should be informed of drills to avoid causing unnecessary alarm.

H-12. Evacuation procedures depend on the circumstances. Prepare, publicize, and rehearse evacuation plans in advance. Address alarm systems, assembly areas, routes to assembly areas, personnel-evacuation responses, building and area clearances, and evacuation drills.

### PERSONNEL-EVACUATION RESPONSE

H-13. The bomb-threat alarm system should be easily distinguished from the fire alarm. When the alarm sounds, personnel should—

- Lock up or secure all classified materials.
- Conduct a quick visual search of their immediate working area.

Instructions: Be calm. Be courteous. Listen, do not interrupt the caller. Notify supervisor/security officer by prearranged signal while caller is on line.

Name of Operator________________________________________Time__________Date_________

**Caller's Identity**

Sex: ❑ Male ❑ Female ❑ Adult ❑ Juvenile Approximate age: Years____________

**Origin of Call**

❑ Local ❑ Booth ❑ Internal (From within bldg)

❑ Long Distance If internal, leave line open for tracing the call.

| Voice Characteristics | | Speech | | Language | |
|---|---|---|---|---|---|
| ❑ Loud | ❑ Soft | ❑ Fast | ❑ Slow | ❑ Excellent | ❑ Good |
| ❑ High Pitch | ❑ Deep | ❑ Distinct | ❑ Distorted | ❑ Fair | ❑ Poor |
| ❑ Raspy | ❑ Pleasant | ❑ Stutter | ❑ Nasal | ❑ Foul | ❑ Other ____ |
| ❑ Intoxicated | ❑ Other____ | ❑ Slurred | ❑ Lisp | | |
| | | | ❑ Other ____ | | |

| Accent | Manner | | Background Noises | |
|---|---|---|---|---|
| ❑ Local | ❑ Calm | ❑ Angry | ❑ Factory Machines | ❑ Trains |
| ❑ Not Local | ❑ Rational | ❑ Irrational | ❑ Bedlam | ❑ Animals |
| Region____ | ❑ Coherent | ❑ Incoherent | ❑ Music | ❑ Quiet |
| ❑ Foreign | ❑ Deliberate | ❑ Emotional | ❑ Office Machines | ❑ Voices |
| Race_____ | ❑ Righteous | ❑ Laughing | ❑ Mixed | ❑ Airplanes |
| | | | ❑ Street Traffic | ❑ Party Atmosphere |

**Bomb Facts**

Pretend difficulty with your hearing. Keep caller talking.

If caller seems agreeable to further conversation, ask questions like —
When will it go off? Certain Hour - Time Remaining - What kind of bomb? - Where are you now?
How do you know so much about the bomb? - What is your name and address?

If building is occupied, inform caller that detonation could cause injury or death.
Did caller appear familiar with plant or building by his description of the bomb location?

Write out the message in its entirety and any other comments on a separate sheet of paper and attach to this checklist.

**Action To Take Immediately After Call**

Notify your supervisor/security officer as instructed. Talk to no one other than as instructed by your supervisor/security officer.

**Figure H-1. Sample Bomb-Threat Data Card**

- Open windows (wherever possible).
- Leave the building, taking only valuable personal belongings.
- Leave doors open and immediately proceed to the assembly area.

H-14. Opening the building will reduce internal damage due to blast effects. It will also somewhat mitigate the extent of debris flying out of or falling from the building should a detonation occur.

## ASSEMBLY AREAS

H-15. Choose the routes to the assembly area so that personnel do not approach the IED at any time. Preselect the routes to the assembly area, but devise a system to inform personnel of the location of the suspected IED and alternate routes. Routes prevent confusion and bunching and avoid potential hazards (such as plate glass, windows, and likely locations of additional IEDs).

H-16. Assembly areas should be preselected and well known to personnel. Establish a clearly defined procedure for controlling, marshaling, and checking personnel within the assembly area. If buildings or establishments are in a public area, coordinate the assembly areas with local police. Assembly areas are selected using the following criteria:

- Locate assembly areas at least 100 meters from the likely target or building (if possible).
- Locate assembly areas in areas where there is little chance of an IED being hidden. Open spaces are best. Avoid parking areas because IEDs can be easily hidden in vehicles.
- Select alternate assembly areas to reduce the likelihood of ambush with a second device or small-arms fire. If possible, search the assembly area before personnel occupy the space.
- Avoid locating assembly areas near expanses of plate glass or windows. Blast effects can cause windows to be sucked outward rather than blown inward.
- Select multiple assembly areas (if possible) to reduce the concentration of key personnel. Drill and exercise personnel to go to different assembly areas to avoid developing an evacuation and emergency pattern that can be used by terrorists to attack identifiable key personnel.

## BUILDING AND AREA CLEARANCE

H-17. Establish procedures to ensure that threatened buildings and areas are cleared. Prevent personnel from reentering the building. Establish a cordon to prevent personnel from entering the danger area. Establish an initial control point (ICP) as the focal point for the PMO and for MP control.

H-18. Cordon suspicious objects to a distance of at least 100 meters, and cordon suspicious vehicles to a distance of at least 200 meters. Ensure that nobody enters the cordoned area. Establish an ICP on the cordon to control access; relinquish ICP responsibility to the PMO or local police upon their arrival. Maintain the cordon until the PMO, security police, security forces, or local police have completed their examination or stated that the cordon may stand down.

## SEARCHING FOR A SUSPECTED IED

H-19. Searches are conducted in response to a telephonic threat or a report of an unidentified object on or near premises occupied by DOD personnel. The following types of searches may be used when searching for a suspected bomb or IED:

- An occupant search is used when the threat's credibility is low. Occupants search their own areas. The search is completed quickly because occupants know their area and are most likely to notice anything unusual.
- A team search is used when the threat's credibility is high. The search is very thorough and places the minimum number of personnel at risk. Evacuate the area completely, and ensure that it remains evacuated until the search is complete. Search teams will make a slow, thorough, systematic search of the area.

H-20. The following procedures should be followed if a search for explosive devices must be conducted before qualified EOD teams arrive:

- Make an audio check, listening for unusual sounds.
- Sweep the area visually up to the waist, then sweep up to the ceiling. Do not forget the tops of cabinets and cupboards.
- Perform a thorough and systematic search in and around containers and fixtures.
- Pass search results as quickly as possible to the leader responsible for controlling the search area. Do not use a radio; it may detonate the explosive.

H-21. Circumstances might arise in the case of a very short warning period. In other instances, a threat of a bomb against some facilities (if true) might necessitate the evacuation of a very large area. In these circumstances, searching for the presence of an explosive device to identify its location, appearance, and possible operating characteristics may be warranted.

H-22. Personnel who have not been trained in IED search and ID techniques should not search for explosive devices. Two types of errors are very common—the false ID of objects as IEDs and the incorrect ID of IEDs as benign objects. Depending on the devices used to arm and trigger an IED, the search process could actually result in an explosion.

## SEARCH ORGANIZATION

H-23. The person controlling the search should have a method of tracking and recording the search results (such as a diagram of the area). Delegate areas of responsibility to the search-team leader, who should report to the person controlling the search when each area has been cleared. Pay particular attention to entrances, toilets, corridors, stairs, unlocked closets, storage spaces, rooms and areas not checked by usual occupants, external building areas, window ledges, ventilators, courtyards, and spaces shielded from normal view.

## DISCOVERY OF A SUSPECTED IED

H-24. When a suspicious object has been found, report its location and general description immediately to the nearest law-enforcement or supervisory person. Do not touch or move a suspicious object. Instead, perform the following steps:

- If an object appears in an area associated with a specific individual or a clearly identified area—
  - Ask the individual/occupant to describe objects they have brought to work in the past few days.
  - Ask for an accounting of objects.
  - Ask for a verbal description/ID of objects.
- If an object's presence remains inexplicable—
  - Evacuate buildings and surrounding areas, including the search team.
  - Ensure that evacuated areas are at least 100 meters from the suspicious object.
  - Establish a cordon and an ICP.
  - Inform personnel at the ICP that an object has been found.
  - Keep the person who located the object at the ICP until questioned.
  - Avoid reentering the facility to identify an object that may or may not be an IED.

## REACTING TO AN EXPLODED IED

H-25. The following procedures should be taken when an explosive/IED detonates at a DOD facility:

- For explosions without casualties—
  - Maintain the cordon. Allow only authorized personnel into the explosion area.
  - Fight any fires threatening undamaged buildings without risking personnel.
  - Report the explosion to the PMO, security police, security forces, or local police if they are not on the scene.
  - Report the explosion to the installation operations center even if an EOD team is on its way. Provide as much detail as possible, such as the time of the explosion, the number of explosions, the color of smoke, and the speed and spread of fire.
  - Ensure that a clear passage for emergency vehicles (fire trucks, ambulances, and so forth) and corresponding personnel is maintained.
  - Refer media inquiries to the PAO.
  - Establish a separate information center to handle inquiries from concerned friends and relatives.
- For explosions with casualties—
  - Select a small number of personnel to help search for casualties.
  - Assign additional personnel the responsibility for maintaining the cordon to keep additional volunteers searching for casualties.

Maintain the cordon until the EOD team verifies no further presence of bombs/IEDs at the site and the fire marshal determines that risk of additional injury to searchers from falling debris is acceptable.

- Prepare a casualty list for notification of next of kin; delay publication of the list until its accuracy is determined.
- Arrange for unaffected personnel to contact their next of kin immediately.

H-26. Civilian management officials and subordinate military commanders continue to have important personal roles to fulfill during a bomb/IED attack on DOD personnel, facilities, and assets. Perform the following procedures when reporting an attack:

- Pass available information to the operations center.
- Avoid delaying reports due to lack of information; report what is known. Do not take risks to obtain information.
- Include the following information in the report:
  - Any warning received and if so, how it was received.
  - The identity of the person who discovered the device.
  - How the device was discovered (casual discovery or organized search).
  - The location of the device (give as much detail as possible).
  - The time of discovery.
  - The estimated length of time the device has been in its location.
  - A description of the device (give as much detail as possible).
  - Safety measures taken.
  - Suggested routes to the scene.
  - Any other pertinent information.

H-27. Perform the following procedures when providing emergency assistance to authorities:

- Ensure that the PMO, security police, security forces, and other emergency-response units from local police, fire and rescue, and EOD teams are not impeded from reaching the ICP. Help maintain crowd control and emergency services' access to the site.
- Evacuate through the doors and windows of buildings.
- Assist the on-scene commander by obtaining a building diagram showing detailed plans of the public-service conduits (gas, electricity, central heating, and so forth), if possible. If unavailable, a sketch can be drawn by someone with detailed knowledge of the building.
- Locate, identify, and make witnesses available to investigative agency representatives when they arrive on the scene. Witnesses include the person who discovered the device, witnessed the explosion, or possesses detailed knowledge of the building or area.

H-28. Performing the above steps will provide substantial assistance to the crisis-management team and give other personnel constructive, supportive actions to take in resolving the crisis. Care must be exercised, however, that additional explosive devices are not concealed for detonation during the midst of rescue operations. These attacks add to the physical damage and emotional devastation of bomb/IED attacks.

H-29. The use of bombs and IEDs during terrorist attacks against DOD personnel, facilities, and assets is a common occurrence. The procedures outlined in this appendix are intended to help a DOD facility respond to an attack before an explosive device detonates. The procedures are also intended to help mitigate the consequences of an attack in case efforts to find an explosive device and render it inoperable are not successful. Incurring the costs to DOD facilities and installations of detecting an explosive device and terminating a terrorist incident before the device can detonate are almost always preferable rather than exercising plans and options to respond to a detonation. Several of the security measures discussed will help reduce the likelihood of a successful bomb/IED attack against DOD assets.

# Appendix I

# Executive Protection

DOD Directive 2000.12 recognizes a need to provide protection to military officers and DOD civilians who are assigned to high-risk billets, who are (by the nature of their work) high-risk personnel, or who are assigned to facilities identified as high-risk targets. The directive defines these terms as follows:

- High-risk billet. Authorized personnel billet (identified and recommended by appropriate authority) that because of grade, assignment, travel itinerary, or symbolic value may make personnel filling them an especially attractive or accessible terrorist target.
- High-risk personnel. US personnel and their family members whose grade, assignment, travel itinerary, or symbolic value may make them an especially attractive or accessible terrorist target.
- High-risk target. US facilities and material resources that, because of mission sensitivity, ease of access, isolation, or symbolic value may be an especially attractive or accessible terrorist target.

**NOTE: For purposes of this appendix, the term executive will be applied to all persons requiring additional security protection who are assigned to high-risk billets, designated as high-risk personnel, or identified as high-risk targets.**

## SUPPLEMENTAL SECURITY MEASURES

I-1. The specific supplemental security measures that may be furnished to executives are subject to a wide range of legal and policy constraints. US law establishes stringent requirements that must be met before certain security measures may be implemented. DOD regulations, instructions, and legal opinions may further constrain the implementation of some protective measures described in this chapter. The SOFAs and MOUs between the US and a foreign government will also limit the use of some supplemental security measures. Leases and other conditions imposed by contract for purchase of land or buildings by the US for DOD use may also limit the application of certain security techniques. All of these constraints should be carefully considered when conducting security surveys, developing plans, and implementing additional security measures to protect high-risk personnel.

## EXECUTIVE PROTECTION GOALS

I-2. In the discussion that follows, several measures are outlined that can afford senior military officers and DOD personnel additional protection against terrorist acts. The purpose underlying these measures is to—

- Increase the interval of time between detecting a threat and the onset of hostile action against executives and their dependents.
- Increase the amount of time required by terrorists to gain physical access to executives from the onset of hostile actions, whether the executives are at home, at the office, or in transit.

I-3. The implementation of supplemental security measures should strive to achieve the following prioritized goals:

- Enhancements should hold the terrorist threat at bay until a response force arrives (delay at a distance).
- Enhancements in physical security should enable executives to flee to safety (delay to permit flight).
- Enhancements should permit the executive to retreat into a safe haven of sufficient strength and survivability. This should enable a response force to wage an effective counterattack to liberate executives and others accompanying them to a safe haven, including family members at home and colleagues and visitors at work (delay, hold, and counterattack).

I-4. The following supplemental measures should be applied with care. There is a clear trade-off between increasing the level of physical security at the office and at home and preserving the anonymity of executives, thereby avoiding telltale signs of activity that point to prominence or criticality. These measures can be expensive. Expense can be measured not just in terms of dollars, but also in terms of changes to organizational routine. Therefore, three questions must be resolved before implementing bold, disruptive, and expensive supplemental security enhancements:

- What are the most cost-effective means of enhancing the security of executives at risk?
- How many changes in organizational routines and personal behaviors will have to be made for security measures to be effective in reducing the risk of terrorist attacks and the vulnerability of executives to such attacks?
- What are the anticipated costs of additional security measures in terms of dollars, organizational functionality, and mission capability?

I-5. Security enhancements can be made to improve the security of executives and can be even more effective if executives and their families take full advantage of and reinforce those measures. If executives do not change their behavior to accommodate additional security and protective measures, then the behaviors can effectively defeat the purpose of additional protection. Additional increments of security can be obtained to defeat virtually any threat. However, there is a point at which it is no longer cost-effective to add layer upon layer of protective measures to defeat a threat.

## RESIDENTIAL SECURITY MEASURES

I-6. While terrorist groups conduct intelligence operations to identify targets, mistakes have been made in the past. DOD personnel should avoid leasing residences previously used by representatives of governments or organizations known to be targets of various terrorist groups. DOD personnel leasing residences formerly used by representatives of such governments may be placing themselves unnecessarily at risk of being attacked as a result of mistaken identity.

I-7. An executive's entire lifestyle should be included in security surveys used to assess the need for supplemental physical-security measures at the office. The executive's home and transportation from home to office and back should also be examined for risk and vulnerability. The same principles used to identify supplemental security improvements in an office environment apply to an executive's home environment as well. The purposes of physical-security enhancements are to—

- Increase the amount of time terrorists need to initiate and complete an attack on executives while at home, thereby giving response forces more time to rescue executives and their dependents.
- Reduce potential harm to executives and their families because of a terrorist assault mounted against the residence.

I-8. The goals of enhanced residential physical-security measures are to—

- Increase the amount of time between detection of a threat and the onset of hostile actions.
- Delay the terrorists as long as possible. Prevent terrorist access to executives and their family members and make it difficult to leave the scene to escape prosecution. These measures should not further jeopardize the lives of executives and their family members.
- Provide a safe haven where executives and their family members may flee for security pending the arrival of a response force on the scene.

I-9. The following measures can be implemented selectively to help security personnel achieve these objectives:

- Increase the time interval between threat detection and the onset of hostile terrorist acts by—
  - Ensuring that all door locks and window clasps are working.
  - Ensuring that all doors and windows are properly secured to their frames and that the frames are properly anchored to the residential structure.
  - Locking driveway gates with a security lock to prevent entry.
  - Installing a through-door viewing device or visitor intercom.
  - Installing security lights to aid in viewing entrances.
- Increase the number of physical barriers between the outer perimeter of the residence and the interior of the residence by—
  - Adding heavy, remotely operated gates to all fences, walls, and perimeter barriers consistent with the penetration resistance of the barrier between the residence, the street, and adjacent neighbors.
  - Creating a vestibule or air lock between living quarters and the exterior of a residence, ensuring that no one can enter the residence directly from the outside.
  - Adding fire doors or security doors or gates between the residence's bedrooms and living areas.

- Increase the time required to penetrate exterior structural walls by explosives, hand-held power tools, and hand tools by—
  - Adding additional armor covered by aesthetically pleasing materials to exterior walls.
  - Adding a separate reinforced masonry wall around the residence.
- Increase the surveillance of the residence and decrease response time by—
  - Installing CCTV systems to permit remote viewing of all doors and windows accessible from the ground, nearby structures, trees, or easily acquired platforms (such as a van parked next to a wall).
  - Installing area IDSs between the residence's perimeter and the residence itself, varying the number and types of sensors, and adding backup communication channels between the IDS and a surveillance assessment/response dispatch center.
- Increase the residence's durability and survivability to a terrorist attack by—
  - Fitting windows with either venetian blinds or thick curtains to reduce the observability of activities within the residence and to reduce hazards of flying glass in case of nearby explosions or gunfire.
  - Installing backup power systems for security devices (surveillance systems, communication systems, and access-control systems).
  - Ensuring that backup communication is available with the installation or embassy's security department via a secure landline or two-way radio.
  - Fitting a panic-alarm bell to the outside of the house with switches on all floor levels. Such an alarm should also annunciate at the local police and cognizant DOD or DOS security office.
  - Installing a safe haven in the home.

## TRANSPORTATION MEASURES

I-10. High-risk personnel are most accessible to terrorists while in transit in official or privately owned vehicles. Specific steps can be taken to reduce the vulnerability of executives in transit.

### SPECIAL TRANSPORTATION IN TRANSIT FROM DOMICILE TO DUTY

I-11. As a rule, Congress has strongly opposed the provision of domicile-to-duty transportation by the federal government to its officers and employees. Only 16 officials are entitled by statute to such assistance. Congress did, however, grant authority to the President and the heads of executive agencies and departments to provide domicile-to-duty transportation under certain circumstances. According to the statute, "a passenger carrier may be used to transport between residence and place of employment an officer or employee with regard to whom the head of a Federal agency makes a determination, [provided] that highly unusual circumstances present a clear and present danger, that an emergency exists, or that compelling operational considerations make such transportation essential to the conduct of official business."

I-12. The phrase "highly unusual circumstances which present a clear and present danger" is understood to mean that—

- The perceived danger is real, not imaginary.
- The perceived danger is immediate or imminent, not merely potential.
- Proof is provided that the use of a government vehicle would provide protection not otherwise available.

I-13. Such a danger would exist where there is an explicit threat of terrorist attacks or riot conditions and such transportation would be the only means of providing safe passage to and from work.

I-14. The phrase "emergency exists" means that there is an immediate, unforeseeable, temporary need to provide home-to-work transportation for an agency's essential employees. The phrase "similarly compelling operational considerations" means that there is an element of gravity or importance for the need of government-furnished transportation comparable to the gravity or importance associated with a clear and present danger or an emergency. Congress suggested further, "in such instances, [it is expected] that home-to-work transportation would be provided only for those employees who are essential to the operation of the government."

I-15. The Secretary of Defense has the statutory authority to allow a CINC to use government-owned or -leased vehicles to provide transportation in an area outside of the US for members of the uniformed services and other DOD personnel under certain circumstances. These circumstances include and are limited to a determination by the CINC that public or private transportation in the area is unsafe or is not available. Under these circumstances, DOD may provide transportation (usually in government buses or passenger vans) to personnel and their family members if it will help the CINC and his subordinate commanders maintain the capability to perform or undertake assigned missions. This transportation is not intended for transporting personnel from their residences to their places of work. The Secretary of Defense and the Service Secretaries also have the statutory authority to provide transportation from home to duty stations and back on a limited basis. This authority is usually implemented by providing a nontactical armored vehicle (NTAV) to protect personnel.

I-16. It is a DOD policy to make NTAVs available where necessary to enhance the security of DOD personnel consistent with the requirements and limitations found in the statute. DOD issuances, service regulations, and CINC guidance stipulate detailed procedures by which DOD manages NTAV programs. The statute also establishes a procedure for Presidential waiver of the "buy American" requirement; DOD and service regulations provide for the delegation of Presidential authority from the President to the Secretary of Defense; to the Director, Defense Security Assistance Agency; and to the Director, DIA. DOD Instruction 5210.84 authorizes DOS acquisition and installation of light vehicle armoring to DOS specifications in local defense-component vehicles on a reimbursable basis. The level of protection provided to the Defense Component Office will comply with approved overseas security policy group armored-vehicle standards.

I-17. The DOD recognizes two classes of NTAVs—heavy and light. Heavy NTAVs are fully armored vehicles intended to protect occupants from attack by bombs; IEDs; grenades; and high-velocity, small-arms projectiles. Light NTAVs are less than fully armored vehicles and are intended to protect occupants from attack by medium-velocity, small-arms projectiles and at least some types of IEDs.

I-18. The dividing lines between heavy and light NTAVs have become less distinct over time as armoring techniques and materials have given greater capability to NTAVs that are not classified as heavy. As a practical matter, add-on vehicle-armoring kits are now in production which (when properly installed in an appropriately powered and suspended vehicle) will provide a level of protection approaching that of the heavy NTAVs.

## Heavy NTAVs

I-19. Heavy NTAVs may be assigned to US personnel upon certification by a Service Secretary only under the following conditions:

- Highly unusual circumstances present a clear and present danger to the health and safety of a nominated protectee.
- Compelling operational considerations make such transportation essential to conducting official business.

I-20. If the physical-security survey concludes that a heavy NTAV is warranted, the nominated protectee's Service Secretary shall, on the advice and recommendation of a combatant commander, determine whether the use of a heavy NTAV is warranted. If so, the Service Secretary shall authorize the use of a vehicle for a renewable 90- to 360-day period. At the end of the period, the requirement will be reexamined and a recertification for the protection shall be issued by the Service Secretary.

I-21. Each of the services manages a portion of the DOD's NTAV program. Each service has issued supplementary mandatory guidance for processing requests for, as well as allocation and use of, these scarce assets.

I-22. Heavy NTAVs are complex systems requiring specialized maintenance and operation. Normally, they will be assigned to DOD personnel with a driver who has been properly trained in the operation and maintenance of the vehicle. The operator is not a chauffeur; he is an integral part of a supplemental security package provided by DOD to meet its obligations in protecting key assets.

## Light NTAVs

I-23. Light NTAVs may also be provided to US employees and officers where highly unusual circumstances present a clear and present danger to the health and safety of a nominated protectee or compelling operational considerations warrant their use. This category of NTAV features add-on armoring. While they are a less-complex armoring system than those used in heavy NTAVs, light NTAVs afford substantial protection to occupants against a variety of threats. New developments in after-manufacture armoring kits for vehicles are occurring at a rapid pace, increasing the number of vehicle manufacturers and models for which other NTAV modifications are suitable. Each service and the DIA have instructions for implementing DOD policy that authorizes the use of other NTAVs to enhance personnel protection of high-risk persons.

### PRIVATELY OWNED VEHICLES

I-24. High-risk personnel may wish to forego the use of POVs during periods of extreme risk. Considerations include selecting measures that—

- Deter secret entry, making undetected placement of IEDs in or under the vehicle difficult for terrorists to accomplish.
- Enhance the vehicle's ability to increase distance between it and pursuers.
- Assist response forces in case of an incident.
- Make the vehicle appear little different than its standard models.

## INDIVIDUAL PROTECTIVE MEASURES

I-25. Executives can enhance their personal security in the office environment by—

- Discouraging staff members who are taking telephone messages from disclosing their whereabouts.
- Ensuring that caution is used when opening mail and being especially careful with letters or packages that might contain IEDs.
- Ensuring that access is strictly limited to their office area.
- Limiting publicity and keeping official biographies short. This includes using outdated photographs if a publicity photograph is essential.
- Ensuring that they are not working alone late at night and on days when the remainder of the staff is absent.
- Working in conference rooms or internal offices where outside observation is not possible if late-night work is necessary. Security officers should be notified of the work so that they can periodically look in.
- Ensuring that office furnishings are not placed directly in front of exterior windows.

### OFFICIAL BUSINESS AWAY FROM THE OFFICE

I-26. The following suggestions reinforce efforts by executives to maintain the high level of security provided in the home or office environment while on official business outside of these locations:

- Discuss security requirements with the person planning the function.
- Travel to and from the function with escorts.
- Choose the route carefully.
- Avoid publicizing planned attendance at official functions (unless required).
- Attempt to sit away from both public areas and windows.
- Encourage the function's sponsor to close the curtains to minimize the likelihood that anyone outside will be able to see inside and determine who is attending the function. This is extremely important for an evening function, when a well-lit interior can be easily viewed from a darkened exterior.
- Request that external floodlights be used to illuminate the area around the building where an evening function will occur.

## LOCAL OFFICIAL AND UNOFFICIAL TRAVEL

I-27. Executives can greatly enhance their personal security when conducting official and unofficial travel by following these general practices:

- Vary daily patterns, such as leaving and returning at different times.
- Consider escorts to and from work or travel with a neighbor.
- Establish a simple oral or visual duress procedure between executives and drivers (for example, a phrase or movement used by the executive or driver only if something is amiss).
- Vary taxi companies. Ensure that the ID photo on the license matches the driver. If uneasy for any reason, take another taxi.
- Attend social functions with others, if possible.
- Examine the car before entering to see if there has been any interference. A small mirror on a rod is a cheap and effective method to inspect underneath cars. Do not touch the vehicle until it has been thoroughly checked (inside, around, and under).
- Avoid leaving personal items exposed in the car (uniform items, service-issued maps, official briefcases, and so forth).

## SECURITY PRACTICES WHILE DRIVING

I-28. Executives can take the following measures to enhance security while driving:

- Keep car doors locked. Do not open windows more than a few inches.
- Avoid overloading a vehicle, and wear seat belts.
- Park vehicles in parking areas that are either locked or monitored. Never park overnight on the street. Before entering vehicles, check for signs of tampering.
- Keep the trunk locked.
- Drive in the inner lanes to keep from being forced to the curb.
- Use defensive and evasive driving techniques. Drill with your driver by watching for suspicious cars and taking evasive action.
- Avoid driving close behind other vehicles (especially service trucks), and be aware of activities and road conditions two to three blocks ahead.
- Beware of minor accidents that could block traffic in suspect areas such as crossroads. Crossroads are preferred areas for terrorist or criminal activities because they offer escape advantages.

I-29. If a terrorist roadblock is encountered, use the shoulder or curb (hit at a 30- to 45-degree angle) of the road to go around it or ram the terrorist's blocking vehicle. Blocking vehicles should be rammed in a nonengine area, at a 45-degree angle, in low gear, and at a constant moderate speed. The goal is to knock the blocking vehicle out of the way. In all cases, do not stop and never allow the executive's vehicle to be boxed in with a loss of maneuverability. Whenever a target vehicle veers away from the terrorist vehicle, it gives adverse maneuvering room and presents a better target to gunfire.

## INTERURBAN, NATIONAL, AND INTERNATIONAL TRAVEL SECURITY PRACTICES AND PROCEDURES

I-30. To enhance security in interurban, national, and international circumstances, executives should—

- Book airline seats at the last moment. Consider using an alias.
- Restrict the use of ranks or titles.
- Avoid allowing unknown visitors in the hotel room or suite.
- Keep staff and family members advised of the itinerary and subsequent changes. Clearly and emphatically restrict this information to those having a need to know.

## HOME SECURITY PRACTICES AND PROCEDURES

I-31. To enhance security at home, executives should—

- Check the ID of persons entering the premises (electricians, plumbers, telephone-maintenance personnel, and so forth). When in doubt, call their office to verify their identity before allowing them in your home.
- Avoid opening the door to a caller at night until he is visually identified through a window or a door viewer.
- Close curtains in a room before turning on lights.
- Consider placing the telephone where you will not be seen from doors or windows when answering.
- Investigate the household staff (especially temporary staff members).
- Stay alert and be on the lookout for the unusual. Ensure that the home is locked and secure whenever the residence is unattended. Be cautious upon return and look for the movement of furniture or the placement of unusual wires.
- Note and report suspicious persons.
- Control house keys strictly.
- Park the car in a locked garage.
- Consider installing a panic-alarm bell to the outside of the house with switches located on all floor levels.
- Clear the area around the house of dense foliage or shrubbery.
- Test duress alarms (if available). Make certain that family members understand how they work as well as the importance of their use.
- Cooperate with law-enforcement personnel, and abide by their security recommendations.

## SECURITY AT SOCIAL AND RECREATIONAL ACTIVITIES

I-32. The risk of terrorist incidents is always present for high-risk personnel or personnel assigned to high-risk billets. The following measures are intended to permit executives to live a close-to-normal life while still remaining mindful of the risks to their security.

- Ensure that the host is aware of and takes appropriate measures for your security.
- Have your personal staff assist a civilian host, if required.
- Arrange for visitors to be subject to adequate security control.

- Screen the invitation list, if possible.
- Vary the times of sporting activities (golfing, jogging, and so forth).

## COMBATING-TERRORISM TRAINING FOR EXECUTIVES

I-33. Combatant commanders annually compile a list of high-risk billets in their AO. These lists are forwarded through the appropriate service personnel channels, enabling each service to identify, plan, and provide resources to meet training requirements. All personnel and adult family members en route to high-risk billets must attend the Individual Terrorism Awareness Course (INTAC) conducted at the US Army John F. Kennedy Special Warfare Center at Fort Bragg, North Carolina. During this one-week course, personnel will receive instruction in defensive-driving techniques and survival shooting as well as individual protective measures and hostage survival. These individuals should also attend the appropriate regional orientation course (Middle East, Asia/Pacific, Latin America, or Africa) offered at the US Air Force Special Operations School at Hurlburt Air Force Base (AFB), Florida. The service member whose duties will require frequent vehicle operation should attend an appropriate evasive-driving course. Information on current offerings may be obtained by contacting the service representative to the DOD Antiterrorism Coordinating Committee or the Combating Terrorism Branch in the Office of the Assistant Secretary of Defense (OASD) Special Operations/ Low-Intensity Conflict (SO/LIC).

## TRAVEL TO POTENTIAL PHYSICAL-THREAT RISK AREAS

I-34. Personnel en route to potential physical-threat risk areas (as identified by the OASDSO/LIC) should attend one of the following courses:

- The Dynamics of International Terrorism Course conducted at the US Air Force Special Operations School at Hurlburt AFB, Florida. During this one-week course, personnel will receive lectures on threats by region (Europe, Middle East, Latin America, Asia/Pacific, and Africa), the history and psychology of terrorism, personnel combating-terrorism measures (vehicle, personal, airline, and physical security), and hostage survival.
- A Regional Orientation Course (Middle East, Latin America, Africa, Asia/Pacific) at the US Air Force Special Operations School at Hurlburt AFB, Florida. This one-week course offers personnel instruction in cultural, political/military, and individual security factors associated with the region.

I-35. Installation security personnel may also receive the above training if they have completed the Antiterrorism Instructor Qualification Course (AIQC) at Fort Bragg, North Carolina.

## PROTECTIVE SECURITY DETAILS

I-36. Each service can provide bodyguards for key senior military officers, DOD civilians, other US officials, or foreign dignitaries requiring personal protection. Each Service Secretary is responsible for assigning protective security details (PSDs) to service members based on the recommendation of their counterintelligence and/or law-enforcement investigation staffs. The PSDs are assigned to DOD personnel who meet requirements established by service regulations. In general, PSDs may be assigned only to those personnel whose position or assignment places them at risk and whose continued availability to the National Command Authorities and the CINCs is vital to DOD's mission execution.

I-37. A PSD provides high levels of security to an executive by establishing a series of protective cordons around him. The establishment of defense in depth often means that the innermost protective layer is in close contact with the executive at all hours of the day and night.

I-38. A PSD is trained to maintain a low profile. It is concerned about the executive's visibility and its ability to blend into his surroundings. There is nothing more damaging to the security of an executive than the obvious, detectable presence of a PSD when all other measures to have him blend into the local environment have been successful. A PSD will strive to keep travel routes and means of transportation from being publicized. If this cannot be accomplished, the PSD may suggest editorial changes to the itinerary scheduled for release to limit details of planned travel from public disclosure. For example, routes to and from announced appointments usually do not need to be revealed.

I-39. During the course of a PSD's mission, its members may be asked to perform several different security functions. They may, for example, perform direct or indirect protection or escort duty. Direct protection is open and obvious; indirect is generally a surveillance measure. The security-guard unit may operate as an interior guard and may consist of one or more men stationed at fixed posts. A PSD's members should know the identity of each individual in the party of a protected official; executives can help by introducing them to each member of the official party.

I-40. The protected person's attitude is critical to the success of the PSD's mission. Executives have a right and a responsibility to make their wishes known with respect to their personal security. They also have an obligation to listen carefully to the head of the PSD who is trained and highly qualified to help make reasonable judgments about manageable risks. A PSD's members understand that their function is inherently intrusive and that executives can easily resent the loss of privacy that accompanies the protection offered. On the other hand, PSDs have jobs to do, not merely to protect executives, but to help safeguard mission-critical assets—senior military and civilian leaders.

I-41. One of the most demanding functions placed on a PSD is to limit the ability of individuals to circulate and approach the executive. This is often very frustrating to executives who wish to shake hands, engage in close conversations with visitors, and move freely and without impediment in a social situation. The PSDs are trained to strictly enforce limitations on the circulation of individuals, carefully checking each person for ID and ascertaining that he is authorized to be present at the occasion.

I-42. DOD personnel who are provided with PSDs and must conduct official business or hold social engagements in large rooms can take steps to minimize the disruptions to such functions. These steps include—

- Providing advance attendee lists to the head of the PSD.
- Having one or more members of the staff who know the attendees stand with PSD members and identify the attendees as they arrive.
- Informing attendees that they will be admitted only at specified entrances.

I-43. The PSD's members are highly trained security specialists. While in the company of executives, they will be accommodating and helpful. Executives should remember, however, that the primary function of the PSD's members is to protect them, not perform errands or carry out personal services. A PSD's members who are performing valet or other chores cannot effectively protect the senior officers or civilian officials to whom they have been assigned.

## EXECUTIVE-PROTECTION SYSTEM INTEGRATION

I-44. This appendix has focused on supplemental security measures used to address terrorist threats to senior high-risk personnel within the DOD. Various methods and measures have been discussed that provide increments of security over and above the base level of security provided to all DOD personnel assigned to an installation, facility, activity, or unit. In making decisions to allocate protective resources to enhance the security of senior officers and senior DOD officials, it is essential to remember that measures must be applied systematically. Additional security measures implemented to protect high-risk personnel in the office environment must be carried over to official functions conducted outside the office. The security measures must also be extended to protected persons' private lives and, depending on the nature of the threat, the lives of their family members.

I-45. The converse is equally true. It makes no sense to provide domicile-to-duty transportation for a high-risk person and make no provision for additional protection at home, at the office, and at official business and social functions. In view of the total costs of security measured in dollars, time, and inconvenience to protected persons, their staffs, colleagues, and families, it may be more prudent to radically alter living and working arrangements than to try to augment security in a piecemeal manner. For example, it might be prudent to house high-risk personnel within a DOD installation rather than to try to secure a detached, private residence at a substantial distance from the operations base of a response force. The key to successful executive protection is to ensure that the level of protection afforded by physical-security measures, operational procedures in the office and at home, and PSDs is constant. The level of protection must be matched to the threat and must be sustainable. Executives have a special responsibility to set a personal example of combating-terrorism awareness; of attention to personal, family, office, information, and OPSEC concerns; and of combating-terrorism security measures implementation. By doing so, they make their colleagues and subordinates more aware and more conscious of their security environment and less likely to be victimized by terrorist attacks.

# Appendix J
# Resource Management

Programs need annual funding to operate. This funding is obtained by devising and documenting a resource program that looks seven years out. Looking ahead one or two years is generally not a problem. Commanders must be able to describe future requirements in the out years so that money will be available when the program arrives in future years.

## FUNDING PROGRAMS

J-1. Physical security is dependent on integrated systems with budgetary constraints. This appendix serves to inform commanders of three basic funding programs contributing to physical security—

- The RJC6, which resources physical-security equipment purchase and maintenance.
- The QLPR, which resources law-enforcement operations (to include security guards and special-reaction teams).
- The VTER, which provides resources for projects and temporary programs that enhance any type of security due to an increase in terrorist threat.

## PROJECTED REQUIREMENTS

J-2. Recurring VTER requirements are usually shifted to QLPR as standard requirements after a period of a few years. Installations send seven-year projected requirements to the MACOMs for submission into the Army's Program Objective Memorandum (POM), which is an annual significant event for resource managers at all levels. Installation requirements should be included in the POM build; otherwise, the installation's program loses visibility from the start of the funding process. Input format is determined locally, but a generic example of the format and the type of information requested is provided at Figure J-1, page J-2.

## OBLIGATION PLAN

J-3. Budget execution deals with the current fiscal year (October through September), with the exception of some types of dollars (appropriations) that are multiyear (such as procurement dollars that are executable for three years). Once the resource manager notifies an installation of its available annual funding, an obligation plan by month or quarter is developed to display how and when the funds will be spent. This obligation plan is also used to forecast when the program will run out of money, which in turn will justify the submission of an unfinanced requirement (UFR) to obtain the proper level of resources. Examples of an annual obligation plan and UFR are located at Figures J-2 and J-3, page J-3.

**Program Objective Memorandum (POM)**

Installation: Fort McClellan, AL
Activity: Military Police School
Appropriation: Operation & Maintenance, Army (OMA)
Army Management Structure (AMS): 321731
Management Decision Package (MDEP): VTER
Element of Resource (EOR): 2500
Issue: Force Protection/Antiterrorism Courses

| Resources | FY00 | FY01 | FY02 | FY03 | FY04 | FY05 | FY06 |
|---|---|---|---|---|---|---|---|
| ($000) | 500 | 510 | 520 | 530 | 540 | 550 | 560 |

Justification/Impact:
Funding is required to continue the eight functional contract training courses conducted by the US Army Military Police School in Force Protection/Antiterrorism for the Army and selected DOD services and agencies. Courses focus on the five pillars of force protection and provide students with a variety of force-protection expertise to prevent, react to, and contain a terrorist event. If funds are not provided, the projected annual student load of 2,500 cannot be trained.

**Figure J-1. Sample POM**

## TYPES OF APPROPRIATIONS

J-4. The most common types of appropriations that managers will be involved with are day-to-day operations or mission (OMA) funds and procurement (OPA) dollars that must be used for major projects or equipment buys over the $100,000 threshold. Army Materiel Command (AMC) uses research, development, and engineering (RD&E) dollars for operations at US Army Test and Evaluation Command (TECOM) facilities and for testing new equipment. Construction projects over $500,000 use military construction (MILCON) funds.

J-5. Resources needed for physical security are the result of planning. Commanders include physical-security programs and improvements as a part of all annual budgets. When the situation changes based on METT-TC, physical-security programs are reviewed, updated and, when necessary, approved. Commanders allocate resources consistent with the threat. Force-protection requirements provide the fundamental reasons for resourcing physical security. Security managers, force-protection officers, and PMs help identify security requirements and prioritize expenditures. When physical-security improvements are not properly planned, integrated, and budgeted for, commanders accept risks for physical-security failures.

**Annual Obligation Plan**
**FY 99**

Installation: Fort McClellan, AL
Activity: Military Police School
Appropriation: Operation & Maintenance, Army (OMA)
Army Management Structure (AMS): 321731
Management Decision Package (MDEP): VTER
Element of Resource (EOR): 2500
Issue: Force Protection/Antiterrorism

Phased Monthly Obligation Plan (Cumulative):

| Month | $000 |
|---|---|
| Oct | 40 |
| Nov | 80 |
| Dec | 120 |
| Jan | 160 |
| Feb | 200 |
| Mar | 240 |
| Apr | 280 |
| May | 320 |
| Jun | 360 |
| Jul | 360 |
| Aug | 360 |
| Sep | 360 |

**Figure J-2. Sample Annual Obligation Plan**

**Unfinanced Requirement (UFR)**
**FY 99**

Installation: Fort McClellan, AL
Activity: Military Police School
Appropriation: Operation & Maintenance, Army (OMA)
Army Management Structure (AMS): 321731
Management Decision Package (MDEP): VTER
Element of Resource (EOR): 2500
Issue: Force Protection/Antiterrorism

Justification/Impact:
Funds in the amount of $360K have been provided to conduct the eight Force-Protection/Antiterrorism contract courses in the US Army Military Police School. This amount will only support training through June 99. An additional $120K is required to conduct training during 4th quarter FY99. Unless additional funds are provided, 625 students will not be trained.

**Figure J-3. Sample UFR**

## Appendix K

# Vulnerability Assessment

After a commander has obtained a threat analysis, he proceeds to complete the analysis by conducting the vulnerability and criticality assessments. (This appendix will discuss only the VA.) This process considers a mission review and analysis of the installation, base, unit, or port in relation to the terrorist threat. The review should assess the cost of antiterrorism measures in terms of lost or reduced mission effectiveness. It should then assess the level of acceptable risk to facilities and personnel given the estimated erosion of mission effectiveness. Often the best operational method and routine may be the worst to counter potential terrorist activities. This review and analysis is performed routinely and particularly for deployment.

## ASSESSMENT CONSIDERATIONS

K-1. The installation, base, unit, or port assessment is derived from the results of the vulnerability and criticality assessments. The assessment provides the staff with the overall vulnerability to terrorist attack. The staff then develops the crisis-management plan, which addresses all terrorist threat levels regardless of the present level. The THREATCONs are then applied according to the local threat. The considerations are—

- **Vulnerability.** The VA is a self-assessment tool used to evaluate its vulnerability to terrorist attack. The more vulnerable an installation, base, unit, or port is, the more attractive it becomes to terrorist attack.
- **Criticality.** The criticality assessment identifies key assets and infrastructures located on and adjacent to the installation, base, unit, or port. These assets are normally symbolic targets that traditionally appeal to a specific terrorist group (such as headquarters buildings and monuments). It addresses the impact of the temporary or permanent loss of key assets or infrastructures to the ability of the installation, base, unit, or port to perform its mission. The staff determines and prioritizes critical assets. The commander approves the prioritized list. The assessment—
  - Selects key assets.
  - Determines whether critical functions can be duplicated under various attack scenarios.
  - Determines the time required to duplicate key assets or infrastructure efforts if temporarily or permanently lost.
  - Determines the vulnerability of key assets or infrastructures to bombs, vehicle crashes, armed assault, and sabotage.
  - Determines the priority of response to key assets and infrastructures in case of fire, multiple bombings, or other terrorist acts.

- **Damage.** The damage assessment determines the ability of the installation, base, unit, or port to plan for and respond to a terrorist attack against key assets and infrastructures.
- **Recovery procedures.** The recovery-procedures assessment determines the capability to recover from the temporary or permanent loss of key assets and infrastructures. Based on this assessment, the staff establishes recovery procedures to ensure the continued ability to perform the mission.

## THREATCON LEVELS

K-2. Specific security measures should be directly linked with THREATCON levels. These considerations are—

- **THREATCON Normal.** This THREATCON level exists when a general threat of possible terrorist activity exists but warrants only a routine security posture.
- **THREATCON Alpha.** This THREATCON applies when there is a general threat of possible terrorist activity against personnel and facilities (the nature and extent of which are unpredictable) and when circumstances do not justify full implementation of THREATCON Bravo measures. It may be necessary to implement measures from higher THREATCONs either resulting from intelligence or as a deterrent. The measures in this THREATCON must be capable of being maintained indefinitely.
- **THREATCON Bravo.** This THREATCON applies when an increased and more predictable threat of terrorist activity exists. The measures in this THREATCON must be capable of being maintained for weeks without causing undue hardship, affecting operational capability, or aggravating relations with local authorities. While in Bravo, the installation should bring manning levels and physical-protection levels to the point where the installation can instantly transition to THREATCON Charlie or Delta.
- **THREATCON Charlie.** The transition to THREATCON Charlie must be done on short notice. It is a result of an incident occurring or the receipt of intelligence indicating that some form of terrorist action against personnel and facilities is imminent. Charlie measures should primarily focus on manning adjustments and procedural changes. Security forces will usually enhance their security presence by acquiring additional manning or by adjusting the work-rest ratio (such as moving from a 3:1 to a 6:1 ratio). At Charlie, off-installation travel should be minimized.
- **THREATCON Delta.** Since the transition to THREATCON Delta is immediate, Delta measures should primarily focus on manning adjustments and procedural changes. THREATCON Delta applies in the immediate area where a terrorist attack has occurred or when intelligence has been received that terrorist action against a specific location or person is likely. The security force's manning level is usually peaked in Charlie; therefore, Delta's additional manning will usually come from an augmentation force. Once in Delta, nonessential operations will cease in order to enhance the security and response

posture. Normally, this THREATCON is declared as a localized condition.

K-3. With exception of THREATCON Normal, all THREATCON levels have certain measures associated with them. These measures are listed in JP 3-07.2, Appendix J, and AR 525-13.

## ASSESSING VULNERABILITY

K-4. A VA addresses the consequences of terrorist attacks in terms of the ability of units, installations, commands, or activities to accomplish their assignments successfully, even if terrorists have inflicted casualties or destroyed or damaged DOD assets. The VA focuses on two broad areas—

- Preventing and, failing that, substantially mitigating the effects of a terrorist act.
- Maintaining emergency preparedness and crisis response.

K-5. The VA provides the commander with a tool to assess the potential vulnerability of an installation, base, unit, or port activity; but it is not a substitute for sound judgment. The VA must stand on its own and be supported by valid considerations. Typically, a small group of knowledgeable individuals develop the VA. The VA team consists of personnel with required areas of expertise. Some of these team members are the—

- Assessment-team chief.
- Physical-security specialist.
- Structural engineer.
- Infrastructure engineer.
- Operations-readiness specialist.
- Intelligence and/or counterintelligence specialist.

K-6. The functions and responsibilities of each team member are outlined in JP 3-07.2. The following paragraphs provide basic information regarding these areas:

- The assessment-team chief's key responsibilities include overseeing the management, training, and performance of the vulnerability-team members; finalizing the assessment-team out briefing; and preparing the population dynamics and risk assessment.
- The physical-security specialist is responsible for the security and safety of the installation, facility, and personnel.
- The structural engineer examines a variety of potential terrorist weapon effects and structural responses. This function serves to better protect personnel from shocks and blasts by reducing damage through the technically appropriate use of standoff measures, hardening, blast shielding, and shatter-resistant window film (such as Mylar®) as described in Chapter 3. The structural engineer's main responsibility is threat and damage assessment from terrorist weapons estimates and suggestions for threat protection or damage-mitigation measures.
- The infrastructure engineer examines protection against the effects of WMD, protection against terrorist-incident induced fires, and utility systems that can be used to minimize terrorist-incident casualties

(including elements of power, environmental control, and life-support systems). The primary responsibilities include infrastructure security and fire, safety, and damage control.

- The operations-readiness specialist examines plans, procedures, and capabilities for crisis response, consequence management, and recovery operations should a terrorist incident occur. The main responsibilities of this position include emergency-medical and individual-readiness assessments.
- The intelligence and/or counterintelligence specialist has the primary responsibility of performing logical analyses and preparing possible conclusions regarding terrorist targets and target vulnerabilities. These are based on processed intelligence information and knowledge of terrorist capabilities and methods in view of US installation, facility, and personnel safety and security practices.

## CONDUCTING THE ASSESSMENT

K-7. Upon its arrival, the assessment team provides an in briefing for the commander, staff, and designated technical point of contact. Site personnel should conduct a site-familiarization briefing and tour. Administrative activities may include establishing the team support area, setting up equipment, scheduling team and technical points of contact meetings and discussions, ensuring classified-material control, establishing a personnel locator, and organizing materials for the out briefing and site folder. Each assessment-team member conducts the assessment based on the specific responsibilities for each function within his area.

## POST-ASSESSMENT ACTIVITIES

K-8. Within 30 days of the visit, a summary narrative report and an annotated briefing should be delivered to the installation commander. Follow-on assistance for the commander may be applicable in areas of technical characteristics of improvement options, cost estimates, and generic sources of materials and equipment.

## DRILLS AND EXERCISES

K-9. Multiechelon war gaming of possible terrorist attacks is the best test, short of an actual incident, to analyze the response of an installation, base, unit, or port. Drills and exercises test suspected vulnerabilities and antiterrorist measures. These exercises and drills also train the staff as well as reaction-force leadership and help maintain a valid threat assessment by identifying and adjusting to changing threat capabilities.

# Glossary

**AA&E** arms, ammunition, and explosives

**ABCS** Army Battle Command System

**AC** alternating current

**admin** administration

**ADP** automated data processing

**AF** Air Force

**AFB** Air Force base

**AFM** Air Force manual

**AFMAN** Air Force manual

**AFOSI** Air Force Office of Special Investigations

**AFR** Air Force regulation

**AIQC** Antiterrorism Instructor Qualification Course

**AIS** automated information system

**AL** Alabama

**AM** amplitude-modulated

**AMC** Army Materiel Command

**AMS** Army management structure

**AO** area of operations

**AP** armor piercing

**Apr** April

**AR** Army regulation

**AR-PERSCOM** Army Reserve Personnel Command

**ASP** ammunition supply point

**AT** antitank

**AT/FP** antiterrorism/force protection

**attn** attention

**Aug** August

**AWG** American wire gauge

**bldg** building

**BMS** balanced magnetic switch

**BTO** barbed-tape obstacle

**BUPERS** Bureau of Naval Personnel

**$C^2$** command and control

**$C^3$** command, control, and communications

**cav** cavalry

**CB** citizen's band

**CCB** Community Counterterrorism Board

**CCIR** commander's critical information requirements

**CCTV** closed-circuit television

**CD-ROM** compact-disk, read-only memory

**cdr** commander

**CG** command guidance

**chap** chapter

**CIA** Central Intelligence Agency

**CID** Criminal Investigation Division

**CINC** commander in chief

**CISO** counterintelligence support officer

**CMU** concrete-masonry unit

**CONEX** container express

**CONPLAN** contingency plan

**CONUS** continental United States

**CP** command post

**CPWG** crime-prevention working group

**CQ** charge of quarters

**CRIMP** Crime Reduction Involving Many People

**CTA** common table of allowance

**DA** Department of the Army

**DARE** Drug Abuse Resistance and Education

**DC** direct current

**DC** District of Columbia

**Dec** December

**DIA** Defense Intelligence Agency

**DOD** Department of Defense

**DOE** Department of Energy

**DOJ** Department of Justice

**DOS** Department of State

**DOT** Department of Transportation

**DS** direct support

**DTM** data-transmission media

**DTOC** division tactical operations center

**EDM** emergency-destruct measures

**EECS** electronic entry-control system

**EOD** explosive-ordnance disposal

**EOR** element of resource

**EPW** enemy prisoner of war

**equip** equipment

**ESS** electronic security system

**FAA** Federal Aviation Administration

**FBI** Federal Bureau of Investigation

**FCC** Federal Communications Commission

**Feb** February

**FIS** foreign-intelligence services

| | |
|---|---|
| **FM** | field manual |
| **FM** | frequency-modulated |
| **ft** | foot, feet |
| **FY** | fiscal year |
| **G2** | Assistant Chief of Staff, G2 (Intelligence) |
| **GHz** | gigahertz |
| **GTA** | graphic training aid |
| **HN** | host nation |
| **HQ** | headquarters |
| **HUD** | Housing and Urban Development Administration |
| **HUMINT** | human intelligence |
| **Hz** | hertz |
| **IAW** | in accordance with |
| **ICP** | initial control point |
| **ID** | identification |
| **IDS** | intrusion-detection system |
| **IED** | improvised explosive device |
| **IG** | inspector general |
| **IID** | improvised incendiary device |
| **in** | inch(es) |
| **INSCOM** | US Army Intelligence and Security Command |

**INTAC** Individual Terrorism Awareness Course

**IPB** intelligence preparation of the battlefield

**IR** infrared

**ISS** information systems security

**J2** Intelligence Directorate (Joint Command)

**Jan** January

**JS** Joint Service

**JSAT** Joint Security Assistance Training

**JSCP** Joint Strategic Capabilities Plan

**JSIIDS** Joint-Service Interior Intrusion-Detection System

**Jul** July

**Jun** June

**K** one thousand

**kHz** kilohertz

**LED** light-emitting diode

**liq** liquid

**LOS** line of sight

**LOTS** logistics over the shore

**LP** listening post

**LRA** local reproduction authorized

**MACOM** major Army command

**maint** maintenance

**Mar** March

**MCO** Marine Corps order

**MDEP** management decision package

**MDMP** military decision-making process

**METT-TC** mission, enemy, terrain, troops, time available, and civilian considerations

**MEVA** mission-essential or vulnerable area

**MI** military intelligence

**MILCON** military construction

**MILPO** military personnel office

**MILVAN** military van

**min** minimum

**mm** millimeter(s)

**MO** modus operandi

**MOU** memorandum of understanding

**MP** military police

**MPACS** Military Police Automated Control System

**MPMIS** Military Police Management Information System

**MPR** military-police report

**MS-DOS** Microsoft®-disk operating system

**MWD** military working dog

**N/A** not applicable

**naut** nautical

**NAVATAC** Navy Antiterrorism Analysis Center

**NBC** nuclear, biological, and chemical

**NCIC** National Crime Information Center

**NCO** noncommissioned officer

**NISCOM** Naval Investigative Service Command

**No.** number

**Nov** November

**NSA** National Security Agency

**NTAV** nontactical armored vehicle

**NVD** night-vision device

**O** official

**OASD** Office of the Assistant Secretary of Defense

**OCOKA** observation and fields of fire, cover and concealment, obstacles, key terrain, and avenues of approach

**OCONUS** outside the continental United States

**Oct** October

**OMA** operations and maintenance, Army

**OPA** operations and procurement, Army

**OPLAN** operation plan

**OPORD** operations order

**OPSEC** operations security

**Pam** pamphlet

**PAO** public affairs office(r)

**PD** probability of detection

**PERSCOM** Personnel Command

**PHOTINT** photographic intelligence

**PI** police intelligence

**PIR** passive infrared

**PM** provost marshal

**PMO** provost marshal office

**POL** petroleum, oil, and lubricants

**POM** Program Objective Memorandum

**POV** privately owned vehicle

**PS** physical security

**PSD** protective security detail

**PSI** physical-security inspector

**PTO** Parent-Teacher Organization

**PX** post exchange

**R&D** research and development

**RD&E** research, development, and engineering

**ref** reference(s)

**RF** radio frequency

**RII** relevant information and intelligence

**ROI** report of investigation

**RORO** roll on/roll off

**RPG** rocket-propelled grenade

**/s/** signed

**S2** Intelligence Officer (US Army)

**SAW** squad automatic weapon

**SDNCO** staff duty noncommissioned officer

**SDO** staff duty officer

**Sep** September

**SIGINT** signal intelligence

**SJA** staff judge advocate

**SO/LIC** Special Operations/Low-Intensity Conflict

**SOFA** status of forces agreement

**SOP** standing operating procedure

**St.** Saint

**STANO** surveillance, target acquisition, and night observation

**stat** statute

**STC** sound-transmission coefficient

**STD** standard

**STU** secure telephone unit

**TAACOM** Theater Army Area Command

**TB** technical bulletin

**TDY** temporary duty

**TECOM** US Army Test and Evaluation Command

**TEMPEST** Terminal Electromagnetic-Pulse Emanation Standard

**THREATCON** threat conditions

**TM** technical manual

**TMDE** test, measurement, and diagnostic equipment

**TNT** trinitrotoluene

**TRADOC** US Army Training and Doctrine Command

**TSC** triple-standard concertina

**UCMJ** Uniform Code of Military Justice

**UFR** unfinanced requirement

**US** United States

**USACE** US Army Corps of Engineers

**USACIDC** US Army Criminal Investigation Command

**USAMPS** US Army Military Police School

**USCG** US Coast Guard

**VA** vulnerability assessment

**VIP** very important person

**vol** volume

**WMD** weapons of mass destruction

# Bibliography


AR 12-15. *Joint Security Assistance Training (JSAT) Regulation (*Navy Instructions 4950.4; AFR 50-29). 28 February 1990.

AR 50-5. *Nuclear and Chemical Weapons and Material—Nuclear Surety.* 3 October 1986.

AR 50-6. *Nuclear and Chemical Weapons and Materiel, Chemical Surety.* 1 February 1995.

AR 190-11. *Physical Security of Arms, Ammunitions, and Explosives.* 30 September 1993.

AR 190-12. *Military Police Working Dogs.* 30 September 1993.

AR 190-13. *The Army Physical Security Program.* 30 September 1993.

AR 190-14. *Carrying of Fire Arms and Use of Force for Law Enforcement and Security Duties.* 12 March 1993.

AR 190-22. *Searches, Seizures and Disposition of Property.* 1 January 1983.

AR 190-27. *Army Participation in National Crime Information Center.* 28 May 1993.

AR 190-51. *Security of Unclassified Army Property (Sensitive and Nonsensitive).* 30 September 1993.

AR 190-56. *The Army Civilian Police and Security Guard Program.* 21 June 1995.

AR 190-59. *Chemical Agent Security Program.* 27 June 1994.

AR 380-5. *Department of the Army Information Security Program.* 25 February 1988.

AR 380-10. *Technology Transfer, Disclosure of Information and Contacts with Foreign Representatives.* 30 December 1994.

AR 380-19. *Information Systems Security.* 27 February 1998.

AR 380-67. *The Department of the Army Personnel Security Program.* 9 September 1988.

AR 381-20. *The Army Counterintelligence Program.* 15 November 1993.

AR 405-20. *Federal Legislative Jurisdiction.* 1 August 1973.

AR 525-13. *Antiterrorism Force Protection (AT/FP): Security of Personnel, Information, and Critical Resources.* 10 September 1998.

AR 600-8-14. *Identification Cards for Members of the Uniformed Services, Their Family Members, and Other Eligible Personnel* (AF 36-3026[I]; BUPERS I 1750.10A; MCO P5512.1B; CG M5512.1; Manual 29.2, Instructions 1 and 2). 1 March 1998.

AR 604-5. *Personnel Security Clearance, Department of the Army Personnel Security Program Regulation* 1 February 1984.

DA Form 2028. *Recommended Changes to Publications and Blank Forms.* 1 February 1974.

DA Form 2806-R. *Physical Security Survey Report.* April 1985.

DA Form 2806-1-R. *Physical Security Inspection Report (LRA).* April 1985.

DA Form 2819. *Law Enforcement and Discipline Report.* May 1988.

DA Form 3975. *Military Police Report.* May 1988.

DA Pam 190-12. *Military Working Dog Program.* 30 September 1993.

DA Pam 190-51. *Risk Analysis for Army Property.* 30 September 1993.

DOD 0-2000.12-H. *Protection of DOD Personnel and Activities Against Acts of Terrorism and Political Turbulence.* February 1993.

DOD 4160.21-M. *Defense Reutilization and Marketing Manual.* August 1997.

DOD Directive 2000.12. *DOD Antiterrorism/Force Protection (AT/FP) Program.* 13 April 1999.

DOD Instruction 2000.16. *DOD Combating Terrorism Program Standards.* 10 May 1999.

DOD Instruction 5210.84. *Security of DOD Personnel at US Missions Abroad.* 22 January 1992.

Federal Specification FF-L-2740. *Locks, Combination.* 12 October 1989.

FM 5-34. *Engineer Field Data.* 30 August 1999.

FM 19-4. *Military Police Battlefield Circulation Control, Area Security, and Enemy Prisoner of War Operations.* 7 May 1993.

FM 55-20. *Army Rail Transport Units and Operations.* 1 June 00.

GTA 19-4-3. *Individual Protective Measures.* July 1997.

JP 3-07.2. *Joint Tactics, Techniques, and Procedures for Antiterrorism.* 17 March 1998.

JS Guide 5260. *Service Member's Personal Protection Guide: A Self-Help Handbook to Combating Terrorism.* July 1996.

TB 9-2300-422-20. *Security of Tactical Wheeled Vehicles.* 27 August 1988.

TM 5-805-8. *Builders' Hardware.* 20 January 1992.

TM 5-820-4. *Drainage for Areas Other Than Airfields* (AFM 88-5, Chap 4). 14 October 1983.

TM 5-853-1. *(O) Security Engineering Project Development* (AFMAN 32-1071, Vol 1). 12 May 1994.

TM 5-853-2. *(O) Security Engineering Concept Design* (AFMAN 32-1071, Vol 2). 12 May 1994.

TM 5-853-3. *(O) Security Engineering Final Design* (AFMAN 32-1071, Vol 3). 12 May 1994.

TM 5-853-4. *Security Engineering Electronic Security Systems.* 12 May 1994.

USACE STD 872-50-01. *Entry Points for US Army Installations.* To be published within the next six months.

USACE STD 872-90-02. *FE5 Chain-Link Security Fence Details.* May 1992.

USACE STD 872-90-03. *FE6 Chain-Link Security Fence Details.* May 1992.

USACE STD 872-90-04. *FE7 Chain-Link Security Fence Details for Non-Sensored Fence.* May 1992.

USACE STD 872-90-05. *FE7 Chain-Link Security Fence Details for Sensored Fence.* May 1992.

USACE STD 872-90-06. *FE8 Chain-Link Security Fence Details for Sensored Fence.* May 1992.

USACE STD 872-90-07. *FE5 Chain-Link Fence Gate Details.* May 1992.

USACE STD 872-90-08. *FE6 Chain-Link Fence Gate Details.* May 1992.

USACE STD 872-90-09. *FE8 Chain-Link Fence Gate Details.* May 1992.

# Index

Made in the USA
Monee, IL
05 November 2026

67811390R00262